8-16-99

Introduction to Digital and Data Communications

Introduction to Digital and Data Communications

MICHAEL A. MILLER

DeVry Institute of Technology, Phoenix

WEST PUBLISHING COMPANY

St. Paul • New York • Los Angeles • San Francisco

Copyediting by Susan Ecklund

Text design by Rick Chafian

Artwork by Rolin Graphics

Proofreading by Robert deFreitas and Luana Richards

Production, Prepress, Printing, and Binding by West Publishing Company

West's Commitment to the Environment

In 1906, West Publishing Company began recycling materials left over from the production of books. This began a tradition of efficient and responsible use of resources. Today, up to 95 percent of our legal books and 70 percent of our college texts are printed on recycled, acid-free stock. West also recycles nearly 22 million pounds of scrap paper annually—the equivalent of 181,717 trees. Since the 1960s, West has devised ways to capture and recycle waste inks, solvents, oils, and vapors created in the printing process. We also recycle plastics of all kinds, wood, glass, corrugated cardboard, and batteries, and have eliminated the use of styrofoam book packaging. We at West are proud of the longevity and the scope of our commitment to our environment.

COPYRIGHT ©1992 By WEST PUBLISHING COMPANY
50 W. Kellogg Boulevard
P.O. Box 64526
St. Paul, MN 55164-0526

Printed in the United States of America
99 98 97 96 95 94 93 92   8 7 6 5 4 3 2 1 0

Library of Congress Cataloging-in-Publication Data

Miller, Michael A.
    Introduction to digital and data communication / Michael A. Miller.
       p.   cm.
    Includes bibliographical references and index.
    ISBN 0-314-93371-9 (hard)
    1. Digital communications. 2. Data transmission systems.
    I. Title.
    TK5103.7.M547   1992
    004.6—dc20   91-37888
    CIP

This book is dedicated to two very important people in my life: my wife, Ann, who had the patience to provide me the freedom and time to write it, and my good friend and colleague, Ivan Plotnick, for supplying me with the incentive and support to undertake this project.

Contents

Seven ● Multiple-Channel Data Communications   **218**

Eight ● Open Systems Network models   **266**

Eleven • Fiber-Optic Communications   357

Twelve • Data Communications Applications   387

Preface

This text on the data and digital communications field is intended to be a junior- or senior-level text in an Electronics Technology curriculum with one or two courses on the topic. A community college or proprietary school with a Communications specialty could easily teach this text to sophomores, after a course in basic communication. In order to do justice to the numerous facets of the data and digital communications field, no one area is treated with the kind of depth necessary to produce telecommunications engineering technologists. It is intended, rather, to provide the electronics technician and technologist with sufficient background in data and digital communications so that a solid, thorough understanding of what is in the field is achieved.

It is expected that the student has had or is familiar with the topics in courses that cover the following material:

1. AM and FM radio
2. Basic communications
3. Basic electronic circuits and devices
4. Basic digital electronic circuits and devices
5. Microprocessors and basic computer architectures.

Chapter 1 starts by giving the student a general overview of the data communications area. It lays the basis for future chapters by establishing some necessary fundamentals.

Chapter 2 deals with the telephone system from the standpoint of data transfer usage. Electrical specifications and impairments that require consideration when using telephone system facilities are also covered.

Chapter 3 provides details on some common error-detection and error-correction methods used in data, digital, and network communications systems.

Chapter 4 introduces the concept of data link protocols which are illustrated by exploring widely used data link protocols.

Chapter 5 is concerned with the hardware of low speed data communications as interfaced through the telephone system. Use of breakout boxes to test communications interfacing is also discussed.

Chapter 6 extends the basics of the data communications system up through phase shift and quadrature amplitude modulation schemes.

Chapter 7 introduces the student to multiplexing communications channels into a single entity. Here the line between data and digital communications begins to be crossed. The concepts of T1 digital lines and the equipment used to monitor and test them are included in this chapter.

Chapter 8 presents the two most common network layer models for data and digital communications networks, open systems interconnection (OSI) and systems network architecture (SNA) models.

Chapter 9 covers a wide variety of communications networks including local area networks (LANs) and private branch exchange (PBX) systems. The IEEE 802 standard is discussed in detail in this chapter.

Chapter 10 deals with the integrated services digital network, or ISDN. ISDN services are provided for combined services such as voice, video, and data to be transmitted simultaneously across the public services network. Protocol analyzers used to test and emulate many of the networks discussed to this point are explored in this chapter.

Chapter 11 discusses fiber optics and fiber-optic networks. In a text of this nature, a single chapter on fiber optics cannot cover this vast subject, which usually requires a separate tome, in its entirety. However, it is the author's intent to provide enough material on the subject so that the student obtains a good background on the subject.

Chapter 12 discusses two data communications applications—facsimile and satellite communications. Both of these services make worldwide communications possible and rapid.

The Appendix provides a list of the abbreviations and acronyms used in this text and in the data and digital communications field.

The author wishes to express his appreciation to the editors at West, Tom Tucker, Chris Conty, Jeff Carpenter, and Tom Hilt, for their guidance during the writing and reviewing of this text, and also thanks copy editor Susan Ecklund for her efforts. Additional thanks are extended to all of the professional instructors who took the time to thoroughly review the manuscript and make the many suggestions that were incorporated into the book to make it more complete and accurate. Specifically, those reviewers are:

Richard Anthony
Cuyahoga Community College
 (Metro. Campus)

Joseph Booker
DeVry Institute of Technology,
 Lombard (Illinois)

Ray Burns
Red River Community College
 (Manitoba)

Shakti Chatterjee
DeVry Institute of Technology,
 Columbus

John Clark
Niagara College (Ontario)

Jerry Cockrell
Indiana State Univ.

Ken Collins
Central Piedmont Community College
    (N.C.)

Elaine Cooley
Indiana/Purdue Univ. at Indianapolis

Peter Edmonson
Mohawk College (Ontario)

Alvis Evans
Tarrant County J. C. (Texas)

Clay Laster
San Antonio College (Texas)

Susan Meardon
Wake Technical Community College
    (N.C.)

Gary Mullett
*Springfield Technical Community
    College* (Massachussetts)

Mark Oliver
Monroe Community College (N.Y.)

James Pearson
DeVry Institute of Technology, Irving
    (Texas)

Willard Peterson
Mohawk Valley Community College
    (N.Y.)

Paul Lin
Indiana/Purdue Univ. at Indianapolis

Susan Garrod
Purdue University

Robert Martin
*Northern Virginia Community
    College*

Margarete Ralston
Metropolitan State Univ. (Colorado)

Carol Richardson
Rochester Institute of Technology

James W. Sadowski
DeVry Technical Institute, Woodbridge
    (N.J.)

Jim Stewart
DeVry Technical Institute, Woodbridge
    (N.J.)

Ronald Tocci
Monroe Community College (N.Y.)

Jon L. Van Donkelaar
Energy Innovations (Ohio)

Robert Weiler
Capitol College (Md.)

Don Weiman
SUNY College of Technology, Alfred

Paul Wojnowiak
Southern College of Technology

Michael A. Miller, Senior Professor
DeVry Institute of Technology, Phoenix, Arizona

One • Introduction to Digital and Data Communications

OBJECTIVES

The purpose of this first chapter is to introduce basic concepts of a communications system involving the transfer of digital information between two stations. The specific topics discussed are

1. Historical background
2. Functional system blocks
3. System and network topologies
4. Character codes
5. Data types and rates
6. Binary data formats

ACRONYMS

ACK—**ACK**nowledge

bps—**B**its **P**er **S**econd

ASCII—**A**merican **S**tandard **C**ode for **I**nformation **I**nterchange

BS—**B**ack **S**pace

CMOS—**C**omplementary **M**etal **O**xide **S**emiconductor

CODEC—**CO**der/**DEC**oder

CR—**C**arriage **R**eturn

CWK—**C**ontinuous **W**ave **K**eying

DC—**D**evice **C**ontrol

1

DCE—Data Communications Equipment or Data Circuit Terminating Equipment

DLE—Data Link Escape

DLE—DeLimitEr

DTE—Data Terminal Equipment

DTFM—Dual Tone Multiple Frequency

EBCDIC—Extended Binary Coded Decimal Interchange Code

EIA—Electronic Industries Association

ENQ—ENQuiry

EOT—End Of Transmission

ETB—End of Transmission Block

ETX—End of TeXt

FAX—FACSimile

FF—Form Feed

IC—Integrated Circuit

ISDN—Integrated Services Digital Network

LAN—Local Area Network

LCU—Line Control Unit

LF—Line Feed

LSB—Least Significant Bit

MAN—Metropolitan Area Network

NAK—Negative Acknowledge

NRZ—Non-Return to Zero

NRZB—NRZ Bipolar

NRZI—NRZ Mark Inversion

PCM—Pulse Coded Modulation

RZ—Return to Zero

RZB—Return to Zero Bipolar

SOH—Start Of Heading

SP—SPace

sps—Symbols Per Second

STACO—STAtion COntroller

STX—Start of TeXt

SYN—SYNchronization

TELCO—TELephone COmpany

TTL—Transistor-Transistor Logic

UART—Universal Asynchronous Receiver Transmitter

USART—Universal **S**ynchronous **A**synchronous **R**eceiver Transmitter

WAN—**W**ide **A**rea **N**etwork

● **1.1 INTRODUCTION**

The sending and receiving of vast amounts of information over vast distances and in short periods of time has been a goal since human beings learned to communicate. Words are continually added to languages to convey new or more complete ideas. Methods of transmitting those words and related information have been and are continually being developed. Along with these methods, codes have arisen to convey letters, words, and data. The beating of drums and sending of smoke signals were early forms of coded data. The coding of information into quantities of sound or smoke were two of the earliest attempts to communicate over long distances.

The transfer of data in digital form began around 1832 with the advent of Morse code, a systematic code that represents the printable characters of a language using a form of binary data. A signal or electrical current is placed onto an interconnecting line between a sender and a receiver when a switch or key is closed by the sender. The longer the key is held closed, the longer the signal or current remains on the line. The form of data modulation using a sine wave signal is called **CONTINUOUS WAVE KEYING (CWK)** because the signal itself is a continuous audio oscillation that is placed on or off the line by use of the key. The switching on and off of a direct current in place of a signal is also referred to as keying. In the latter case, an electromagnetic relay is energized in the presence of the current and released when the key or switch opens and removes current from the line. The length of time the signal or current is present on the line defines the signal as a dot (short time) or a dash (long time). Each combination of dots and dashes represents a letter, number, or punctuation mark. These readable characters are called **ALPHANUMERIC CHARACTERS.**

Morse code served as a means of communicating information for a considerable amount of time. With the invention of the telephone and the creation of the **TELEPHONE COMPANY** system **(TELCO),** quicker means of transferring data evolved. An early type of printer/keyboard system, developed by the Teletype Corporation, employed electromechanical relays to replace the action of the key. Instead of continuing with dots and dashes, these machines used the presence or absence of 20 ma of current to represent binary data. The presence of the 20 ma current signifies a logic high or "one" state often called a **MARK** and the absence of current, a logic low or "zero" state called a **SPACE**. These Teletype machines, combined with equipment that allowed them to be interfaced to the telephone system, provided the machinery for the development of Teletype and Telex systems in the 1930s.

Teletype machines were slow, noisy, and consumed large amounts of power. Additionally, since these machines were electromechanical in nature, they were highly prone to mechanical failures and constant adjustments and maintenance. As technology improved, so did the equipment that benefited from the technology. Systems have gone beyond transferring basic data to handling pictures, sharing production and test information, managing large quantities of data transfers, and allowing everyone to do basic banking at any time through automatic teller machines (ATMs). Communications networks have advanced from simple two-station communications, through multistation interconnections, to multichannel fiber-optic cable, satellite, and microwave communications.

Within the production work and business office environments, requirements for quick access to large amounts of information have led to the development of localized communications systems, such as **LOCAL AREA NETWORKS (LANs), METROPOLITAN AREA NETWORKS (MANs),** and **WIDE AREA NETWORKS (WANs).** These networks are formed with computer systems, communications equipment, and various terminals and peripheral devices connected to a common communications network. Data and programs can be easily moved between the equipment connected to the network. The differences in the three types are the size of the network and, ultimately, how the communications systems operate. Local area networks usually occupy one building interconnecting various offices and departments; another common practice is to interconnect product work and quality-testing workstations onto a common network. Metropolitan area networks encompass numerous locations within a medium-size area such as a city or state. Automatic teller machines used by a local bank or a number of drugstores with a computerized prescription service are examples of applications of metropolitan area networks. Wide area networks extend coverage to national and international applications. Extending the automatic teller metropolitan area network example, a bank that does business in the western half of the United States would utilize a wide area network. Companies such as Motorola and Intel, which manufacture a good many of the integrated circuits used in communications and computers, use wide area networks to interconnect plants and offices dispersed throughout the United States and the world.

Still, the boundaries of communications are being extended. Multiple channels of communication are carried simultaneously through networks spanning the globe. More communication satellites circle the earth, while fiber-optic transatlantic cables are being laid. The fiber-optic cables will supply many times the number of communication lines presently available on the existing transatlantic lines.

Integrated communication services have been developed and are being established that provide a way to send and receive voice, video, and data information on a single network. The telephone company (AT&T) is incorporating this service, called **INTEGRATED SERVICES DIGITAL NETWORK (ISDN),** using the common carrier telephone system. Many communications networks are finding application in the business and manufacturing environments. It is important for today's technician as well as today's business and production men and women to have a grasp on the use and inner workings of these networks. Before these types of networks can be explored, a good basic knowledge of a communications system, starting with two points of communication, must be established.

● **1.2 A FUNDAMENTAL COMMUNICATIONS SYSTEM**

Figure 1.1 illustrates a simple system connection between two locations. This two-point network begins on the left with the controlling station in the system called the **PRIMARY STATION.** Data to be sent from this station to the **REMOTE** or **SECONDARY STATION** at the other end of the communications **LINK** usually originate in parallel form from one or more peripheral devices. These peripherals include, but are not limited to, computer terminals, printers, keyboards, **FACSIM-ILE (FAX)** machines, and data display terminals. Note that the information supplied by these devices could be anything from key presses to pictures. The information is converted from its natural form into digital form by the peripheral and is presented as groups of parallel data to the system. **PARALLEL DATA** are a group of digital bits that are all available at the same time. An individual communications path is required for each bit, allowing data to move quickly, transferring several bits of information at one time. However, the need for multiple data paths is impractical and costly for long-distance transfers. Instead, it is preferable to send data along a single data path between two stations. In order to do this, the parallel data need to be converted into **SERIAL** form, with one bit of data following another. While this method of moving information is slower, it uses only a single data path to move the data.

Example 1.1   Illustrate the difference between sending the following two sets of numerical values in parallel and in serial:

2C3B in hexadecimal, which is 0010110000111011 in binary,

and

BADF00D5 or 10111010110111110000000011010101

Solution: Assume that data in parallel are grouped in 8 binary bit groups. With this stipulation, the first two hexadecimal digits (2C or 00101100 in the first number, and BA or 10111010 in the second) would be moved first as a binary group of the 8 bits, with the least significant bit of each character located to the right. After this group of bits is transferred to a destination, the next pair of digits is moved as 3B or 00111011 in the first example and DF or 11011111 in the second. If it took one millisecond (ms) for each move, the complete transfer of the first number (2C3B) would take 2 ms. Eight data lines would be required to facilitate the move.

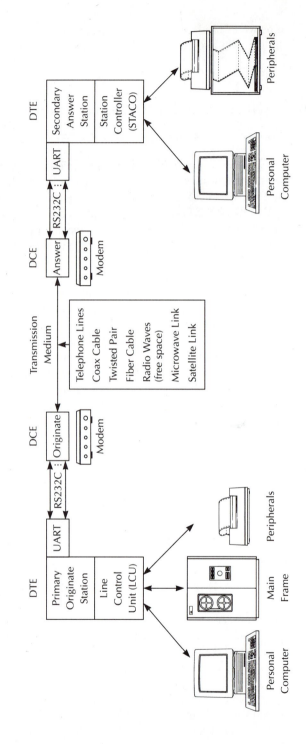

Figure 1.1 Communications Link

As a serial stream of data, 2C3B would be moved on a single line, 1 bit at a time, starting with the least significant bit of the first character and continuing through to the most significant bit of the last character. As a binary stream of data bits, the transmission follows this sequence:

$$2 \quad C \quad 3 \quad B$$
$$0100001111001101$$

Since there are a total of 16 digital bits to be moved, moving them at 1 ms per move requires 16 ms to complete the move. However, only one transmission path is required for the bits to travel on from source to destination.

BADF00D5 transferred on the 8-bit parallel path is sent in the following sequence: BA = 10111010 followed by DF = 11011111, 00 = 00000000, and D5 = 11010101. Four transfers at 1 ms per transfer takes a total time of 4 ms. To transmit serially, once again start with the least significant bit of the first digit and send the entire value out as a steady stream of bits, each requiring 1 ms to send:

$$B \quad A \quad D \quad F \quad 0 \quad 0 \quad D \quad 5$$
$$1101 \quad 0101 \quad 1011 \quad 1111 \quad 0000 \quad 0000 \quad 1011 \quad 1010$$

Gaps are left for clarification. The total time required to send the serial stream of bits is 32 ms. Keep in mind that the 1-ms transfer time is maintained for comparison purposes. Parallel bits and serial bits would actually be sent at different rates depending on the system doing the sending and receiving.

Devices that perform parallel-to-serial conversion (and vice versa at the receiving station) are the **UNIVERSAL ASYNCHRONOUS RECEIVER TRANSMITTER (UART)** and the **UNIVERSAL SYNCHRONOUS/ASYNCHRONOUS RECEIVER TRANSMITTER (USART).** The conversion process and the rates at which parallel data are sent to the UART or USART and the rate at which serial data are sent and received are controlled by the computer system to which the UART or USART is connected. UARTs and USARTs, which are discussed in detail in chapter 5, are produced in medium-scale integrated circuit packages. They perform other required tasks in addition to the serial-parallel conversion. The basic application for the conversion process is to take the data supplied on a computer system's parallel data paths and convert those data into serial data to be placed, ultimately, onto a single transmission path like a telephone line. For long distances the single transmission path is more practical to use than a parallel path consisting of many data lines. The size and cost of the parallel lines would be too prohibitive for long-distance connections.

A **LINE CONTROL UNIT (LCU)** located as part of the primary station directs the interface and interconnection of the peripherals to the communications portion of the station. Additionally, the LCU contains the software needed to establish and control the data communications link between the primary and secondary stations. An applications software program used by the LCU, called a

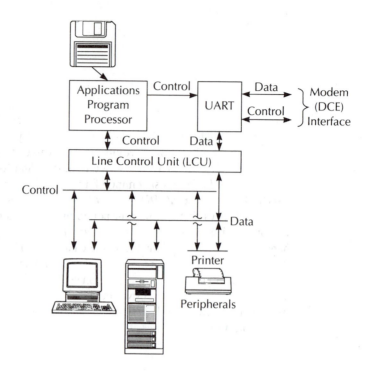

Figure 1.2 Data Terminal Equipment (DTE)

PROTOCOL, defines a set of rules that determine the requirements for the successful establishment of a data link and the transfer of actual information between the stations. Other protocol software and hardware applications control higher-level communications interfacing in various types of networks. Protocols at the lowest level, used to establish communications linkages, are discussed in chapter 4, while higher-level protocols are presented in chapters 8 and 9, which deal with networking.

The units discussed thus far are grouped together, functionally, into a single unit called **DATA TERMINAL EQUIPMENT (DTE)** (Figure 1.2). The communications section of the DTE includes an applications program processor, the LCU, and the UART. Stations are required to maintain communications channels in the advent of failure of individual peripherals at the station's site. The illustration shows how the communications section is isolated and interfaced with the peripherals using the line control unit. It is important to note that under the general blanket of peripherals there could also be included mainframe computer systems as well as simple keyboard terminals or personal computers. Software programs are written to select the order in which the peripherals connected to the line control unit are accessed and the format used for interpreting the data stream into useful information. This information includes control, message, and error-detection data.

Applications programs also direct control information to the line control unit and UART to allow data to flow from the peripheral currently serviced by the line control unit to the UART and out to **DATA COMMUNICATIONS EQUIPMENT** or **DATA CIRCUIT TERMINATING EQUIPMENT,** both known by the acronym **DCE.** Additional applications programs interpret the data to determine its destination and usage. Data communications equipment devices are those that accept the serial data stream from the data terminal equipment, convert that data into a form usable by the transmitting **MEDIUM,** usually an analog signal, and send the converted data out onto that medium. At the receive end the process is reversed. The received analog signals are converted back to digital format and fed to the receive side of a UART. The most common form of data communications equipment is the **MODULATOR/DEMODULATOR (MODEM).** This unit converts the serial data into audio tones that can be transported using telephone lines designed for voice communications. Modems come in many types based on the data rate they are to handle, the type of data, and how the data are to be transmitted and received. An originating modem at the primary end of a two-point system like that in Figure 1.1 converts the digital data into two originate frequency tones, one for a logic 1 level and one for a logic 0. The answer modem at the secondary end of the data link must be capable of receiving those tones and converting back to digital signals. Equally important, when the secondary sends a reply back to the primary, the answer modem must convert the digital data into two different sets of tones that the receive side of the originate modem must be able to receive. Other modem types perform these functions using a single set of tones under strict control for who is sending and who is receiving. Details on modems and how they do their job are covered in chapters 5 and 6. Other types of data communications equipment may code digital data into other digital formats or codes before converting the serial data into analog signals. An example of such a device is the **CODER/DECODER (CODEC).** These devices and the processes that are used to create the codes are discussed in chapter 7. One of the most common of these processes is called pulse-coded modulation (PCM).

The medium between the primary and secondary stations can be as simple as a coaxial or twisted-pair cable in the case of local area communications networks. Another common medium type is the existing telephone lines, which use twisted pairs and/or fiber-optic cables. The telephone company system of interconnecting local and long-distance lines is a popular medium for use simply because it is already there. Early work in data communications centered around adopting the telephone system for handling data transfers. Additional types of media include radio, microwave, and satellite communications using the airwaves.

Interconnecting data terminal and data communications equipment so they will work harmoniously is complicated because different manufacturers produce varied types of devices. A need for a standard interface between these DTE and DCE units is crucial. One example of a commonly used standard, the **RS232C,** was written by communications engineers for the **Electronic Industries Association (EIA),** which is one of many organizations responsible for establishing standards for a variety of electronics and electrical applications. RS stands for recommended standard, which means that there is no enforcing authority to assure the proper use or complete compliance with the specifications of the standard. Enforcement of its use falls to the consumer, who benefits by not purchasing those systems that do not include the standard and buying those that do include it. Encompassed in the standard are electrical and physical specifications for users wishing to connect DTE equipment to DCE equipment. This standard and others are examined in chapter 5 in detail.

At the right side of Figure 1.1 is the remote or slave station, more commonly called the secondary station. A **STATION CONTROLLER (STACO)** at the secondary station performs functions similar to those of the primary station's line control unit. However, instead of controlling the data link, the purpose of the software at the STACO is to respond to the commands and protocol rules sent by the primary. The STACO interprets these commands and the protocol so that it can route the incoming information to the correct peripheral at the secondary station and return any responses that are required.

● 1.3 COMMUNICATION LINKS

Communication links are configured to satisfy particular requirements for a given system. The simplest link is the one used in the previous discussion. It contains a single primary and a single secondary station. The primary station initiates (or originates) the communication link and maintains control over that link until the data transfers are completed. The answering station is the secondary. Both stations must use the same protocol, data rates, and data codes for data to be correctly sent and received. The actual method of sending and receiving data is further divided into three types—simplex, half duplex, and full duplex.

A system can be configured to send and receive data in one direction only (from primary to secondary, for example). This transmission is referred to as a **SIMPLEX** transmission. This type of data transfer link can be performed by the system of Figure 1.1 if, for example, the primary would always transmit data to the secondary and the secondary would not be required to respond or send anything in return. Simplex transmission is useful in an environment where large quantities of data are to be sent without acknowledgment of the reception. The data can be sent fast and continuously. Of course, if an error should occur, there is no way for the secondary to inform the primary that it happened. A separate communication, such as a voice telephone call, would have to be used between operators at both stations. Prime examples of half-duplex operation are the transfer of inventory data from a supply warehouse to an accounting office and the sending of a large volume of test results from an environmental lab to a quality-control office for evaluation.

There are two basic methods of bidirectional data transfer, **HALF DUPLEX** and **FULL DUPLEX.** Half duplex allows transmission in both directions (primary to secondary and secondary to primary) but restricts these transfers to one direction at a time. Referring again to Figure 1.1, the primary would begin by sending a message to the secondary. After the message is completely sent, the secondary can reply with a message of its own. The second bidirectional method, full duplex, is the simultaneous transmission and reception of data between the two stations, primary to secondary and secondary to primary. In this case the primary sends messages to the secondary using a selected pair of audio tones (each at its own frequency) to represent logic 1s and 0s. At the same time, the secondary also transmits data to the primary using a different set of audio tones to represent logic 1s and 0s. These different tone pairs allow both transmitters to use the medium at the same time to send their messages. Duplex systems are commonly used in teleconferencing, in which meetings between persons at different physical locations are accomplished through a network that allows communications to be sent and received by all members of the conference at the same time. Thus, a question asked by a participant in New Jersey can be answered and discussed by the moderator in Washington.

When additional secondaries share the same communications system, the network is called a **MULTIPOINT SYSTEM.** There are numerous ways to achieve a multipoint network, but whatever method is used, there will be one and only one primary in the system. This station controls the communication linkage, determining which station it communicates with at any given time.

● 1.4 COMMUNICATIONS SYSTEM FORMATS

The term **TOPOLOGY** describes the physical configuration of a multipoint system or network. Three common topologies are **BUS, RING,** and **STAR** or **HUB.** The bus topology (Figure 1.3) uses a common transmission line between the primary and all the secondaries on the line. The primary, as controlling station, initiates all data communications between it and the secondaries. Individual secondaries cannot communicate directly with each other; the primary establishes a communication (or data) link with one of the secondaries by addressing that secondary. For example, if the primary wished to send data to secondary 2 in Figure 1.3a, it would send an addressing message to it and wait for a response. The response would indicate whether that secondary is ready to receive the primary's messages. After the primary is finished with one secondary, it goes to the next one and establishes a link with it. Besides sending messages to each secondary, the primary can send a querying message to find out whether a secondary has messages to be relayed to the primary. An extension of this topology, shown in Figure 1.3b, is a system with a secondary acting as a primary to a sub-multipoint branch. Traffic destined for one of the secondaries in the subsystem is routed through the subprimary, which appears to the primary as just another secondary. At other times the subprimary can act as a primary to control the secondaries connected to it, while the main primary communicates with other secondaries in the system.

The **STAR** or **HUB** (Figure 1.4a) topology places the primary in the center or hub, with each secondary connected to an individual communication line (or spoke) to the primary. As with the bus system, secondaries cannot communicate directly

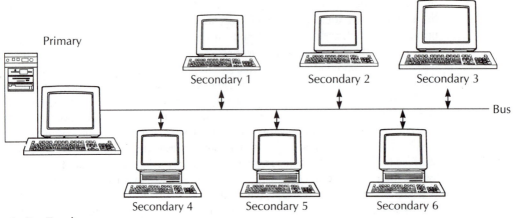

a) Bus Topology

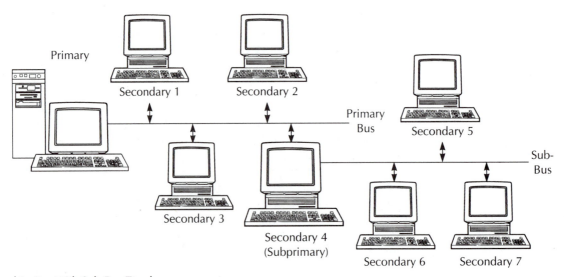

b) Bus With Sub-Bus Topology

Figure 1.3 Bus Topology

with each other in the star system. Secondaries in this system can also act as subhubs for substar systems as shown in Figure 1.4b. Once again, the subhub and primary can communicate simultaneously with other secondaries as long as they are not directly communicating with each other.

A third common topology is the **RING** system (Figure 1.5). This topology operates without a designated primary. Messages are passed around the ring in a specified direction. Messages destined for a particular station are read by that station, while the remaining messages are passed on to the next station in the ring. Each station may add additional messages to the stream and pass them on to the next station. Control of the system relies on a separate message called a **TOKEN,** which is passed from station to station in the ring. The station that currently holds the token can read messages destined for it and append messages it wishes to send on to other stations in the ring. That station then sends the token on to the next station in the ring. Details on token passing are discussed in chapter 9. One of the stations in the ring can be used to interconnect a second ring to the system, as shown in Figure 1.5b. A device at that secondary, called a **BRIDGE,** performs the function of

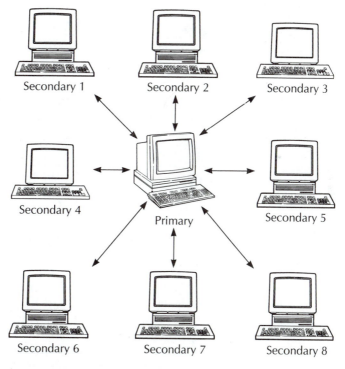

a) Star or Hub Topology

Figure 1.4a STAR or HUB Topology

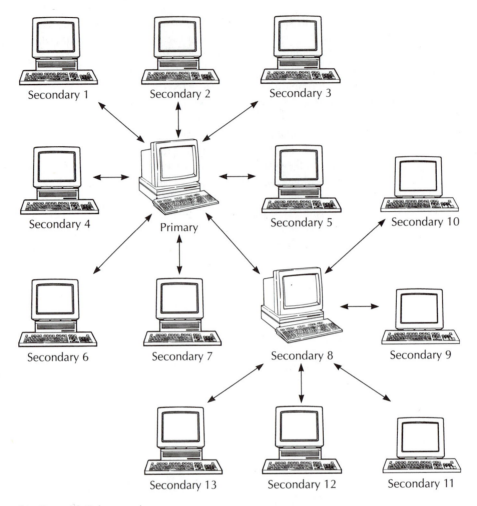

Secondary 1　　　Secondary 2　　　Secondary 3

Secondary 4　　　Primary　　　Secondary 5　　　Secondary 10

Secondary 6　　　Secondary 7　　　Secondary 8　　　Secondary 9

Secondary 13　　　Secondary 12　　　Secondary 11

b) Star with Subnetwork

Figure 1.4b Star with Sub Network

determining when data held in one ring are meant for the other ring. Upon sensing that condition, the bridge passes the messages from one ring to the next, while the remaining messages continue around the original ring. Bridges and other interconnecting devices and systems are discussed in chapter 9.

Topologies may be mixed in an overall system, as illustrated in Figure 1.6. The key point is that all of the subsystems act independently as long as they are not being addressed by the main primary. Messages can be handled within each subsystem and then routed to the main primary when the subprimary is accessed by the main primary as another secondary on the main line.

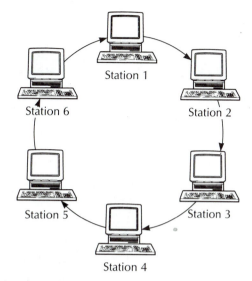

a) Ring Topology

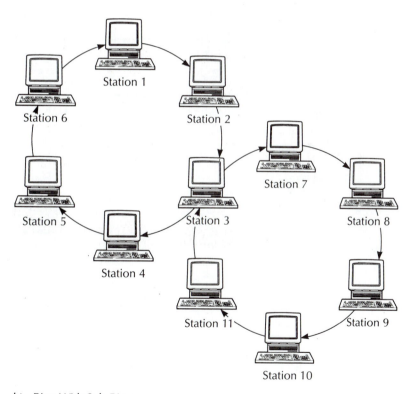

b) Ring With Sub-Ring

Figure 1.5 Ring Topology

15

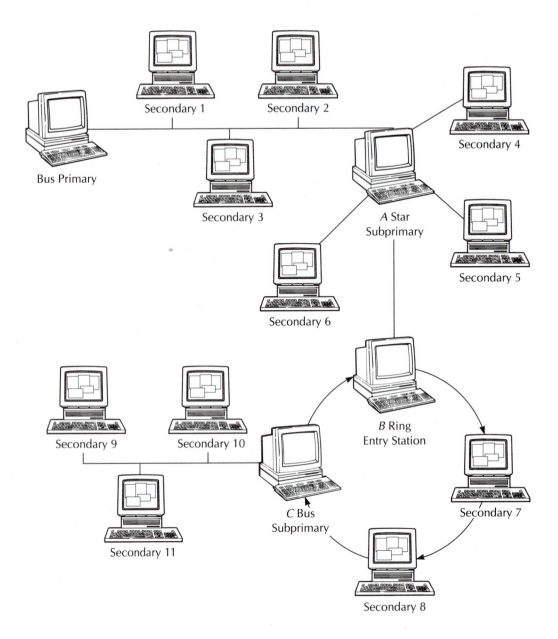

Figure 1.6 Mixed Topology System

In Figure 1.6, station *A* appears to the bus primary as a secondary station. All traffic destined for the remainder of the system, except stations 1, 2, and 3, is routed through station *A*. Station *A,* acting as the hub station for a star network, has three standard secondaries (stations 4, 5, and 6) and one ring entry or bridge station (station *B*). Messages destined for the stations in the ring or lower bus network all must be routed through station *B*. Lastly, station *C* on the ring performs the primary tasks for the bus network connected to stations 9, 10, and 11. It is necessary to note at this point that this is an introductory concept and that a lot more is involved in interconnecting different networks to have them work as a cohesive system.

The systems described thus far are fundamental and represent the basic form that larger, more sophisticated networks take. Local area networks, for instance, utilize all three topology classes, sometimes intermixing topologies within a well-defined network. Specialized interfacing devices and software manage appropriate protocol and message-format conversions between various subsystems within the entire network. These systems are discussed in succeeding chapters, beginning with a basic two-point communications system in chapter 5 and concluding with the numerous forms of networking in place today discussed in the chapters following chapter 5. Basic to every system developed and employed today is the transference of information via a digital code. These **CHARACTER CODES** use selected combinations of 1s and 0s to represent different characters. Besides data communications applications, binary character codes are used wherever a digital processor or computer is used to interpret letters, numbers, and directed functions. A common example most students are familiar with is the personal computer keyboard and display screen. As the user types on the keyboard, the character code for each key is sent from the keyboard to the computer's processor or memory. The user understands the characters printed on the keys. The computer's system relates only to digital bits, in this case the character codes. Similarly, to display readable characters on the screen, the computer system interprets the codes and controls the placement of dots to form each character. For a communications system, the inputting of data starts in the same way—characters are entered from a keyboard or other input peripheral (disk, facsimile machine, scanner, etc.), which has the job of creating the character codes for the system. The communications circuitry takes these character codes and causes them to be sent to some destination where the process is reversed. The character codes are used by the receiving station to create the readable messages, which can be displayed on a screen, sent to a printer, or stored in an output peripheral storage unit such as a floppy diskette.

● 1.5 CHARACTER CODES

Coding became more sophisticated by replacing the series of dots and dashes used in the Morse code with combinations of binary data bits formulated into a character code. One of the earliest of these character codes is called the Baudot code, in which groups of 5 binary bits are used to represent a single character. Thus each different 5-bit binary word is recognized as a different character as listed in Table 1.1. The binary codes for each character are shown with their equivalent hexadecimal value to the left of the chart. Do not lose sight of the fact that 5 bits is insufficient to fill two hexadecimal digits (4 bits per digit) and contains more than can be represented by one digit. The three missing leading bits of the upper hexadecimal digit are set to 0 so that each binary combination can be illustrated in hex.

Using 5 bits per **CHARACTER WORD** limits the total number of characters that can be directly represented by this code to 32 (base 2 raised to the fifth power). This hardly covers 26 alphabet and 10 number characters, yet Table 1.1 shows that there are many more characters. To achieve this the Baudot code uses the shift character to switch from letters to numbers (and the associated character set of each). Pressing the shift key on the keyboard generates the shift character and tells the receiving end to adjust the interpretation of the characters that follow the shift character.

Example 1.2   What is the code sequence to send the message N.Y. GIANTS 17, L.A. RAMS 15?

Solution: The message begins with the letter character N. The first character sent must assure that the receiving communications equipment is capable of interpreting letter characters. To do this, the letter-shift character (11011) is sent first. This is followed by the N character (00110). A period is required next, which is found in the number-shift column. Before the period can be sent, the receiving device has to be "shifted" to the numbers column. This is accomplished by sending the number-shift character next (11111). Now comes the period (00111), letter-shift (11011), Y character (10101), number-shift (11111), a period (00111) again, a space (00100), and so forth through the remaining message.

As can be seen in the example, the main drawback to using the Baudot code is the need to shift each time numbers or punctuation was used in a message. The chief benefit of the code is its size. It takes less time to send 5 bits serially than it takes to send a larger number of bit codes. The Baudot code was used chiefly in the early European Telex system using Teletype machines. These machines, which used electromechanical relays to generate digital 1s and 0s, were slow and consumed a lot of electricity. A different model of Teletype machine used in the United States Teletype system recognized a 7-bit code called the **AMERICAN STANDARD CODE FOR INFORMATION INTERCHANGE (ASCII).** The 7-bit code, shown in Table 1.2, provided for direct coding for 128 different characters. This eliminated the need for the shift character to change from one set

Table 1.1 Baudot 5-Bit Character Code

Hex Code	Binary Code	Letter Shift	Number Shift
00	00000	NULL	NULL
01	00001	T	5
02	00010	Return	Return
03	00011	O	9
04	00100	Space	Space
05	00101	H	#
06	00110	N	,
07	00111	M	.
08	01000	Line Feed	Line Feed
09	01001	L	)
0A	01010	R	4
0B	01011	G	&
0C	01100	I	8
0D	01101	P	0
0E	01110	C	:
0F	01111	V	;
10	10000	E	3
11	10001	Z	"
12	10010	D	$
13	10011	B	?
14	10100	S	BELL
15	10101	Y	6
16	10110	F	!
17	10111	X	/
18	11000	A	—
19	11001	W	2
1A	11010	J	'
1B	11011	none	Letter Shift
1C	11100	U	7
1D	11101	Q	1
1E	11110	K	(
1F	11111	Number Shift	None

Table 1.2 ASCII Character Code

Binary codes are shown in their hexadecimal (HEX) equivalent

HEX	ASCII	HEX	ASCII	HEX	ASCII	HEX	ASCII
00	NULL	20	Space	40	@	60	'
01	SOH	21	!	41	A	61	a
02	STX	22	"	42	B	62	b
03	ETX	23	#	43	C	63	c
04	EOT	24	$	44	D	64	d
05	ENQ	25	%	45	E	65	e
06	ACK	26	&	46	F	66	f
07	BELL	27	'	47	G	67	g
08	BKSP	28	(	48	H	68	h
09	HTAB	29	)	49	I	69	i
0A	LNFD	2A	*	4A	J	6A	j
0B	VTAB	2B	+	4B	K	6B	k
0C	FMFD	2C	,	4C	L	6C	l
0D	CRET	2D	−	4D	M	6D	m
0E	SHOUT	2E	.	4E	N	6E	n
0F	SHIN	2F	/	4F	O	6F	o
10	DLE	30	0	50	P	70	p
11	DC1	31	1	51	Q	71	q
12	DC2	32	2	52	R	72	r
13	DC3	33	3	53	S	73	s
14	DC4	34	4	54	T	74	t
15	NACK	35	5	55	U	75	u
16	SYNC	36	6	56	V	76	v
17	ETB	37	7	57	W	77	w
18	CAN	38	8	58	X	78	x
19	ENDM	39	9	59	Y	79	y
1A	SUB	3A	:	5A	Z	7A	z
1B	ESC	3B	;	5B	[	7B	{
1C	FLSP	3C	<	5C	\	7C	¦
1D	GPSP	3D	=	5D	]	7D	}
1E	RDSP	3E	>	5E	∧	7E	~
1F	UNSP	3F	?	5F	___	7F	DEL

of characters to another. In the table, hexadecimal representations of the binary codes are shown. However, in this case, only the most significant bit of the hexadecimal digit pairs is missing and held low. The ASCII code has been expanded to 8 bits to allow many special graphics and mathematical characters to be added in what is termed the extended ASCII character code. In this text we will use the basic 7-bit ASCII code set presented in Table 1.2. Besides communications applications, the ASCII code is widely used in personal computers as the prime character code.

Example 1.3   What sequence of binary information is sent for the message of example 1.2 using the ASCII code?

Solution: This time the message is straightforward, without the need for shift characters. The letter N (4E or 1001110 in binary) is followed by the period character (2E or 0101110), Y (59 or 1011001), another period (2E), a space (20 or 0100000), and so forth. Although it takes 2 more bits per character to use the ASCII code, unnecessary shift characters are eliminated.

Character codes contain the alphanumeric characters, which include all the printable characters, such as letters, numbers, and punctuation. Additional characters provide **GRAPHICS** and **DATA LINK CONTROL** information used by the communications network. Graphics characters are those that help define how text will appear on a data terminal screen or printer page. They include carriage return, line feed, tabs, and so on. Data link control characters, on the other hand, establish information about the interpretation of a message itself. Examples are start of text (STX), end of transmission (EOT), and device control (DC). STX and EOT are stand-alone control characters since their meaning is defined by the character itself. STX signifies when text begins, and EOT designates the end of a data transmission. On the other hand, the DC character requires a second following character to define its use. The DC character usually alerts the receiver that it should be prepared to respond to some data link command signified by a following character, most often a number, appearing as DC1, DC2, DC3, or DC4. DC functions are defined by the protocol in use, with the DC character signaling the program to interpret the following character (1–4) as having a specific data link control application.

Translating each character code from Table 1.2 into its hexadecimal equivalent, the ASCII code can be grouped as follows:

1. Codes 30–39 represent numbers 0–9.
2. Codes 41–5A represent the 26 uppercase alpha characters.
3. Codes 61–7A represent the 26 lowercase alpha characters.
4. The remaining codes represent punctuation, graphics, and data link characters.

The graphics characters consist of the following:

1. **LINE FEED (LF),** which moves the printed page or display screen up one line;
2. **CARRIAGE RETURN (CR),** which returns the print head or screen cursor to the leftmost position;
3. **TAB,** which allows the print head or cursor to move several spaces along a line;
4. **FORM FEED (FF),** which causes an entire page to be moved up or scrolled until the beginning of the next page is sensed;
5. **BACK SPACE (BS),** which causes the head or cursor to back up one space;
6. **SPACE (SP),** which causes the print head or cursor to move right one space.

The functions of data link characters, which are used to establish the communications link once the hardware link is set, are defined by protocol applications software. This group includes but is not limited to **START OF TEXT (STX), END OF TEXT (ETX), END OF TRANSMISSION (EOT), START OF HEADING (SOH), ACKNOWLEDGE (ACK), DATA LINK ESCAPE (DLE), ENQUIRY (ENQ), SYNCHRONIZATION (SYN), NEGATIVE ACKNOWL-EDGE (NAK), DEVICE CONTROL (DC),** and **END OF TRANSMISSION BLOCK (ETB).**

A third popularly used code was developed by International Business Machines (IBM) Corporation for use with its mainframe computers. It is an 8-bit-per-character code called **Extended Binary Coded Decimal Interchange Code (EBCDIC),** shown in Table 1.3. Eight bits per character allows for 256 possible codes. Notice in Table 1.3, however, that quite a few codes are not used. These gaps in the code occasionally lead to difficulty for users, which may be one reason that ASCII has emerged as a more popular code for personal computer and communications usage. The number of bits (8) was selected to allow the code to be easily expanded in the future. That is, the gaps can be filled in with new characters at any time. In contrast, to expand the ASCII character code set required the code to be redefined as an 8-bit code.

Example 1.4   What is the binary equivalent of the Baudot, ASCII, and EBCDIC codes for the capital letter E? Assume that Baudot is set to interpret the correct set of characters.

Solution: From Table 1.1, the Baudot code for E is 10000. From Table 1.2, the ASCII code for E is 1000101. Lastly, from Table 1.3, the EBCDIC code for E is 11000101. Each of these can be represented as hexadecimal digits by using leading 0s to replace missing binary bits. Thus, the Baudot code for E becomes 10 hex, for ASCII, 45, and for EBCDIC, C5.

NOTE: Throughout the rest of the text binary numbers will be written in groups of four with a space between each group. This is only to aid in reading them. As an example, the binary code for an EBCDIC E would appear as 1100 0101.

Table 1.3 EBCDIC Character Code

Binary codes are shown as their hexadecimal (HEX) equivalents.

HEX	EBCDIC	HEX	EBCDIC	HEX	EBCDIC	HEX	EBCDIC	HEX	EBCDIC
00	NULL	20	DIGSEL	50	&	91	j	D0	}
01	SOH	21	STSIG	5A	!	92	k	D1	J
02	STX	22	FLSEP	5B	$	93	l	D2	K
03	ETX	24	BYPASS	5C	*	94	m	D3	L
04	PNOFF	25	LNFD	5D	)	95	n	D4	M
05	HZTAB	26	ENDBLK	5E	;	96	o	D5	N
06	LWRCASE	27	ESC	5F		97	p	D6	O
07	DELETE	2A	STMESS	60	−	98	q	D7	P
09	RLF	2D	ENQR	61	/	99	r	D8	Q
0A	Repeat	2E	ACK			A1	~	D9	R
0B	VERTAB	2F	BELL			A2	s	E2	S
0C	FMFD	32	SYNC	6B	,	A3	t	E3	T
0D	CARET	34	PNON	6C	%	A4	u	E4	U
0E	SHFTOUT	35	RCDSEP	6D	−	A5	v	E5	V
0F	SHFTIN	36	UPCASE	6E	>	A6	w	E6	W
10	DLE	37	EOT	6F	?	A7	x	E7	X
11	DC1	3C	DC4	7A	:	A8	y	E8	Y
12	DC2	3D	NACK	7B	#	A9	z	E9	Z
13	DC3	3F	SUB	7C	@	C0	{	F0	0
14	Restore	40	Space	7D	'	C1	A	F1	1
15	NewLine	4A		7E	=	C2	B	F2	2
16	BckSp	4B	.	7F	"	C3	C	F3	3
17	Idle	4C	<	81	a	C4	D	F4	4
18	Cancel	4D	(	82	b	C5	E	F5	5
19	EndMed	4E	+	83	c	C6	F	F6	6
1A	UnitBkSp	4F			84	d	C7	G	F7	7
1C	IntF1Sep			85	e	C8	H	F8	8
1D	IntGpSep			86	f	C9	I	F9	9
1E	IntRcSep			87	g				
1F	IntUnSep			88	h				
				89	i				

Note: Unused codes have been omitted from this chart.

Digital information in serial form moves at a distinct data rate. It takes time to send information, one bit at a time, from one place to another. Data are sent serially to reduce the number of transmitting lines to a single pair (electrically active and a return line). The rate at which digital information is sent or received is called the BIT RATE, whose unit is **BITS PER SECOND (bps).** Each character code uses a fixed number of binary bits called a **WORD,**[1] whose combinations of highs and lows represent the various characters in the code. Baudot, which uses 5 bits to represent each character, has a data word of 5 bits. ASCII's word is 7 bits wide, and EBCDIC uses an 8-bit word for each character.

Basic data rate, in bits per second, defines the rate at which the binary bits move from source to destination. It does not take into consideration any factors of bit grouping or other methods used to speed up overall transmission rates. Later in the text there are discussions involving the grouping of binary data bits into **SYMBOLS.** A symbol is basically any single measurable electrical occurrence. It can be a voltage, a number of cycles of a sine wave at a given frequency, a sine wave sent with a particular phase compared to a given reference, a current, or a resistance. What is most significant for the field of communications is that a symbol can be created through various processes from a grouping of binary bits. As an example, the ASCII code, instead of being represented as a 7-bit code, could be set up to use a different voltage level for each character. In other words, 1.2113 V could indicate a letter C and 1.2114 V a letter D. We are not concerned with the practicality of such a system, but since there is a unique 7-bit binary code for each of those characters, we can also say there would be a correlation between the ASCII code and the voltage level—that is, 100 0011 (43H)[2] for a C is equivalent to 1.2113 V, and 100 0100 (44H) for a D is equivalent to a 1.2114 V. The bottom line is that 7 bits can be used to generate a single symbol (in this case a voltage level). The single symbol can be sent as one entity, which is faster than sending the 7 individual bits that are its equivalent.

A system that converts the 7 bits of an ASCII code in the previous discussion into a single symbol would be sending data out at a given bit rate (say 4900 bps)

---

[1]The conventional definition of *word* as applied to digital information is a fixed number of binary bits indicating a selected size of information. Thus a computer system using an 8-bit bus operates with a byte-size data word of information. Other systems using, say, an 11-bit bus, operate using 11-bit words. With the advent of 16- and 32-bit microprocessors, manufacturers have taken liberties with the term *word*. For example, Motorola's user manuals for the 68000 family microprocessor specifically define a *word* as having 16 bits of data and a *long word* as containing 32 bits. However, as it later develops, Motorola, like other makers of microprocessors, is not consistent. In the user manual for its newer processors, *word* is redefined as containing 32 bits! This is mentioned so that students who may be studying these processors concurrently with data and digital communications are made aware of the conflicting use of the term *word*. In this text the term refers to its original meaning, that is, binary data in a fixed number of bits.

[2]An H placed after a number indicates that number is hexadecimal.

while actually sending the message at a rate of 700 symbols per second (bit rate divided by the number of bits per symbol). Why do this? System transmission rates are limited by the system's bandwidth. For example, the practical bandwidth of the telephone system is approximately 3 kHz and lies between 300 and 3 kHz. Data bit rates are limited by the 3,000-Hz upper end of that bandwidth. Once the 7-bit codes are converted to symbols, though, the symbols can be sent up to the 3,000-Hz rate. At a rate of 7 bits per symbol, that would allow the system to send data at a rate of 3,000 symbols per second $\times$ 7 bits per symbol or 21,000 bps.

A more common term for symbols per second is the **BAUD** for the system. A broader interpretation of baud is the rate of information transfer. This definition applies because much can happen to the original data before they reach the form at which they are actually transmitted.

The tendency in studying data communications and, to some extent, its application in the field is to dismiss the differences in baud and bit rate. Many authors and, to a lesser extent, persons working in the field use the terms *bit rate* and *baud* to mean the same thing. For many low-speed applications the differences between the two are insignificant. Thus a 300- or 1200-bps modem used with many personal computers is often termed a 300- or 1200-baud modem. There is no problem here, since at these rates one symbol is produced for each data bit, resulting in an equivalent baud and bit rate. Until the distinction becomes significant in a particular area under discussion, this text will consider baud and bit rate as being similar. This is not to minimize the differences, but a good many concepts are unaffected by them. Analysis of data transfer efficiency, rates, and bandwidth limitations is specified using baud; in that context the difference is critical.

● 1.7 ASYNCHRONOUS AND SYNCHRONOUS DATA

Serial data are transmitted in one of two forms, synchronous or asynchronous. Synchronous data require a coherent clocking signal between transmitter and receiver called a **DATA CLOCK** to synchronize the interpretation of the data sent and received. The data clock is extracted from the serial data stream at the receiver by special circuits called **CLOCK RECOVERY CIRCUITS.** Once the clock is recovered at the receiving end, bit and character synchronization can be established. Bit synchronization requires that the high and low condition of the binary data sent matches that received, and is not in an inverted state. Character synchronization implies that the beginning and end of a character word is established so that these characters can be decoded and defined. Overall synchronization is maintained by the clock, recovered from the message data stream itself.

Asynchronous data incorporate the use of extra **FRAMING BITS** to establish the start and ending (stop) of a data character word. A clocking signal is not recovered from the data stream, although the internal clocks of the transmitter and receiver must be the same frequency for data to be correctly received. The receiver

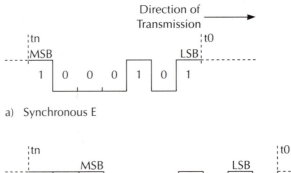

a) Synchronous E

b) Asynchronous E

Figure 1.7 Comparing Data Formats for ASCII E

responds to the data stream when it detects a **START BIT.** A data character is decoded and defined after the **STOP BIT** is received and confirmed. Asynchronous data are easier to detect and synchronize, but the efficiency of data transmissions is reduced by the addition of framing bits as overhead (no message data) bits. Although a clock signal is not recovered from the data stream, internal clocks at the transmitter and receiver must be very close in frequency so that the received data are sampled at the same rate as they were transmitted. This sampling rate will establish bit synchronization, while the detection of the start and stop framing bits establishes character synchronization.

A comparison of a single character using the two data types is shown in Figure 1.7. For purposes of illustration, the ASCII code for the letter E (100 0101, or 45 in hexadecimal) is used. The usual order of transmission is to send the **LEAST SIGNIFICANT BIT (LSB)** first, as in this illustration, which shows the character as it appears sent as a serial stream with respect to transmission direction. The number of synchronous bits used is 7, the actual code bits for the character itself. The number of framing bits used for asynchronous data varies depending on the stations in the communication link. For this example, 1 start and 2 stop bits have been selected. This adds 3 additional bits to the character word. Ten total bits are required to send the letter E using asynchronous data. Efficiency of transmission is defined as the ratio of the number of message bits to the total number of transmitted bits:

$$\text{efficiency} = \frac{\text{data bits}}{\text{total bits}} \times 100\% \qquad (1.1)$$

Example 1.5   Compare the efficiency between sending the character ASCII E as asynchronous and synchronous data.

Solution: The efficiency of sending E as synchronous data is 100% since all of the bits are used as message bits. The asynchronous character uses 7 of the total 10 bits as message bits. The efficiency of that transmission is 7/10 × 100%, or 70%.

In example 1.5 it takes more time to send the asynchronous character than the synchronous because of the additional overhead bits. When considering the efficiency of transmission, more than one character of a message must be involved. The overall efficiency of a transmitted message uses the total number of data bits in the message divided by the total number of bits of that message. Messages include error-detection characters, to determine whether an error in the message has occurred, and other overhead characters. For synchronous transmissions, these overhead characters are used to establish the initial clock recovery and synchronization before data can be properly received. They are called **SYN** characters for synchronization characters. Since they are not actual data characters, they are not included in the data bit total but are included in the message total count, reducing the overall transmission efficiency. For large messages this reduction in efficiency is minimal.

Example 1.6   Compare the overall efficiency of sending a 600-character message using ASCII code as a synchronous data stream with two SYN characters and an asynchronous data stream using 1 start and 1 stop bit. Both transmissions include a single error-detection character.

Solution: The total number of message bits for both transmissions is 600 × 7 = 4,200. For the synchronous data, three additional characters are added (two SYN and one error-detection character), for a total of 603 × 7 = 4,221 bits. The efficiency for this transmission is:

$$\text{efficiency} = \frac{4,200}{4,221} \times 100\% = 99.5\%$$

The asynchronous data stream includes only one additional character for error detection, but each character contains 9 bits (1 start, 7 data, and 1 stop bit each), for a total bit count of 601 × 9 = 5,409 bits. The efficiency for that transmission is:

$$\text{efficiency} = \frac{4,200}{5,409} \times 100\% = 77.6\%$$

While it appears that the use of synchronous data is preferable because of the improved efficiency, other factors need to be considered before deciding which type of data to use. For instance, synchronous data require the implementation of additional circuitry for clock recovery. In general, the less expensive asynchronous data are used for lower data rates when time is not crucial. The most common application of asynchronous data is for home computers and modems that operate at 300 or 1200 bps. For higher data rates, the more efficient synchronous data are used. Systems employing synchronous data are very time conscious. Most local area and other multipoint networks use synchronous data.

● 1.8 BINARY DATA SIGNAL FORMATS

Besides different character codes and data types (asynchronous and synchronous), digital data can be transmitted or coded into different electrical signal formats. Each of the following forms has its own advantage and/or use. They are illustrated in Figure 1.8, and each will be discussed in the sequence shown. Each data signal format, sent as a serial stream of data, may be interpreted as generating a square wave signal whose frequency varies according to the changing bit pattern. Depending on the signal format type, the frequency of the "square wave" usually gets lower as the number of consecutive 1s or 0s increases. This is not the case for all of the signal formats, as we will see, but this serves as an example of the meaning of a fundamental sine wave frequency. Suppose that there is a serial stream of alternating 1s and 0s. Drawn out, this would appear as a signal whose level constantly switches between two voltages. In essence, this signal appears as a square wave. Now suppose there is a serial stream that alternates pairs of 1s and 0s. The square wave signal this creates is one-half the frequency of the first signal.

Square waves are actually a composite of a fundamental sine wave and all of its odd harmonics. The maximum power of the square wave is in its fundamental sine wave. Systems transmitting data in any one of the following signal formats must have a bandwidth high enough to accommodate the fundamental sine wave so that the data stream can be sent through that system and the bit rate replicated at the receiving end. The highest fundamental sine wave cycle is shown as a dotted line in each data signal format of Figure 1.8, occurring during the portion of the data stream that generates it. Notice that for other 1 and 0 combinations in any given format, the sine wave representing the change of data states is either the same or at a lower frequency than the highest fundamental rate.

The following discussion centers on the relationship between a system's bandwidth and the bit rate. Other factors are dependent on the system's bandwidth besides the bit rate. For some types of modulation, changing from digital- to analog-type signals requires the use of one or more carrier signals. The bandwidth must be wide enough to accommodate those signal frequencies and the rate at which switching occurs (usually based on the data or bit rate). These issues are discussed

DIGITAL & DATA COMMUNICATIONS

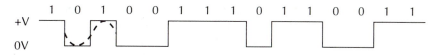

1 0 1 0 0 1 1 1 0 1 1 0 0 1 1

a) Non-Return to Zero (NRZ). [*Note*: Non-Return to Zero Bipolar (NRZB) —Same as NRZ Except Logic 0 is –V Instead of 0V.]

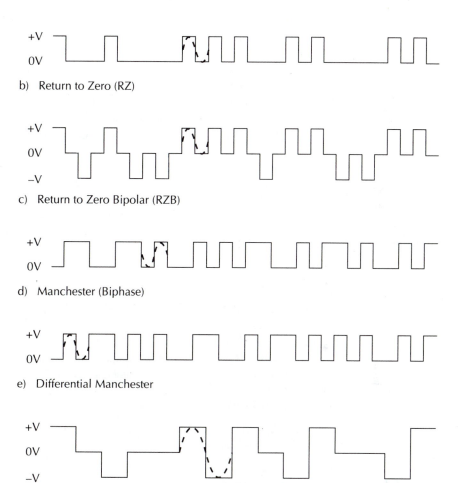

b) Return to Zero (RZ)

c) Return to Zero Bipolar (RZB)

d) Manchester (Biphase)

e) Differential Manchester

f) Non-Return to Zero Inverted (NRZI) Bipolar

Figure 1.8 Binary Data Forms

in chapters 5 and 6 as we explore the modulation techniques used in data communications. Our concern at this point is the relationship between bit rate and system bandwidth.

● 1.8.1 Non-Return to Zero

The **NON-RETURN TO ZERO (NRZ)** signal form is the basic two-level binary form. A logic 1 is at one level (+V in Figure 1.8a), and logic 0 is at a ground reference voltage level (0V in Figure 1.8). This is the form that most students are familiar with from preceding digital courses. This is also the type of square wave signal discussed in the preceding section to illustrate the relationship between the square wave frequency and the sequence of alternating data bits. Also illustrated in Figure 1.8 is the fundamental sine wave for the highest switching data rate used for each data form. For NRZ this rate occurs when the data consist of alternating 1s and 0s. This fundamental sine wave is the highest rate of change to be used by a system employing a given data type and must not exceed the upper limit of that system's bandwidth. Generally, this binary data signal format is the easiest to produce since it requires that a device is either on, generating a high level for a logic 1, or off, generating ground or 0V for a logic 0.

Example 1.7   A system with a bandwidth of 8,000 Hz uses non-return to zero data. What is the highest bit rate at which data can be sent on this system?

Solution: The highest fundamental frequency that can be used in the sample system is 8,000 Hz. Note in Figure 1.8 that 2 bits of NRZ data alternating states between 1 and 0 generate the highest fundamental frequency. Since 2 bits occupy one cycle of that fundamental sine wave, the data rate is twice that of the fundamental frequency, or 16000 bps.

● 1.8.2 Non-Return to Zero Bipolar

A very similar digital signal form that is not pictured in Figure 1.8 is the **NON-RETURN TO ZERO BIPOLAR (NRZB).** Its form is identical to that of the NRZ form, except that the logic 0 level is at a −V instead of 0V. This form is used instead of the NRZ form when recording data onto a magnetic surface. The opposing voltages for 1s and 0s cause the magnetic material to become magnetized in opposite directions (that is, the magnetic lines of flux are in one direction for a current passing through the coil in a recording head in one direction and in an opposite direction for current generated through the coil in the head in the opposite direction). The frequency for the fundamental sine wave of a NRZB is the same as the NRZ form, so the system bandwidth requirements to handle a specific bit rate are the same. That is, the maximum bit rate is twice the bandwidth.

● 1.8.3 Return to Zero

The **RETURN TO ZERO (RZ)** signal format (Figure 1.8b) uses the 0V level for a logic 0 and the +V level for a logic 1. This form, however, causes the logic 1 data bit to return to zero midway through its bit time. This return to zero transition is incorporated to prevent the data stream from having long periods of time at the +V level in the event that a continuous string of 1s is required to be sent. The fundamental sine wave frequency occurs with two consecutive 1s since the transition at midbit and the return to +V for the next bit frame the sine wave cycle. This gives a one-to-one relationship between bit rate and the fundamental sine wave rate. For the system bandwidth of example 1.7, this would limit the data rate to a maximum of 8000 bps, which is one-half of the non-return to zero format. This signaling format is frequently used to aid in a system that recovers clock timing signals from the digital serial data stream. The recovering of the clock signal is aided by the additional level transition that occurs in the center of each data bit when using the return to zero signal format.

● 1.8.4 Return to Zero Bipolar

The next signal format (Figure 1.8c) is that of **RETURN TO ZERO BIPOLAR (RZB)** digital data. As with the NRZB, logic 1s and 0s are represented by opposing voltages (+V and −V). The change this time is that at the midpoint of each data bit, the voltage is dropped to 0V, hence the name return to zero. This form not only supplies the characteristic of opposing voltages but also includes the generation of a transition (change of level) in the middle of each data period. Synchronous systems find this advantageous to assist in recovering a clock from the data stream. Because there is a regular change of level, a clock recovery circuit can synchronize the clock to the center of each data bit in the data stream.

Example 1.8   What is the maximum bit rate for a return to zero bipolar transmission on a system whose maximum bandwidth is 8,000 Hz?

Solution: The highest fundamental sine wave frequency for the return to zero bipolar format occurs with each data bit whenever two consecutive 1s or 0s are sent (see Figure 1.8c). The transition to zero level followed by a return to the previous logic level frames in a complete cycle. This makes the fundamental frequency have the same value as the bit rate. For this example, the maximum bit rate is 8000 bps. (Compare this with example 1.7 for the non-return to zero format.) The price for improving clock recovery is that for the same bandwidth system, the maximum bit rate that data can be sent is halved when compared with the non-return to zero format.

● 1.8.5 Manchester Encoding or Biphase

MANCHESTER ENCODING (Figure 1.8d) is a coding signal form rather than an actual digital signal format. The data stream is fed through a circuit that complements (inverts) the first half of the data bit. The second half is left uncomplemented. Note the first logic 1 appears with a low level during the first half of the first bit period and a high level during the second half. The purpose in doing this is similar to the RZ form: to create a regular transition at the center of each data bit to aid in clock recovery. Note that the clock is synchronized to the transition at the center of each bit and that the data is detected toward the trailing half (the uncomplemented portion) following the clock transition of the data period.

● 1.8.6 Differential Manchester Encoding

DIFFERENTIAL MANCHESTER ENCODING (Figure 1.8e) was developed to move the detection of the actual data level from the second half of the data period to the beginning of the data period. As with Manchester encoding, each data bit has a transition at the middle of the bit period. For differential Manchester encoding, however, logic levels are formed by comparing adjacent data bits. If the second bit in the comparison is a logic 1, the level of the first half of its bit time is the same as the preceding bit's second-half level. If the second bit is a 0, the first half of the second bit time is inverted (the complement) of the first bit's second-half level. In both cases the second half of the second bit's time period is the complement of the first half.

Example 1.9   Show differential Manchester encoding for the bit pattern 101001110110011 as shown in Figure 1.8e.

DIGITAL & DATA COMMUNICATIONS

Solution: Look at the differential Manchester data stream in Figure 1.8e to follow this example. The first bit (logic 1) is assumed to start low and then is complemented during its second-half time period. This assumption is made arbitrarily to establish a starting point. The second bit is a logic 0. Since the second half of the first data bit is high, the first half of the second (logic 0) bit's time is set low (complement of the first bit's second-half level). The second half of the second bit is the complement of its first half (high level). The third bit is a logic 1, which means that its first-half level will be unchanged from the preceding bit's (second bit—logic 0) second-half level. Its second half is complemented from its first half as before, and the process is continued throughout the data stream. Whenever a logic 0 appears as the following bit, a level change is made. When the following bit is a logic 1, no level change is made. All bits, after the first-half level is determined, are complemented during their second-half time period. The fully encoded message is shown in Figure 1.8e.

● **1.8.7 Non-Return to Zero Mark Inversion**

The final data signal format shown is the **NON-RETURN TO ZERO MARK INVERSION (NRZI)** bipolar form of Figure 1.8f. The zero level is reserved for logic 0 data bits. Logic level 1 data bits appear as alternating voltage levels. The first logic 1 is $+V$, for instance. Then the second is $-V$, the third one will be $+V$ again, and so forth. This format adds the ability for error detection. Two consecutive $+V$ or $-V$ levels signal the receiver that one or more logic 1 bits are incorrect, since each logic 1 bit must have the opposite voltage level. The fundamental sine wave frequency for this form is produced by two consecutive 1s and uses the full time period of both data bits. Like the non-return to zero format, the maximum bit rate is twice the system bandwidth.

SUMMARY

Data communications systems, no matter how complex, require fundamental functional units—terminal equipment, which includes application program processors and computers, line control units that direct traffic to a specific peripheral at a station, UARTs for converting computer parallel data to serial data, and standard interface connections. The other overall category of equipment is the data communications equipment used to send out the terminal's data onto the transmitting medium. This equipment includes devices such as modems, which convert the serial digital information into analog signals or tones to be used on media such as telephone lines, and CODECs, which code binary data before converting those codes into analog signals.

To transmit information as digital data, binary codes are required to represent each of the message's characters. Three of the most popular codes in use today are Baudot, ASCII, and EBCDIC. The signal types of digital information can take one of several forms depending on the system's bandwidth and application requirements. These are the basic digital signal form, called return to zero, those that are used to aid clock recovery for synchronous data, such as bipolar return to zero and Manchester encoding, and signal types used to incorporate a measure of error detection within the form (non-return to zero inverted).

GLOSSARY

Alphanumeric characters—Any printable characters, including letters, numbers, and punctuation symbols.

Asynchronous data—Digital data that are not synchronized with a clocking signal.

Baud—The rate at which digital information is transferred from one location to another.

Bit rate (or **bits per second** [bps])—Rate at which binary bits are transferred.

Bridge—Electronic connection between two similar networks.

Bus—Network topology in which all the stations are connected to a common line.

Character code—Codes that define groups of binary bits as representing particular characters.

Character word—A fixed group of binary bits that represent a character.

Clock recovery circuits—Electronic circuits in synchronous data systems used to extract clock synchronization form the data stream.

Data clock—Clock signal used to synchronize data transfers between transmitter and receiver.

Data communications—Communication systems in which the source of information is in digital form.

Data link or link—Actual physical or logical connection between two communicating stations.

Data link control characters—Those characters in a character code that are used to establish a data link.

Digital communications—Communications systems in which the form of information sent on the transmitting medium is in digital form.

Duplex—Defines the type of interactive communication system in use, which can be simplex, half duplex, or full duplex.

Framing bits—Additional bits in an asynchronous data stream that define the beginning and end of character word.

Full duplex—Denotes simultaneous two way communication.

Graphics characters—Characters defined in a character code that are used to govern how and where a message appears on a screen or paper. Examples are carriage return, tab, and line feed.

Half duplex—Denotes communication between two stations in two directions but not at the same time.

Hub or star—Network topology in which each station is connected to a central, controlling station.

Line control unit (LCU)—Circuitry that determines the source device of a message to be sent by a station.

Link—Communication connection between two stations.

Manchester and differential Manchester encoding—Encoding schemes for digital data in which a transition is created in the center of each data bit time. This is to assist in clock recovery by providing a regular change of state with which to synchronize clocking.

Mark—Logic 1 level.

Medium—The physical carrier of a message between stations.

Modem (modulator/demodulator)—Converts between digital and analog forms of the data being sent and received.

Multipoint system—Communications network using more than two stations.

Originating station—The station that initiates the communications link.

Parallel data—Data word in which all of its binary bits are transferred simultaneously.

Primary station—The controlling station in a data link.

Protocol—Set of rules to establish a successful transfer and interpretation of data.

Remote station (also **answering station**)—The nonoriginating station in a communications link.

Ring—Communications network in which stations are linked together sequentially with the last station returned to the first, thereby creating a ring configuration.

RS232C—Recommended interfacing standard between terminals and modems.

Secondary station—The slave or noncontrolling station in a communications link.

Serial data—Sequential transfer (one at a time) of the binary bits of a data word.

Simplex—Communications between two stations in one direction only.

Space—Logic 0 level.

Start and stop bits—Used for framing the beginning and end of an asynchronous character.

Station controller (STACO)—Circuitry that determines the final device destination of received data.

Symbols—Electrical quantity that represents a group of data bits.

SYN—Character used in a synchronous data system for timing synchronization between transmitted and received data.

Token—Short message used to pass communication control from one station to the next in a ring network.

Topology—Defines the configuration of a communications system. Examples are bus, star, and ring.

Word—A specific number of binary bits grouped to represent information.

QUESTIONS

1.1. What is the difference between originating and answering stations? Page 139

1.2. Define primary and secondary stations.

1.3. How do line controllers and station controllers differ?

1.4. What do DTE and DCE stand for? Explain the differences between the functions of each.
1.5. What are the principal jobs of UARTs and modems?
1.6. Define and give three examples of a medium.
1.7. What are the three types of multipoint topologies? How do they differ, functionally?
1.8. What is the purpose of the RS232C interface standard?
1.9. Which of the following examples use local, wide, or metropolitan area networks?
    a. Four veterinary clinics in Gainsborough
    b. Hanson manufacturing plant
    c. Stock exchange
    d. Delphi University campus
    e. Arizona department of motor vehicles
    f. Internal Revenue Service
1.10. Define character code.
1.11. What are the three types of characters represented by a character code such as ASCII? Give an example of each.
1.12. What is the advantage and disadvantage of using the Baudot code?
1.13. Give two differences between asynchronous and synchronous data. What are the prime advantages of each data type?
1.14. Which bits of an asynchronous data word are used for framing?
1.15. Why are synchronous data transmissions more efficient than asynchronous?
1.16. What type of system uses transmission of data in one direction?
1.17. What is the difference between half duplex and full duplex?
1.18. Which binary data formats specify at least one change of state for each data bit? What is the significance of this guaranteed regular transition?
1.19. Which binary data formats have fundamental sine waves whose frequencies match the data rate?
1.20. What is the prime difference between Manchester and differential Manchester encoding schemes?
1.21. In what way does the non-return to zero mark inversion binary data format differ from the other formats?

PROBLEMS

1.1. What are the binary codes for each character below as used in the Baudot, ASCII and EBCDIC character codes?
    a. B   b. f   c. !   d. line feed   e. STX   f. NULL
    g. 5   h. "   i. shift   j. EOT
1.2. Show the binary data stream for the asynchronous ASCII message Do It! using 1 start and 1 stop bit per character for framing.
1.3. Show the synchronous EBCDIC message for This time? using two SYNC characters for synchronization.
1.4. What is the efficiency of a 256 ASCII character message using asynchronous data with 2 stop bits and a single error-detection character?

1.5. What is the efficiency of a 256 EBCDIC character message using synchronous data, a single SYNC, and a single error-detection character?

1.6. A system converts EBCDIC characters into symbols and transmits the symbols at 8848 baud. Each symbol represents a single EBCDIC character. What is the bit rate of the system? What is the efficiency of transmission if two SYNC characters are required for synchronization and two for error detection?

1.7. The bit rate for a system is 1200 bps. What is the fundamental sine wave frequency for each of the data format types of Figure 1.8?

1.8. Show how the ASCII character G would appear in each of the binary data formats of Figure 1.8. Start with the least significant bit on the left as used in the figure.

1.9. Compare the time it takes to send 256 bytes of data using an 8-bit parallel system versus a serial system. Both transfer data at a rate of 1200 transfers per second.

DESIGN PROBLEMS

1.1. Another type of character code is the **AUTOMATIC REQUEST FOR RETRANSMISSION (ARQ),** which uses 7 bits per character. What is unique about this code is that every 7-bit code has exactly three logic 1 bits and four logic 0 bits. Explain how a code like this is used to detect errors in transmission without requiring an additional error-detection character. How does the ability to detect characters affect the efficiency of data transmission?

1.2. Design a bus topology using one primary and four secondaries. Include the functional blocks used in Figures 1.1 and 1.2. Set up a station-addressing scheme that allows the primary to send messages to the secondaries in one of three fashions: all stations at once, pairs of stations at a single time, or each station individually. In which functional block would this addressing be found?

1.3. Illustrate a communications network that uses a star topology as its main core. One of the secondaries in the star serves as a subprimary for a bus topology. Another secondary is part of a ring, while four other secondaries in the star are individual stations.

1.4. Design a data modulator that converts a serial data stream into 3-bit-per-symbol data stream. The input data rate is to be 3000 bps, and the output symbol rate is to be 1000 symbols per second (sps). The design is to be a functional block diagram indicating the types of circuits used in each block. As an optional lab project, extend the design to actual circuits, build them, and confirm the circuit operation.

1.5. It is desirable to transmit messages at a rate of 9600 bps onto a system whose bandwidth is 50 to 2,700 Hz. Select the parameters for such a system. They should include the type of binary data format; asynchronous or synchronous transmission; number of bits per symbol; simplex, half duplex, or full duplex; and any other information you feel is pertinent. Justify or explain each of your choices.

Answers to Odd-Numbered Questions

1.1. The originating station initiates the communications link. The answering station responds to the originator's call.

1.3. Both line controllers and station controllers select the peripheral device being accessed at a particular station. Line controllers, however, also control the establishment and use of a data link.

1.5. UARTs convert parallel digital data to serial form, and modems convert digital data to analog form.

1.7. Bus, star, and ring. A bus has all the stations connected to a common data line. The primary controls who it is communicating with. In the star the primary is a central station communicating with secondaries located on the spokes of the star. In a ring, data is passed from one station to the next until it completes the circuit of the ring back to the originating station.

1.9. a. MAN   b. LAN   c. WAN   d. LAN or MAN, depending on size of the campus   e. MAN   f. WAN

1.11. Alphanumeric (examples are any letter, number, or punctuation mark), graphics (examples are line feed, carriage return, tab, etc.), and data link control (examples are STX, EOT, ETX, DC, etc.)

1.13. Synchronous data requires clocking and is transferred at higher data rates. Asynchronous data requires start and stop framing bits. The circuitry for bit and character synchronization is simpler since it does not have to recover a clock for synchronization.

1.15. Synchronous data transmissions are more efficient because it takes fewer total bits to send a synchronous message compared with an asynchronous one.

1.17. Both allow two-way data transmission. Half duplex is one way at a time, while full duplex is both ways simultaneously.

1.19. Return to zero, return to zero bipolar, Manchester, differential Manchester.

1.21. The voltage levels of the logic 1 states are inverted for each logic 1 data bit.

Solutions to Odd-Numbered Problems

1.1.	Character	Baudot	ASCII	EBCDIC
a.	B	14H	42H	C2H
b.	f	16H	66H	86H
c.	!	16H	21H	5AH
d.	line feed	08H	0AH	25H
e.	STX	none	02H	02H
f.	NULL	00H	00H	00H
g.	5	01H	35H	F5H
h.	"	11H	22H	6FH
i.	shift	1FH 1BH	0EH 0FH	0EH 0FH
j.	EOT	none	04H	37H

1.3. SYNC   SYNC   T   h   i   s   space

01001100010011001100111000010001100100010100010100000010 . . .

t   i   m   e   ?

11000101100100010010100110100000111110110

1.5. Total message characters = 8 × 256 = 2048. Total bits = 8 × 258 = 2,064. Efficiency = 2,048/2,064 × 100% = 99.22%

1.7. Non-return to zero and non-return to zero inverted, 600 Hz; all others, 1,200 Hz

1.9. 1200 transfers per second converts to one transfer every .833 ms. Transferring 256 bytes, 8 bits at a time, requires 256 transfers, or 256 × .833 ms = .2133 seconds. To transfer this data serially, 1 bit at a time, requires 256 × 8 or 2,048 transfers for a total transfer time of 2,048 × .833 ms = 1.707 seconds.

Solutions to Odd-Numbered Design Problems

1.1. ARQ has error detection built in. All characters must contain exactly 3 logic 1 bits. Any character detected by the receiver as containing other than 3 bits has an error in it. These errors are corrected by requesting the sender to retransmit the bad data.

1.3. See solution, Figure 1.9.

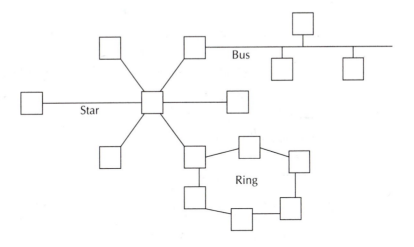

Figure 1.9 Solution to Design Problem 1.3.

Two • The Telephone System

OBJECTIVES

After reading this chapter, the student will be familiar with the basic parts of the telephone system, its specifications, and its impairments. Topics in this chapter include the following:

1. General operation of the **TELEPHONE COMPANY** system
2. Telephone system terminology and specifications
3. Line conditioning and its uses
4. Telephone line and equipment impairments that result in data errors

ACRONYMS

bps—**B**its **P**er **S**econd

BW—**B**and**W**idth or bandstop

CMOS—**C**omplementary **M**etal **O**xide **S**emiconductor

dB—**D**eci**B**el

DDD—**D**irect **D**istance **D**ialing

DTMF—**D**ual **T**one **M**ultiple **F**requency

IC—**I**ntegrated **C**ircuit

LSB—**L**east **S**ignificant **B**it

SNR—**S**ignal-to-**N**oise **R**atio

sps—**S**ymbols **P**er **S**econd

TELCO—**TEL**ephone **CO**mpany

TTL—**T**ransister-**T**ransistor **L**ogic

WATS—**W**ide **A**rea **T**elephone **S**ervice

40

● 2.1 INTRODUCTION

One of the first problems that faced the designer of digital and data communications systems was that of developing a medium for the physical handling of information transfers. Information in this sense includes text, numerical values, graphics, and data link control data, each of whose characters can be represented using one of the character codes discussed in chapter 1. The actual electronics required to handle the transfer was largely dependent on the medium used to carry that information. The system or method selected had to be cost-effective as well as practical to use. Requirements to interconnect numerous locations with one another, many of which were spread over large distances, became a priority consideration. The telephone system was selected primarily because it was there. That sounds simplistic and direct—it is and was. The choice was to either design circuits to work on the existing telephone system or develop a complete and new interconnecting network to handle the transfer of digital information. Time and money dictated using the existing telephone network, although this decision was not without problems—the telephone company was pretty autonomous in those days. It could afford to dictate requirements and costs for the use of its lines and equipment. On the other hand, those in control of the telephone company wisely detected a source of revenue that could only grow.

Once the choice of using the telephone network was made, telephone line parameters, specifications, and limitations had to be studied and incorporated into the data communications equipment design. Telephone lines are being used to carry a large portion of data communications traffic today. It is as important for the data communications student and user to be aware of telephone system specifications, limitations, and impairments as it was to the original system designers. These details affect the use and upkeep of numerous communications links and systems.

● 2.2 THE TELEPHONE SYSTEM IN GENERAL

Telephone calls are initiated by **SUBSCRIBERS** (users) by going **OFF HOOK**—that is, by lifting the handset of the telephone off its cradle or hook. Physically, to go off hook is the act of lifting the handset from its cradle, causing a switch to close. This switch completes a direct current (DC) path (**LOCAL LOOP**) between the telephone set and the local switch office (Figure 2.1). The local station detects the off-hook condition by sensing a drop in voltage from a -48 V battery source to less than -10 V on the subscriber's line caused by the change from an unloaded line (switch open) to a loaded line (switch closing the current loop). The switch station then returns a dial tone, which is a mixture of two sine wave signals, one at 350 Hz and the other at 440 Hz. Upon hearing the dial tone, the user

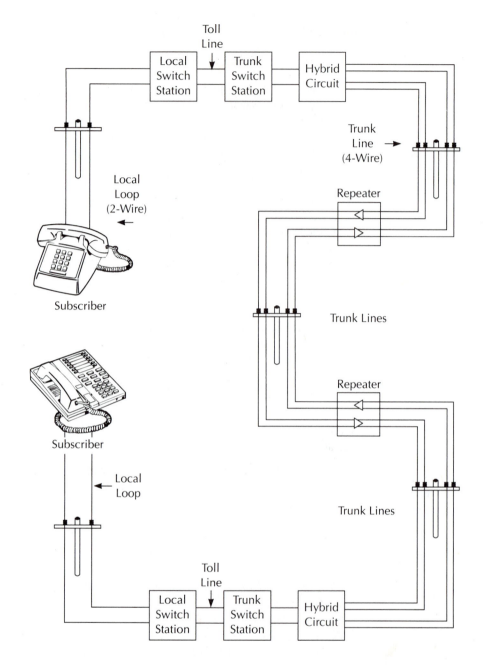

Figure 2.1 Telephone System Elements

can begin to dial the desired number. One way this is done is by using a rotary dial. A number is selected and the dial rotated to the stop position. Upon release, the dial returns to a rest condition. As the dial is returning to this position, a switch connected to the dial is opened and closed as each finger hole in the dial passes the stop position. This opening and closing of the switch causes a current path to be made and broken, generating current pulses at a rate of ten pulses per second (pps). The number of pulses generated is determined by the number selected by the user. The rate at which the dial returns, and hence the pulse rate, is physically fixed. Try forcing the dial to return faster, and you will find that there is a significant force resisting you. Figure 2.2 shows the general specifications and timing diagram for rotary dial. It begins with the on-hook condition (no current), followed by an off-hook (high level, indicating the presence of circuit current). Next on the diagram is an example of a digit (3) set of pulses. There is an interdigit period between each digit dialed. Finally, after the last number is dialed, the line is active and the switch station begins to connect the call. The caller hears the ringing tone, and the called subscriber hears his or her telephone ring. These do not occur at the same time. For instance, a caller may be startled to hear the called party's voice without having heard a ringing tone. This is because there are two different circuits generating the ring signals. One generates a low-power tone back to the caller, while the other sends a large signal ringing voltage to activate the handset ringer.

The newer **TOUCH TONE** (registered trademark of AT&T) dialers use a keypad that generates two simultaneous tones for each number pressed. The name for the method used to send and detect these two tones is called **DUAL TONE MULTIPLE FREQUENCY (DTMF).** Figure 2.3 illustrates the Touch Tone pad and the frequencies of the tones associated with each button. Pressing a particular number on the pad causes a tone at the frequency to the right of the button in Figure 2.3 and one at the frequency indicated below the button to be sent to the local switch station. At the local switch station, circuits detect the tone and decipher the numbers that generated them.

Example 2.1   What two tones are sent on the telephone lines when button 9 is pressed?

Solution: Pressing button 9 causes an 852-Hz (to the right in Figure 2.2) and a 1,477-Hz (at the bottom in Figure 2.2) tone to be sent to the switch station.

Specifications for the power level, interdigit delay, and frequency deviation of the DTMF tones are listed in Table 2.1. The required parameters sent by the transmitter (caller) to the receiver (switch station) are shown in the left and right columns, respectively. The parameter name is listed in the center.

If the call is a local one, it is completed by being connected to another local loop at the local switch station. The switch station senses whether the called party

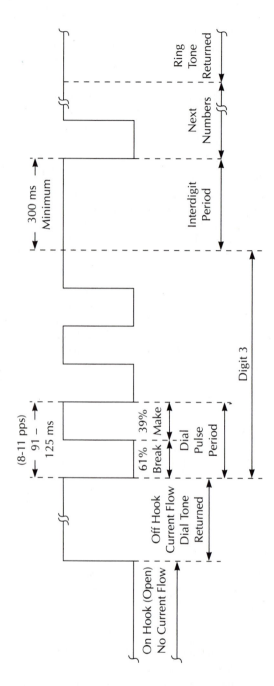

Figure 2.2 Line Characteristic During Dialing

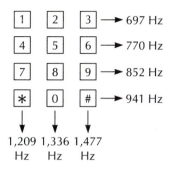

Figure 2.3 Dual Tone Multiple Frequency (DTMF) Touch Tone Pad and Associated Frequencies.

is in use (the phone is off hook) or is free. Upon detecting an off-hook condition, a busy tone, which is a mixture of 480-Hz and 620-Hz tones, is returned to the originating caller. It is sent altering half a second on and half a second silence. If the line is free, a ring signal is sent to the called station. It consists of a 20-Hz sine wave at a voltage between 90 and 120 V. This voltage is needed to operate the electromechanical ringing relay (also known as a ringer coil) that strikes the bell in

Table 2.1 DTMF Specifications

Transmitter	Parameter	Receiver
−10 dBm	Minimum Power Level (single frequency)	−25 dBm −55 dBm Optional
+2 dBm	Maximum Power Level (frequency pair)	0 dBm
4 dB	Maximum Power Difference Between Frequencies in a Pair	+4 dB −8 dB
50 ms	Minimum Duration of Digit	40 ms 23 ms Optional
45 ms	Minimum Interdigit Period	40 ms 23 ms Optional
3 s	Maximum Interdigit Period	3 s
±1.5%	Frequency Deviation Maximum	±1.5%
N/A	Maximum Echo Level Below Primary Frequency Level	−10 dB
N/A	Maximum Echo Delay	20 ms

the telephone set. The ring is on for 2 seconds and off for 4 seconds. The ring signal that is heard by the calling party is not the same one sent to the ringer of the called subscriber. Instead, it is a 480 Hz tone returned by the switch station at the same on and off times to inform the caller that the phone should be ringing at the distant end. Bear in mind that the ring signal you hear is not the same one causing the telephone set ringer to operate. This is why sometimes it seems that the party you were calling picked up the phone before you detected that it was supposed to ring. They got the first ring and answered the phone before the switch station sent you the 480 Hz tone. This is also why you think a call is being placed even though the phone at the other end is not connected to its wall plug. You hear the 480 Hz ring tone, but the 20 Hz tone is not driving any ringer.

For long-distance or message-unit calls, the call is routed through **TRUNK LINES** via **TOLL STATIONS.** Toll stations are switch stations used to select which long-distance trunk lines are to be used to route your call. The ends of the trunk lines are connected via other toll stations to local switch stations, which route the call to its final destination. Longer calls may require longer trunk lines and additional intermediate switching stations. The hierarchy of the telephone company switching stations, shown in Figure 2.4, begins with the local switch station, which has the direct lines to the end user, the subscriber. Many calls can be completed by connecting subscribers at the local switch station. Others require connections through high levels of switching. Local switch stations are classified as class 5 stations. They are connected between each other and in clusters to a **TANDEM SWITCH,** which is also classified as a class 5 station.

Tandem switches, in turn, may connect the incoming call to other tandem switch stations, which pass the call on to a local station to be connected to the called subscriber. Tandem stations are usually the beginning of the long-distance or toll or trunk line network. They are connected through toll stations (class 4), which attempt to route the call to another toll station. In the event that the call is not within a specific primary area, it is routed on up through the class 3 **PRIMARY** station. Several of these primary stations are further interconnected through class 2 **SECTIONAL** stations and finally to class 1 **REGIONAL** stations. The hierarchy takes the form of a switch tree. There are twelve regional centers in the United States (ten) and Canada (two). The number of stations for each level in descending order increases until the bottom of the tree is reached, at the local office level. There are approximately 20,000 local switch stations, and that number increases as the number of subscribers keeps expanding. It should be noted that the switching procedure is such that the minimal number of connections is made to complete a call. As shown in Figure 2.4, there are parallel-level connections as well as cross-connections between levels to meet this purpose.

Most of the upper-level switching stations incorporate **REPEATERS,** which are amplifiers and signal shapers, to regenerate information that may have deteriorated as it traveled along the long trunk line.

The interfacing of the stations for purposes of connecting lines involves a number of characteristics classified under the term **BORSHT,** an acronym for the following:

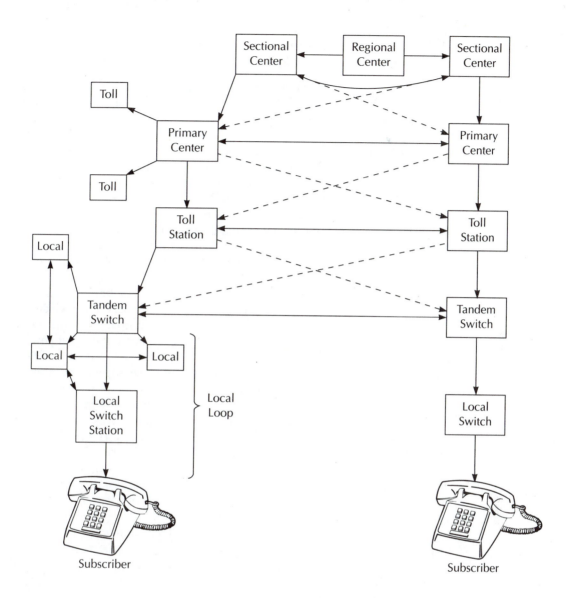

Figure 2.4 Switching Hierarchy

Battery—signifying the −48 V battery used to supply the DC current for on/off hook and dial detection

Overvoltage protection—built-in protection against line surges and induced voltage from electrical storms and lightning strikes

Ringing—designating the 20-Hz, 88 to 105 V ringing signal

Supervision control—responses to on/off hook, etc.

Hybrid circuit to interface 2-wire local lines to 4-wire toll lines

Testing—allowing line access for testing purposes

Two-wire cabling is used for local loops to carry voice and data between two stations. These two wires are the active line and the signal return line required to complete an electrical circuit. Conversations using the local loop are done in a full-duplex manner. That is, the parties at both ends of the line can talk to each other at the same time. Long-distance trunk lines (shown in Figure 2.1 between two hybrid blocks) employ four-wire cables to reduce signal degradation. One pair of wires carries signals in one direction, say from station A to station B. A second pair carries traffic in the opposite direction, from B back to A. In essence, each pair of lines is operating in a simplex manner, that is, allowing communication in one direction. However, the use of both pairs allows communication to occur in a duplex (two-way) manner. A circuit used to convert from the local 2-wire system to the long-distance four-wire is called a **HYBRID CIRCUIT**. An example of a hybrid transformer network is illustrated in Figure 2.5.

A signal introduced on the active lead of the 2-wire interface travels around the loop into the return side of the 2-wire interface. The arrangement of the windings of transformers T1 and T2 are such that the signals sent out from the 2-wire lines (Figure 2.5a) are coupled through them in aiding fashion, that is the signal power is added and non-canceling, amplified through amplifier 1 and sent out on the transmit side of the 4-wire interface. Those same signals at the output of the receive amplifier 2 are identical in level and phase. Since they are identical, there is no voltage potential between them, so amplifier 2 does not sense their presence. This causes very little signal to be present on the receive side of the four-wire interface.

A signal coming from the trunk line on the receive side (Figure 2.5b) is amplified by amplifier 2 and travels through transformers T1 and T2 in opposite directions. These opposing signals coupled through the same transformers as above now cancel instead of aid each other. As a result, the data received from the 4-wire receive side are not coupled back through the transmit pair of wires. The received signal continues through the circuit and is presented to the two-wire lines. Additional components to the right of the diagram are used for impedance matching purposes.

● 2.3 TELEPHONE SYSTEM SPECIFICATIONS AND PARAMETERS

One of the most important telephone system parameters to be considered is the bandwidth of the telephone lines. The bandwidth limits the rate at which data can be transferred using those lines. Ideally, the voice bandwidth of the telephone lines is 0 to 4 kHz. More realistically, the operating bandwidth is 300 Hz to 3 kHz. Over this range, the band response is fairly flat, with some roll-off at either end as shown in Figure 2.6. This range results in some loss of audio at the high end (the human

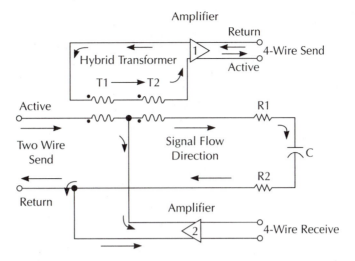

a) Send Loops

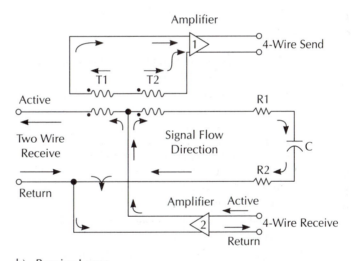

b) Receive Loops

Figure 2.5 Hybrid Two-wire/Four-wire Interface

ear being able to detect, in some cases, as high as 10 kHz or better), but the telephone system is designed for voice communications and not high-fidelity applications.

Numerous studies concerning sound and noise have produced interesting results that led to the establishment of the telephone system bandwidth. One such test to determine the useful frequency band required to carry the greatest percentage

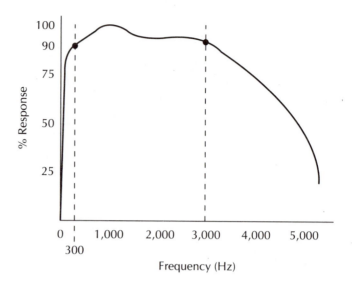

Figure 2.6 Telephone System C-Message Response Bandwidth

of voice calls occurred in the early 1900s. A number of participants were fitted with headsets connected to an audio sine wave generator. They were asked to signify when they heard a sound in the head set. The time duration and frequencies from the generator were set randomly, and resulting data were accumulated and graphed into a **C-MESSAGE** curve similar to the one in Figure 2.6. The heaviest responses occurred within the 300-Hz to 3-kHz range that the telephone system designers observed in fashioning the telephone system.

Additional tests and inputs from psychologists suggested a fascinating phenomenon about silence and the human mind. Human beings do *not* like silence, particularly on the telephone line. When we pick up the phone and hear a ''dead'' or quiet line, our first impulse is to believe that there is something wrong. This assumption is absolutely correct, even if the silence occurs after the call has been connected. Because of our dislike of pure silence, the telephone system is designed so that a small amount of noise is always present on the line, even when no one is talking.

Another important factor is the line impedance used for telephone line terminations. The telephone company specifies line impedance as 600 ohms. Some telephone lines have been strung and in existence as long as the telephone company itself. They have been subjected to every kind of weather and stress or strain known to humans. The physical stresses placed on these lines have altered their characteristics considerably. Telephone line impedances have been known to vary from below 100 ohms to above 5,000 ohms at extremes. Complications concerning line impedance arise as a result of the routing of a call between two telephone sets

DIGITAL & DATA COMMUNICATIONS

using the dial-up network. Physical routes between two sites are established at the time of the call and are dependent on the lines that local and trunk stations find available for use. These lines are then connected through various switching methods. Each time the same call is made, a different route is used to complete the call. Changes in routing result in changes of overall impedance and DC line resistance between the sites connected.

Example 2.2   Illustrate the possible routing of a call between Phoenix and Tucson, Arizona.

Solution: An individual in Phoenix, Arizona, desires to place a call to Tucson, Arizona. These two cities are used for illustration because of their close proximity to each other. There are also few inhabited areas between the two cities. One would expect calls from one to the other to be direct and simple. Most of the time they are (Figure 2.7a). There are long-distance trunk lines between local switch stations that connect these two cities. However, during peak hours (middle of a workday, for instance), these lines may become busy due to heavy use. If this were to prevent calls from being completed, the telephone company would never have reached the level of success that it has. The telephone switch network will try to route the call along the shortest and most direct route. If that cannot be done, a different route will be sought. The call, in this example, might first be routed to a switching station in Albuquerque, New Mexico, then down to El Paso, Texas, and finally on to Tucson before the connection is completed (Figure 2.7b). The caller does not detect anything, making the whole process **TRANSPARENT** to the user. Furthermore, the caller probably is not concerned about the route taken as long as the connection is made and is billed only for the distance between Phoenix and Tucson.

The scenario in example 2.2 is a common occurrence and is not limited in scope to minor rerouting. Calls have been known to take some long, roundabout routes to be completed. It is only on certain days, such as Mother's Day, that the telephone system becomes so overcrowded that a call cannot be completed through some form of routing. Additionally, a call placed between the same two numbers within minutes of each other may not take the same path to be completed. For voice communications, these routings are inconsequential. For data communications, this can be a serious determent. Specifications, parameters, and impairments vary from one set of routing to another. This discussion began by addressing the effects of line impedance. Considering that different cables along the telephone network experience different impedances, changing the routing of calls results in different circuit impedances each time the call is placed.

a) Direct Connection

b) Indirect Routing

Figure 2.7 Long-Distance Routing Examples

• 2.4 DDD AND LEASED LINES

The telephone company provides two kinds of communication links. One is the **DIRECT DISTANCE DIALING (DDD)** network, which is the standard switched network as discussed in the previous section. This is the system that we all use for our normal telephone calls. The other service provided is called **LEASED LINES**, in which a subscriber can lease a **DEDICATED LINE** to interconnect between two

DIGITAL & DATA COMMUNICATIONS

locations. When a subscriber leases one of these lines, it becomes a permanent connection between the two points, providing the same telephone lines and equipment each time a call is placed from one to the other. These lines are sometimes referred to as **PRIVATE LINES** since they are leased to an individual (or company) for private, nonshared use. These lines are useful for high-volume data transfers between designated locations where a consistent set of line parameters is required each time the communication link (connection) is established. Additionally, these lines can be **CONDITIONED** to meet specifications that are more stringent than general telephone line specifications. An example of a communications network that employs lease lines is WATS, for Wide-Area Telephone Service. Long-distance lines designated to this network are used by companies for regular voice and data communications. The same set of conditioned lines is used throughout the network, presenting a constant set of line conditions to the user each time a communications link is established. Use of the WATS lines is restricted to subscribers who lease its use from the telephone company.

● **2.5 LINE CONDITIONING**

Dedicated lines may be conditioned to limit a number of telephone line parameters, including gain or attenuation fluctuations, propagation time variables, **SIGNAL-TO-NOISE RATIO (SNR),** and harmonic distortion. General specifications for these parameters are listed by the telephone company as **BASIC CONDITIONING.** Capacitors and inductors are added to the lines to force them to meet other levels of conditioning. Before looking at the actual conditioning specifications, it is important to know how lines are tested to determine what their condition levels are.

Gain or attenuation variations throughout the bandwidth of the telephone system are measured in reference to values detected at a standard signal frequency of 1,004 Hz applied to the telephone line. The power of the signal received at the other end of the line is measured and recorded as a 0-dB reference. The actual power level is not important as long as it is reasonable. This level is the reference by which the levels of all other signals will be compared. Figure 2.8 illustrates the basic test setup with a graph showing the use of the 1,004-Hz reference level. Generally, the output of the sending station is held steady regardless of the frequency of the signal placed on the line. The source frequency is changed in steps, starting with 300 Hz and up to 3 kHz, producing measurable results throughout the line bandwidth. To meet basic conditioning specifications, changes in the level of signals across the entire bandwidth cannot vary by an increase of 3 dB or a drop of more than -12 dB from the 0-dB reference level at 1,004 Hz. As shown in Figure 2.8, the original level of the 1,004 Hz, as an example, could be $+12$ dBm. Recall that a dBm is a measure of power level using 1 mw as a reference level. Decibels are measures of power ratios between two levels. To say that the output throughout the band shall

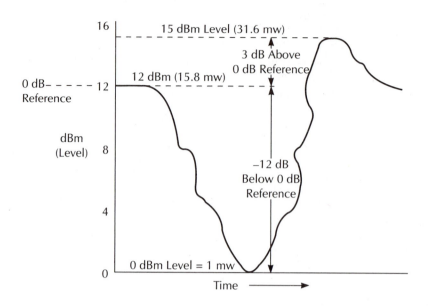

Figure 2.8 Relationship of Level (dBm) and Gain/Attenuation (dB)

be no greater than 3 dB from the 1,004-Hz reference says that the actual level should not increase beyond +12 dBm +3 dB, or +15 dBm for the example shown in Figure 2.8.

Additionally, within the frequency range of 500 to 2,500 Hz, the variation is reduced to +2-dB and −8-dB variations. A graph of the bandwidth response in terms of this variation is shown as Figure 2.9. Note that this is not a standard frequency response curve. The standard curve would show a specific level, and the bandwidth would be determined by looking for the 3-dB roll-off points at the high and low ends. This graph shows allowable variations from a given reference. The uneven line through the graph illustrates an actual bandwidth response curve whose power level at 1,004 Hz is used as the reference. As long as the levels measured at all other frequencies remain within the bounds of the attenuation graph, the circuit that has that response meets basic conditioning.

Basic conditioning, as expressed in the previous paragraph, is one of the specifications supplied by the telephone company to describe its system. These specifications are for voice-grade quality, meaning that they are sufficient to guarantee successful voice communications. Other levels of specification that outline more stringent guidelines for gain variations are provided by the telephone company under the several levels of conditioning. These are C1, C2, C3, C4, and C5. Each has applications for different types of communication interconnections.

In addition to specifying tighter gain fluctuation parameters, C conditioning also details requirements concerning fluctuations in propagation delay throughout the 300-Hz to 3-kHz bandwidth. To grasp the implication of propagation delay,

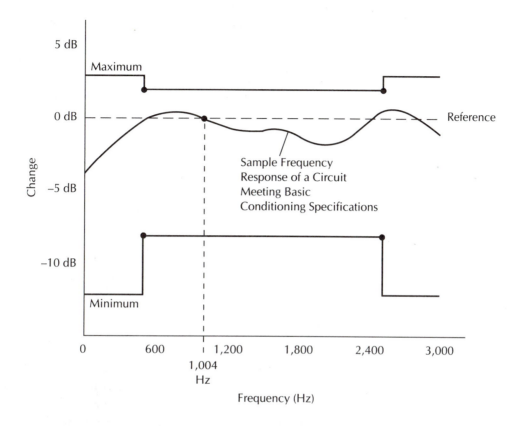

Figure 2.9 Basic Conditioning Limits Graph

consider a coaxial cable connected between two points. The cable has specific DC characteristics such as resistance and specific AC characteristics such as line impedance. The DC characteristics are those that are the same regardless of the frequency of the signal on the line. In contrast, AC characteristics are dependent on and distinctly different for each signal frequency value. Line impedance is the effective resistance and capacitive or inductive reactive load presented to the transmitter by the cable in response to the signal frequency. These effects are created by the inherent reactive quality of the cable created by its physical makeup. For instance, two conductors (center lead and outer shield return) separated by insulating material (acting like a dielectric) appear to have capacitive characteristics at a given frequency. Capacitances tend to oppose changes in voltages such as presented by analog (AC) signals. This opposition causes signals to be slowed down, arriving at the far end of the cable at some time after it was sent from the transmitter. The delay caused by the opposition to signal change is called **PROPAGATION DELAY,** which is the amount of time it takes for a signal to propagate or move through the cable. Since these capacitive and other reactive

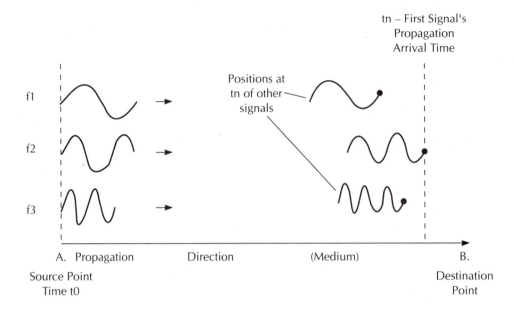

Figure 2.10 Effects of Propagation Delay on Three Different Signals

characteristics of a cable vary with the frequency of the signal on the cable, propagation times will also vary. Signals sent at the same time but at different frequencies will arrive at the far end of the cable at different times. An illustration of three signals and their relative arrival times at the receive station is shown in Figure 2.10. Note that the longest delay is not necessarily related directly to comparative frequency values. Since the characteristics of the line vary with each signal's frequency, the effect on each signal is different and independent of other signals. The first signal arrives at the destination at time tn. At that precise moment, the other two signals are at different physical points within the cable. They both arrive at the destination at some time after tn.

One effect of propagation delay is **PULSE SPREADING.** Consider the signal shown in Figure 2.11 possibly created by sending a number of alternating 1s and 0s using non-return to zero digital format. This signal appears as a square wave to the line (Figure 2.11a). A square wave is composed of many sine waves, starting with the fundamental wave (equal to one-half the bit rate) and including all the fundamental's odd harmonics (frequencies that are an odd multiple of the fundamental). As this square wave propagates down the line (Figure 2.11b), some of those harmonic waves are delayed more than others. The ones that experience shorter delays arrive at the receiving end first, while the ones having longer delays arrive measurably later. The effect is to spread the shape of the square wave (Figure 2.11c). Another effect of the spreading is to reduce the additive effects of the sine waves, causing the amplitude to reduce. If the spreading is severe enough, part of the wave shape can become totally lost (Figure 2.11d), which for digital data can result in the loss of one or more bits.

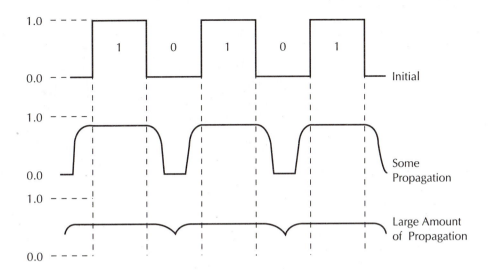

Figure 2.11 Pulse Spreading Effect of Propagation Delay

Measuring propagation time is not an easy task. It would be convenient if we could start a signal down a line, signal the other end that we did so, and measure the amount of time it took for the signal to reach the far end. The problem is that it also takes time for the start signal to get to the other end. Another way might be to echo the signal back—that is, to send it down the cable and then back to the originating source. Divide the round-trip time by 2 to obtain the propagation time. There are two problems here: First, whatever is used to echo the signal takes time to perform the turnaround. Second, how long do you send the signal before stopping and beginning to measure the return signal? This type of propagation measurement, if it could be made, would be an absolute propagation time measurement. The value obtained would be the actual propagation time.

Fortunately, absolute propagation time is not crucial unless it is extreme. What is more important is the relative **PROPAGATION TIME** between different signals on the same line. For instance, in the example of the square wave mentioned in the preceding paragraph, the relative differences in propagation times of the fundamental and the odd harmonics cause the square wave to be distorted. Large variations in propagation time tend to distort signals, causing losses in amplitude and changes in periodic signal time periods. The problem is how to measure relative propagation time, and the answer lies in a method called **ENVELOPE DELAY DISTORTION.**

A sweep generator is used to supply carrier signals sent on the line to be measured. These carrier signals are first modulated by a very low sine wave (Figure 2.12a). The frequency and amplitude of the modulating sine wave remain the same for each signal sent. As each carrier signal is modulated and sent, the resulting time between envelope peaks (tpk) is measured at the receiving end (Figure 2.12b). After all desired carrier signal frequencies are used, the time periods between the peaks

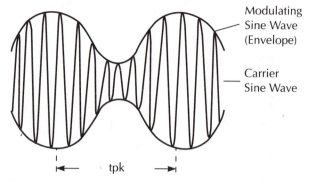

a) Modulated Carrier

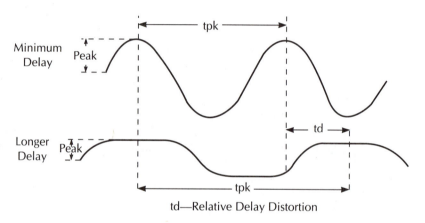

td—Relative Delay Distortion

b) Comparison of Envelope Signals Affected By Delay of Carrier Signals

Figure 2.12 Propagation Delay Distortion

are compared. As the modulating sine wave frequency is not varied, changes in the time between peaks must be caused by propagation effects of the carrier signal that was modulated. Notice that the measurements of these propagation effects are not absolute. The measurement does not tell how much delay (td) of the carrier caused the actual measured envelope peak time. However, because there are *differences* in propagation delays for each modulated carrier signal, there will be *differences* in the measured envelope peak times when those peak time measurements are compared with one another. These differences are the values for envelope delay distortion specifications. For basic and C conditioning the carrier frequency that results in the shortest time between envelope peaks is used as a reference (usually occurring at 1,800 Hz for telephone lines). All remaining en-

Table 2.2 Telephone Line Conditioning Parameters

Condition Level	Gain Limitations			Propagation Delay	
	FREQUENCY (Hz)	LIMITS (dB) MAX	MIN	FREQUENCY (Hz)	LIMIT (ms) MAX
Basic Voice Grade	500–2,500 300–499 2,501–3,000	+2 +3 +3	−8 −12 −12	800–2,600	1.75
C1 Station to One or More Stations	1,000–2,400 300–999 2,401–2,700	+1 +2 +2	−1 −6 −6	1,000–2,400 800–999 2,401–2,600	1.00 1.75 1.75
C2 Same as C1	500–2,800 300–499 2,801–3,000	+3 +2 +2	−3 −6 −6	1,000–2,600 600–999 500–599 2,601–2,800	0.50 1.50 3.00 3.00
C3 Local Private Line	500–2,800 300–499 2,801–3,000	+.5 +.8 +.8	−1.5 −3 −3	1,000–2,600 600–999 500–599 2,601–1,800	0.11 0.30 0.65 0.65
C3 Trunk Private Line	500–2,800 300–499 2,801–3,000	+.5 +.8 +.8	−1 −2 −2	1,000–2,600 600–999 500–599 2,601–2,800	0.08 0.26 0.50 0.50
C4 Maximum of Four Interconnected Stations	500–3,000 300–499 3,001–3,200	+2 +2 +2	−3 −6 −6	1,000–2,600 800–999 2,601–2,800 600–799 2,801–3,000 500–599	0.30 0.50 0.50 1.50 1.50 3.00
C5 Station-to- Station Only	500–2,800 300–499 2,801–3,000	+.5 +1 +1	−1.5 −3 −3	1,000–2,600 600–999 500–599 2,601–2,800	0.10 0.30 0.60 0.60

velope peak period times cannot vary by specified amounts given for each level of conditioning.

Table 2.2 summarizes the specifications for basic and five levels of C conditioning for both **ATTENUATION DISTORTION** and envelope delay distortion. Attenuation distortion is the measure of gain and attenuation variations, and envelope delay distortion is the measure of propagation delay variations.

Column 1 in the table denotes the type of conditioning. Column 2, the frequency range in hertz (Hz) for the limits shown. Column 3 is the gain/attenuation variations allowed in dB. Column 4 is the frequency range for envelope delay measurements, and column 5 contains the actual delay variations in milliseconds (ms) using a modulating sine wave of 100 Hz.

Example 2.3

What are the parameters for a communications system that interconnects three stations? Illustrate a bandwidth response for attenuation distortion and envelope delay distortion for these parameters.

Solution: According to Table 2.2, C4 conditioning is required to interconnect three stations. C4 has the following gain and attenuation variation limits for C4 conditioning:

1. Between 500 and 3,000 Hz, no more than $+2$ dB or less than -3 dB compared with 1,004 Hz.
2. From 300 to 499 Hz and from 3,001 to 3,200 Hz, the limits are $+2$ dB and -6 dB.

Envelope delay distortion for C4 has the following limits:

1. From 1,000 to 2,600 Hz compared with the shortest peak period, other times should not exceed 0.30 ms.
2. From 800 to 999 and 2,601 to 2,800 Hz, the limit is increased to 0.50 ms.
3. Further, between 600 and 799 Hz and between 2,801 and 3,000 Hz the limit is raised to 1.5 ms.
4. Lastly, between the frequencies of 500 and 599 Hz, allowable variation from the reference is 3.0 ms.

The graph for the C4 conditioning response is shown in Figure 2.13. The horizontal axis on each graph is dimensioned in 600-Hz increments of frequency. The vertical axis of the attenuation/gain graph is marked in relative attenuation or gain in 5-dB steps. Signals to meet C4 conditioning must lie between the two solid lines of the graph. The propagation delay graph is scaled in relative delay periods of 1 ms each. Signals placed on a C4 line must have delays that do not exceed the times above the graph line.

Data transfers at higher data rates (9600 bps, for example) are done by grouping binary data bits together to form data **SYMBOLS.** Recall from chapter 1 that any measurable electrical quantity can be used as a symbol. Symbols represent groups of binary data bits used to generate them. These signals may vary in amplitude, frequency, or phase. The circuits and systems used to create these

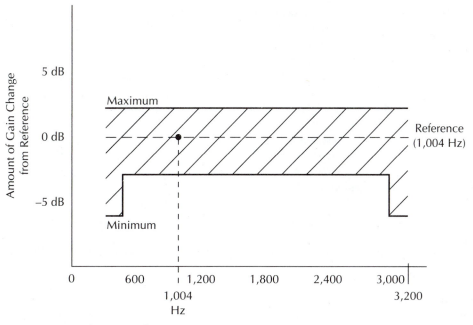

a) Attenuation/Gain Graph

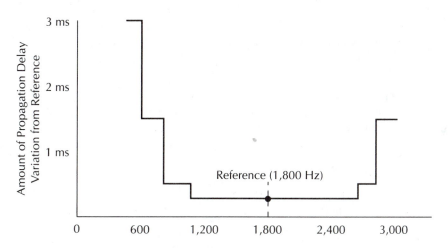

b) Propagation Delay Variations Graph

Figure 2.13 C4 Conditioning Graphs

symbols are discussed in future chapters dealing with various types of modems. The simplest example is a one-to-one relationship—that is, 1 binary bit associated with one symbol. In such a scheme, there would be two different symbols used, one for a logic 1 and the other for a logic 0. This is the essence of frequency shift keying, where the logic state of the data bits causes one of two signals that vary in frequency to be sent on the telephone lines. For instance, every logic 1 data bit sent would cause a signal at 2,200 Hz to be sent, while each logic 0 data bit might generate a symbol of 1,900 Hz.

This idea could be extended to a scheme that partitions a digital stream of data in groups of 3 bits each. Each combination of 3 bits causes a symbol of different frequency to be sent. This scheme formulates into eight different symbols for each combination of 3 binary bits (000–111).

Example 2.4

A particular system develops eight different symbols for binary bits 000–111 based on this chart:

binary	frequency	binary	frequency
000	750 Hz	100	4,000 Hz
001	1,000 Hz	101	5,000 Hz
010	2,000 Hz	110	6,000 Hz
011	3,000 Hz	111	7,000 Hz

For the following data stream (least significant bit—LSB on the right), what are the frequency and order of the symbols sent by the transmitter?

111101100000101010100010101011

Solution: Starting at the right, the first group of 3 bits is 011. This would cause a symbol of 3,000 Hz to be sent. The next group, 101, sends one at 5,000 Hz. The rest follow as shown below:

$$010-2,000 \text{ Hz}$$
$$100-4,000 \text{ Hz}$$
$$010-2,000 \text{ Hz}$$
$$101-5,000 \text{ Hz}$$
$$000-750 \text{ Hz}$$
$$100-4,000 \text{ Hz}$$
$$101-5,000 \text{ Hz}$$
$$111-7,000 \text{ Hz}$$

The symbols are transmitted one following the other at the symbol rate in symbols per second (sps) determined by the system used.

C conditioning specifications can be applied for higher data rates as a result of the reduction of the transmission rate for sending data as symbols versus bits. A data stream at 4800 bps is at a rate beyond the telephone system bandwidth. By converting groups of 3 bits into symbols, the transmission rate can be reduced to a symbol rate of 1600 sps. These higher data rates and the methods used to perform the conversions to symbols make them more susceptible to transmission line noise and harmonic interference. A usable measure of noise as it relates to signal is called the signal-to-noise ratio (SNR), which is computed by dividing the signal power by the noise power. The larger the SNR, the less effect noise has on the signal. C conditioning does not specify SNR limits. Nor does it address harmonic distortion, which is the presence of signals whose frequencies are exactly a whole number times the desired frequency value.

For data rates that approach and pass 9600 bps, dedicated lines are required to have D conditioning. D conditioning specifies guidelines for SNR and harmonic distortion but does not address bandwidth concerns that are covered by C conditioning. Telephone lines are specified to have an SNR of 24 dB. D conditioning requires improvement of this ratio by 4 dB to 28 dB. The presence of a second harmonic signal (one that is twice the frequency of the desired signal) must not exceed a level of −35 dB below the level of the desired signal to meet D conditioning. Third harmonic signals (three times the desired frequency) are required to be an additional −5 dB lower.

Signal-to-noise measurements can be made in two ways. One method is to measure the noise present on an **UNWEIGHTED** line, which is a line without any signal applied. This noise is known as the **INHERENT** or **WHITE NOISE** of a circuit or system. Since the level of the signal applied later is known, a signal-to-noise ratio can be formulated by dividing the signal power by the noise power. Unfortunately, this form of measurement does not give a complete picture of the relationship between signal and noise in a system. This is because **UNWEIGHTED NOISE** is present without the aid of the activity caused by an applied signal. When a circuit is fed a signal, the semiconductor and resistive devices, which are part of the circuit, become agitated. there is more electron activity within these devices because of the energy brought by the applied signal. As a result, these semiconductors and resistors generate additional noise above their inherent white noise. Remove the signal and after a short period, these devices settle back to their ambient state and the noise is reduced. The additional noise generated by applying a signal to a circuit or system is called **WEIGHTED NOISE.** The problem is how to measure weighted noise.

A block diagram for a C-notched test system is shown in Figure 2.14. The sweep generator supplies a signal input whose frequency (fc) constantly changes. A **NOTCH FILTER** between the sweep generator and the circuit under test

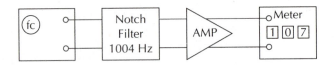

Figure 2.14 Weighted Noise Measurement

(represented by the amplifier in Figure 2.14) allows all the signals generated to pass through to the circuit except one. The signal that is absent is gone for a very short period of time. This is because the notch filter has a very narrow bandwidth and the sweep generator frequency keeps automatically changing. As a result, for a brief instant, the signal is removed from the circuit under test. The time that the signal is not present is too short to allow the components of the circuit to return to their ambient condition. A power meter monitors the output from the circuit. For most of the time, it shows the power level of the signal from the circuit. When the notch frequency is applied and stopped by the notch filter, the power meter measures the noise present minus the signal. Since there is not sufficient time for the weighted noise to reduce, the noise measured is the actual total noise of the circuit. This noise is its inherent white noise plus the weighted noise produced by the signal. Since the level of the signal is known, the actual circuit signal-to-noise ratio can now be calculated by dividing the signal power by the *total* noise power.

Figure 2.15 is a schematic for an active notch filter with its frequency response curve. Notch filters are characterized by a high Q, resulting in a narrow bandstop. Equations 2.1 through 2.3 summarize the relationship between notch filter components, notch frequency (fn), Q, and bandstop (**BW**).

$$\text{fn} = \frac{1}{2\pi C \sqrt{(R1)(R2)}} \tag{2.1}$$

$$Q = 0.5\sqrt{R2/R1} \tag{2.2}$$

$$BW = \frac{\text{fn}}{Q} \tag{2.3}$$

Example 2.5 Select reasonable values for R and C for a notch filter with an fn of 1,004 Hz and bandstop (notch bandwidth) of approximately 50 Hz.

Solution: To find values for R and C given the requirements of the problem, find Q first:

$$Q = \text{fn/BW} = 1{,}004/50 = 20.08$$

Next use equation 2.2 to find the ratio of R2 to R1:

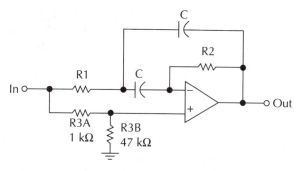

a) Schematic

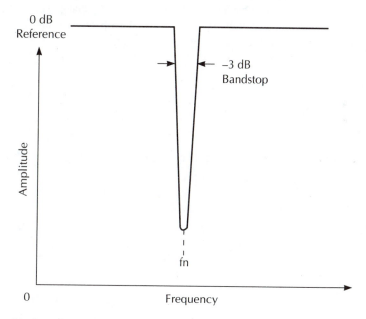

b) Frequency Response

Figure 2.15 Active Notch Filter

$$Q = 0.5\sqrt{R2/R1} = 20.08$$
$$R2/R1 = 1{,}613$$

Then select a value for C, say 0.01 μf, and solve for one of the resistor values. Substitute back into the ratio to find the other resistor value. If any of the component values are not reasonable, then change the value of C and try again.

$$fn = \frac{1}{2(3.14)\,(0.01\ \mu f)\,(\sqrt{1{,}613\ R1 \cdot R1})} = 1{,}004$$

$$R1 = 395\ \text{ohms}$$
$$R2 = 1{,}613\ R_1 = 637\ \text{Kohms}$$

Notch filters also come in integrated circuit (IC) packages that allow you to select the notch frequency by using external components and/or connections. One example of such a device is Motorola's MC145433 shown in Figure 2.16. This IC contains both a notch bandstop and a bandpass filter. The analog signal is applied to pin 4 (notch in), and the notch filtered output is generated at pin 2 (A0). The output is buffered by an internal op amp that can drive a 600-ohm impedance easily. The negative input (A−) of the op amp is brought out on pin 3. To provide unity gain buffering for this output amplifier, A− needs to be connected directly to A0.

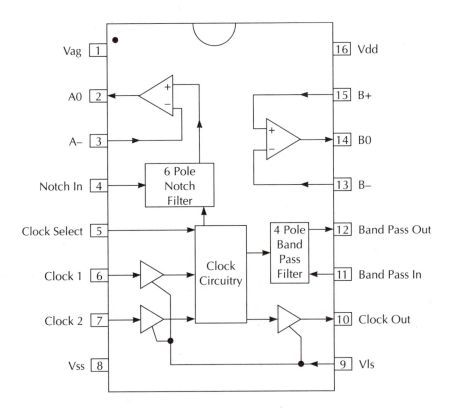

Figure 2.16 MC145433 Tunable Notch Filter Courtesy of Motorola, Inc.

A second op amp is provided on the chip. It has no direct connections to other circuitry within the chip and may be used as the circuit designer sees fit. The bandpass filter has a separate input (pin 11) and output (pin 12). Both filters are tunable to the same frequency (fc). Calculation of the notch (bandstop) or resonant frequency (bandpass) is dependent on the type of clock input or crystal selected to generate a clock at C1 and C2. This chip is usable with different logic family types, primarily transistor-transistor logic (TTL) and complementary metaloxide semiconductor (CMOS.) The voltage level applied to voltage level select (Vls), pin 9 determines logic family voltage levels expected, while Vss and Vdd are the actual source voltage input pins. By proper calculation and application of input signals, this device and ones similar to it provide notch filtering with the narrow bandstop required.

● 2.7 LINE IMPAIRMENTS

Once a line is conditioned, normal signal variations due to telephone equipment are minimized. However, sudden and unexpected changes in signal strength are caused by external factors. These changes, which are random and usually last for very short periods of time, are termed **IMPULSE IMPAIRMENTS** because of their sudden and short occurrences. Examples of the causes of these impairments are induced currents into the lines due to lightning strikes, a power surge from the power station, or magnetic disturbances from a solar flare. For regular voice communications, they are a source of annoyance and not much more, but for data communications they can cause data errors to occur.

The telephone company specifies a maximum allowable impulse gain variation (Figure 2.17) to be a change in signal strength within +/−3 dB lasting no more than 4 ms. These variations are not considered part of the conditioning specifications discussed in the previous section but are in addition to those specifications. The telephone company considers changes within the 4-ms time allowed to be random and not in violation of conditioning limits.

Electrical disturbances, as from electrical storms, also can cause changes in the phase of a signal for a short duration. Recall that a symbol is a single electrical parameter that represents a group of binary bits, and that variations in symbols can be based on amplitude, frequency, or the phase of the signals. On a system that bases the formation of symbols on the signal's phase, sudden changes in the phase, even for a short time period, can result in data errors. To avoid this designers of systems utilizing symbol phases must adhere to telephone system specifications for impulse phase changes. These specifications limit impulse phase shift to 20° in phase (positive or negative) as permissible on telephone lines (Figure 2.18). Again, this is for a short period of time, but it is a concern for reliable communications, which depend on phase differences between signals to supply digital information.

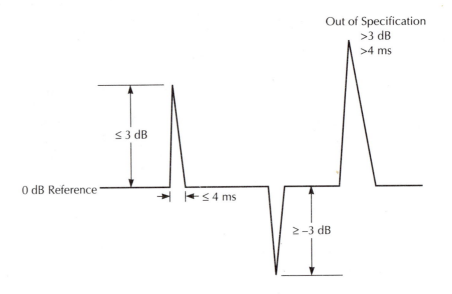

Figure 2.17 Impulse Gain Hits

An additional difficulty that can cause problems in using telephone lines is **CROSS TALK** between neighboring lines. Most of us have experienced this problem on long-distance calls at one time or another. While talking with your party, you can distinctly hear, albeit at a very low volume, another phone conversation going on. What you are experiencing is cross talk, in which one line is inducing current into the other in the same manner as a transformer inducing current from its primary to its secondary winding. Once again, for voice conversations this is a small annoyance, but for data communications it is a possible source of data errors. There are no telephone company specifications addressing limits to cross talk.

Because telephone lines and equipment are updated, repaired, or replaced periodically, impedance mismatches between older portions of the system and newer ones occur. One result of this mismatch is degradation of signals on telephone lines. Amplitudes are attenuated, and the shapes of signals become slightly distorted. Both of these events are minimized by **LINE CONDITIONING.** Once a line is conditioned (and the subscriber continues to pay to have the conditioning maintained), it is the responsibility of the telephone company to assure that the conditioning is retained regardless of changes made to existing equipment or lines.

Echoes or standing waves are another result of line mismatches. Some of the signal is lost because of attenuation; some of the signal reaches its destination as designed; another portion of the signal is reflected back down the line because of the mismatch. This reflected wave is called an **ECHO** or **STANDING WAVE** because

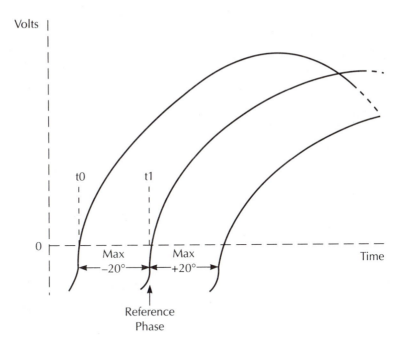

Figure 2.18 Phase Impulse Hits

it appears to echo back what is sent and while it is returning, it appears to "stand" on the line instead of being absorbed by the receive end. The problem of echoing is increased by the use of repeaters that inadvertently amplify the echo signal. Echoes can also be produced by cross talk effect. Some of the signal being sent is induced into the wires carrying response signals from the called station. The induced signal returns to the originating station as an echo of the information that the station sent. To reduce the effects of echoing for voice calls, repeaters are equipped with **ECHO SUPPRESSORS.** On four-wire trunk lines there are repeaters placed at regular intervals to reamplify and reshape attenuated signals. Echo suppressors are used to shut off one repeater while its opposite member is on. As an example, on a long-distance call, you are speaking with a person who does not know exactly when to cease talking and let you add something to the conversation. You yell, trying get his or her attention, but it seems to no avail. Is this person that insensitive? No! It's the echo suppressor that is preventing your pleas for attention from going through while the other party is still talking. Ah, but let them pause for even a short breath and you can reverse the situation.

Echo suppressors prevent full-duplex data transmissions for the same reason that you could not interrupt the person on the other end of the line. To defeat·this problem, echo suppressors have a circuit built into them that detects a tone between

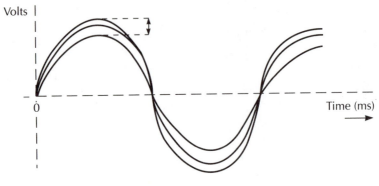

a) Amplitude Jitter

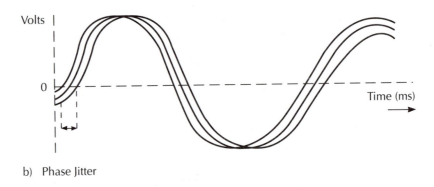

b) Phase Jitter

Figure 2.19 Examples of Jitter

2,010 and 2,240 Hz. If this tone is maintained for 400 ms, the echo suppressor is defeated and full-duplex operation can commence. The suppressors will remain disabled for the duration of the call as long as signals within the voice bandwidth (300 Hz to 3 kHz) remain on the line. The loss of a signal for as short a time as 5 ms will allow the suppressors to be enabled again.

Interference and electrical activity in telephone circuits can also cause phase and amplitude **JITTER** (Figure 2.19). Jitter is a small, constantly changing swing in phase or amplitude or both. It differs from impulse-type impairments by its constant nature. Because of its usually small size and predictability, this type of problem is easy to resolve or compensate for. Other problems are caused by equipment failure, aging, or large swings in the environment. They may result in frequency or phase translations. In effect, a constant error occurs in all frequencies or phases generated on the affected circuits. Some repair or adjustment is required to correct these problems.

SUMMARY

The accessibility of the telephone system for communications use led early designers of data communications systems to adapt the telephone system for their use. Impairments and telephone system specifications were studied, since the systems being designed had to operate in the specified telephone environment. These impairments and specifications discussed in this chapter do not cover every aspect of the telephone system. However, most of the significant concerns and considerations involved in using the telephone system to carry the traffic of data communications have been explored.

GLOSSARY

Attenuation distortion—Measure of the change in signal gain on a communication line for signals present in the telephone line bandwidth.

Basic conditioning—Standard telephone company parameter specifications for its lines.

Conditioned lines—Lines which have been electrically altered to meet more stringent requirements.

C message—Voice bandwidth between 300 Hz and 3 VHz.

Cross talk—Coupling of an electrical signal from one line to another.

Direct distance dialing (DDD)—Common telephone service.

Dedicated line (also called a **private** or **lease line**)—Specific telephone line between two stations.

Echo—A portion of the transmitted signal returned to the sender.

Echo suppressor—Circuit that disables line amplifiers in trunk circuits to defeat echoes.

Envelope delay distortion—A measure of the relative variances in propagation delay of signals on the telephone lines.

Hybrid circuit—Circuit used to interface 2-wire to 4-wire telephone lines.

Impulse impairment—Sudden short change in signal amplitude or phase.

Inherent or **White noise**—Noise generated by the nature of a system's physical properties without the presence of a signal.

Jitter—Small, constant change in signal amplitude or phase.

Leased line—Private line whose use is leased from the telephone company.

Line conditioning—Process by which telephone line parameter specifications are tightened.

Local loop—Part of the telephone system between the subscriber and the local switch station.

Notch filter—A narrow bandstop filter.

On and **Off hook**—Respectively, disconnecting and connecting the subscriber telephone set to the telephone lines.

Primary station—Station with the highest level in a telephone system hierarchy.

Private line—Dedicated telephone line assigned to a single user.

Propagation delay—Time it takes a signal to travel from one end of a communication line to another.

Pulse spreading—Signal distortion resulting from propagation delay.

Regional station—Switch station with the lowest level in the telephone system hierarchy.

Repeater—Long-distance line amplifier used to regenerate telephone signals.

Sectional station—Mid-level switch station within the telephone hierarchy, used to interconnect primary stations with sectional stations.

Standing wave—See echo.

Subscriber—Telephone user.

Switch station—Telephone station used to connect one subscriber with another. Also called toll stations for long-distance applications.

Symbols—An electrical signal that represents one or more data bits.

Tandem switch—Telephone switch station that interconnects several local switch stations and toll stations.

Toll station—Long-distance switch station.

Touch tone—Dialer which generates dual multiple frequency tone signals whenever a number is pressed.

Transparent—Any part of a communication link that aids in transferring data without user's awareness of its existence.

Trunk line—Long-distance four-wire telephone lines

Unweighted—Circuit condition or noise without the presence of a signal.

Weighted line—Circuit which has a signal applied to it.

Weighted noise—Noise generated by circuit activity caused by the application of a signal.

QUESTIONS

2.1. Calls connected between two subscribers on the same city block are routed through a _____.

2.2. What type of lines carry long-distance calls?

2.3. What type of circuit converts two-wire lines to four-wire?

2.4. Four-wire lines are used for _____ lines.

2.5. What is the purpose of a repeater in a long-distance line?

2.6. What is the usable bandwidth for a telephone line?

2.7. Why is it impractical to use the DDD network for high-speed data communications?

2.8. What frequency is "notched" for C-notched noise measurements?

2.9. Explain the differences between weighted and unweighted noise.

2.10. What two parameters are specified by C conditioning?

2.11. What parameters are specified by D conditioning?

2.12. What is an impulse gain hit?

2.13. What is the reference frequency for envelope delay distortion measurements?

2.14. What is the purpose of an echo suppressor? Why is it necessary for successful data communications?

2.15. Which form of data communications, simplex, half duplex, or full duplex, is not usable with an enabled echo suppressor?

2.16. Which types of impairments are caused by the following activity:
   a. Electrical thunderstorm
   b. "Ghost" voice on the line
   c. Constant line "static"
   d. Gain set too high on a repeater
   e. Line impedance mismatch

PROBLEMS

2.1. Describe the limits for a C2-conditioned line. Draw the response graphs for attenuation and envelope delay distortion.

2.2. Using equations 2.1 to 2.3, design an active notch filter for a notch frequency of 2,200 Hz and a bandwidth of 100 Hz.

DESIGN PROBLEMS

2.1. A system is experiencing a phase jitter of $+/-5$ degrees. Design a circuit that can be used to remove or minimize this jitter. The input and output impedance of the system in question is 600 ohms.

2.2. Design a circuit to decode a Touch Tone (DTMF) keypad. Use either two sine wave generators to generate the tones or Touch Tone telephone set and test your circuit in a lab environment.

2.3. Design a trunk line repeater circuit that has the following capabilities:
   a. Amplify four-wire signals.
   b. When a signal in one direction is being amplified, the responding line should be disabled to suppress echoes.
   c. Option: add a circuit to defeat the echo suppression function in response to a steady 1,200-Hz signal. Echo suppression resumes when the line is quiet for 1 second.

   Build and test your circuit in the lab.

Answers to Odd-Numbered Questions

2.1. Local loop or local switch station

2.3. Hybrid transformer

2.5. Regenerate signals on trunk lines

2.7. Different data link path established for each call

2.9. Weighted noise is the noise generated in an electrical circuit by the application of signal power. Unweighted noise is the inherent noise present in a circuit without a signal.

2.11. Signal-to-noise ratio and harmonic distortion

2.13. 1,800 Hz

2.15. Full duplex

Solution to Odd-Numbered Problem

2.1. Gain variations: +1 to −3 dB from 500 to 2,800 Hz; +2 to −6 dB from 300 to 499 Hz and 2,801 to 3,000 Hz. Envelope delay distortion: .5 ms from 1,000 to 2,600 Hz; 1.5 ms from 600 to 999 Hz; 3.0 ms from 500 to 599 Hz and 2,601 to 2,800 Hz. (See Figure 2.20).

Solutions to Odd-Numbered Design Problems

2.1. Actual design of jitter filter will depend on student background.

2.3. This circuit requires two amplifiers. When one amplifier is on, the other one is disabled. This can be done with a carrier detector circuit (bridge rectifier and latch). The detector circuit is disabled by a 1,200-Hz tone. A filter followed by a rectifier can be used to supply the control signal to do the disabling. A one-shot circuit (555) can be incorporated to supply a 1-second timer to reenable the suppressor with a loss of signal through both amplifiers.

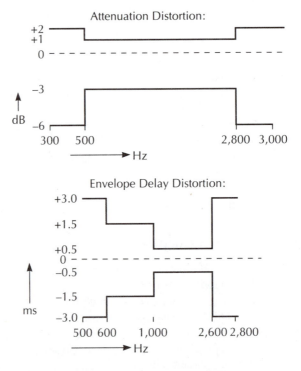

Figure 2.20 Solution to Design Problem 2.1

Three • Error Detection and Correction

OBJECTIVES

After studying this chapter, the student will be familiar with various methods of error detection and correction and how they are used. Included in this chapter is material on the following:

1. Parity and longitudinal redundancy check methods of error detection and correction generally used for asynchronous data communications

2. Hamming code and cyclic redundancy check used for synchronous data error detection and correction

3. Automatic request for retransmission code for error detection

ACRONYMS

ARQ—**A**utomatic **ReQ**uest for retransmission or **A**utomatic **R**epeat **ReQ**uest

ASCII—**A**merican **S**tandard **C**ode for **I**nformation **I**nterchange

BERT—**B**it **E**rror **R**ate **T**ester

CRC—**C**yclic **R**edundancy **C**heck

DEF—**D**ual-**E**rror **F**lag

EFS—**E**rror **F**ree **S**econds

FCMOS—**F**ast **C**omplementary **M**etal **O**xide **S**emiconductor

FEC—**F**orward **E**rror **C**orrection

IC—**I**ntegrated **C**ircuit

LRC—Longitudinal Redundancy Check

MSB—Most Significant Bit

SEF—Single-Error Flag

VRC—Vertical Redundancy Check

3.1 INTRODUCTION

Bit error rates for large quantities of transmitted digital data are extremely low. Error occurrences are rare and usually involve either a single bit error out of thousands or a burst of errors within a small number of consecutive bits. The burst error event is the more common because electrical interferences in communications links affect a series of data bits rather than a single isolated bit. Methods developed to detect and/or correct these errors are aimed at both types of errors. Some processes are used for single-bit detection and/or correction, while others detect occurrences of any number of errors within a transmitted message. Because of the specific application of each error detection/correction process, most methods are extremely reliable.

3.2 PARITY: ASYNCHRONOUS DATA ERROR DETECTION

PARITY has been used for a long time to detect errors in asynchronous data streams where burst-type problems either do not exist or are extremely rare. Both the Telex and Teletype networks employed this method. The parity bit is an added error-detection bit included with each character of the asynchronous data stream. In one respect this is a disadvantage of using parity. It adds an additional bit to be transmitted with each character, reducing the efficiency of the data transfer. In other words, it takes more time to transmit a data character with a parity bit than to send one without a parity bit.

Example 3.1   What is the efficiency of asynchronously transmitting an ASCII character with 1 start and 2 stop bits? Give two answers, one including the parity bit and one without parity for comparison purposes.

Solution: Recall from chapter 1 that the efficiency of transmitting a data stream is found by dividing the number of actual information bits by the total number of bits (information, overhead, and error detection) in the message stream. An ASCII character contains 7 bits of character information. Transmitting an ASCII character asynchronously as described above includes 3 overhead bits (1 start and 2 stop) as well as a parity bit. Total bits transmitted is 11 with parity and 10 without parity. The efficiency of transmitting an ASCII character message without parity is 7/10, or 70%, and with parity is 7/11, or 63.7%

Parity works by counting the number of 1s in a character and causing the parity bit itself to be set or reset to make the total sum of logic 1 bits (character plus parity) even or odd depending on the type of parity system used. An odd-parity system requires the total to be odd and an even one, even.

Example 3.2

What is the state of the parity bit for an ASCII R using both even- and odd-parity systems?

Solution: An ASCII R is 101 0010 in binary. The total number of 1s in the character is three, an odd number. For an odd-parity system the parity bit is a 0 to keep the count odd, while in an even-parity system, the parity bit is set to make the total even (3 + 1 = 4).

A number of exclusive OR gates form the circuit used to generate a parity bit as shown in Figure 3.1. The exclusive ORs (XORs) perform the function of "adding" the character bits together. Table 3.1 is the truth table for an exclusive OR. Considering inputs A and B as two binary numbers, the output, C, is the sum (ignoring any carry information) of these two numbers (0 + 0 = 0, 0 + 1 = 1, 1 + 0 = 1, and 1 + 1 = 10, or just 0 if carry is ignored). After each bit pair is added, their sums are added to the other sums until a final single sum is reached. Any carries generated by these additions are ignored. The final sum is fed to an exclusive OR being used as a controlled inverter. Again, observe the exclusive OR's truth table. Consider input A as a controlling input. Input B is data in. When the control is low, output C is the same state as input B. When control A is high, output C is the opposite or inverse of input B. The control input of the last exclusive OR in the parity generator is connected to a line that selects an even-parity (noninverted output) or odd-parity (inverted output) system.

The initial output of the exclusive OR adders is the state of the parity bit using an even parity system. To maintain this condition, the control input to the inverting exclusive OR is low. To make the output function as an odd-parity system, the control input is made high to cause the parity state to be inverted from the even-parity condition. An 8-bit parity generator chip, the 8262, is supplied by

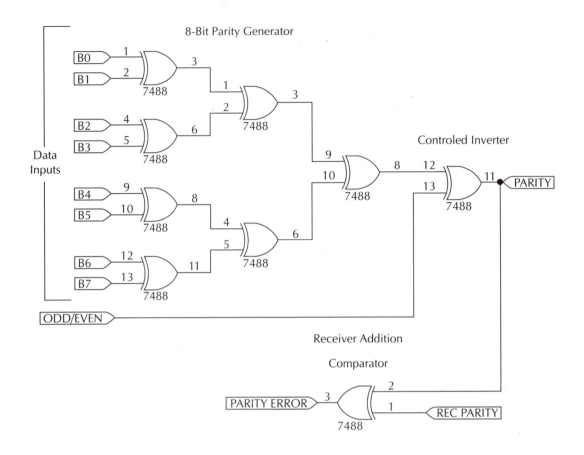

Figure 3.1 Parity Generator

Signetics Incorporated and is shown in Figure 3.2. Both odd- and even-parity outputs are present at pins on the chip. An inhibit input is included to allow the choice of not using parity. With this option both parity outputs are held low.

A parity generator is included at the sending station to produce a parity bit for each character sent. The parity bit is inserted into the serial data stream preceding each character's first stop bit. As an example, an ASCII C is 100 0011 (43 in hexadecimal). Since the least significant bit of the character (shown on the right in this example) is transmitted first following a start bit, the serial sequence for the letter C is 0 (start bit) followed by the 7 character bits. (0110 0001 with the least significant bit on the left since it is sent first). The next bit in the stream is the parity bit (0 for odd parity and 1 for even parity for the letter C), which is then followed by the stop bit (a logic 1). At the receiver a second parity bit is generated for each character received. The parity bit accompanying each character is then compared with the parity bit generated at the receiver. If they match there is no error. A

Table 3.1 Exclusive OR Truth Table

A	B	Sum (C) = A + B
0	0	0
0	1	1
1	0	1
1	1	0

Exclusive OR Truth Table Used as Sum

Control	Data Input	Output	
0	0	0	Noninverted Output
0	1	1	
1	0	1	Inverted Output
1	1	0	

Exclusive OR Truth Table Controlled Inverter Application

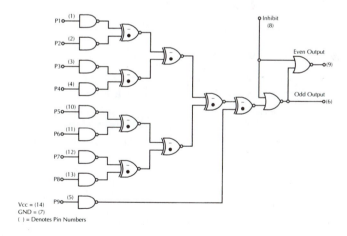

Figure 3.2 IC Parity Checker

mismatch generates a parity error. This comparison is achieved by using another exclusive OR gate. Once again, observe the truth table for the exclusive OR and notice that when both inputs are the same state, the output is low. If the inputs are different, the output is high. The high state indicates a parity error and the low state, the absence of an error.

Example 3.3   Which parity (odd or even) is used with the ASCII asynchronous data stream shown below? Two stop and 1 start bits are being used with each character. Which character has an error in it? The data appear as they would on a data scope, which is an oscilloscope designed to show serial data. The horizontal sweep is synchronized to the data rate. Furthermore, the device is designed to hold a stream of data on the screen until the user releases it. Data are shown starting with the first received bit (usually the least significant bit) at the left side of the screen where the sweep begins. Succeeding bits follow in correct sequence until the last bit is displayed. What is the intended message?

0000 1001 0110 1010 0110 1100 0010 1101 1000 0011 1111 0100 0010 011

Solution: A number of things to remember:

1. A data scope shows the bits as they are sent and displays them left (LSB) to right. The message starts with the first character sent on the left.

2. Each character starts with 1 start bit and ends with 2 stop bits. Adding the parity bit and 7 code bits, each transmitted character contains 11 bits. The first step is to separate the message into characters:

    0000 1001 011   0101 0011 011   0000 1011 011   0000 0111 111
    0100 0010 011

Next strip out the start and stop bits:

    0001 0010   1010 0110   0001 0110   0000 1111   1000 0100

Then determine the parity used and the character that has a bad parity bit. Most of the characters in the message use even parity with a parity error in the middle character.

Remove the parity bits and reverse the order of each character to place the LSB of each on the right to facilitate their interpretation:

    100 1000   110 0101   110 1000   111 0000   010 0001

Convert each binary group to its ASCII equivalent using the ASCII table from chapter 1: H e h p ! The error shows up in the middle character.

What if 2 bits in a single character are bad? A 2-bit change in a single character causes the parity condition to be the same as if an error did not occur. The sum of

1s in both cases would result in the same state of the parity bit. On the surface this may seem like a limitation of parity checking for error detection. Essentially, it is—parity detection is reliable only if a single error occurs in a character. Thus, parity checking is limited to environments that experience infrequent errors in data transmissions.

Example 3.4

Illustrate how two errors in a single character (ASCII B) would not be detected using (even) parity.

Solution: An ASCII B is 100 0010. The state of the even-parity bit is low, so the complete ASCII character with parity is

0100 0010

To illustrate what happens when 2 bits are bad, suppose the 2 least significant bits are wrong (any 2 bits could be selected and the resulting parity indication would be the same). These errors make the character appear as 100 0001, which would be interpreted as an ASCII A instead of a B. The even-parity bit would still be low since the sum of 1s in the character is still even, making the character become 0100 0001. Parity detection by a receiver would not consider this character to be in error, since the received parity and the computed parity at the receiver would both match (both would be low in this example).

To correct errors detected using parity, the receiving station can only request that the message containing the error be retransmitted. A system that is set to request retransmission automatically in response to detecting an error includes **AUTO-MATIC REQUEST FOR RETRANSMISSION,** also known as **AUTOMATIC REPEAT REQUEST (ARQ)** processing within its communications software. ARQ was originally designed to be used with a special type of character code, which used a 7-bit character size. The uniqueness of that code was that each character code contained 3 bits that were high and 4 that were low. If any character received is detected with more or less than 3 high bits in it, it is flagged as bad and the receive station automatically requests that the character be retransmitted, thus the name for the code. The automatic request process has been incorporated into other error-detection software such as that which responds to parity errors.

Not all systems using parity have ARQ. Some systems cause a parity error flag to be set in a status register, which, when read using applicable software, causes a message to be sent to a terminal to inform the user that a data error has occurred. It is then up to the user to request a message to be re-sent or not. The advantage to doing this is that some errors are less crucial. A user can easily figure what the correct character should have been. For instance, if the word "today" had an error in the last character, so that the word read as "todax," it can be readily understood and corrected without the need to retransmit the message containing the word. This saves transmission time, since the message does not have to be re-sent. It does require attention by the user to note the error and to make the decision to request retransmission or not.

● 3.3 DATA CORRECTION USING PARITY

Use of parity described in section 3.3 is for the detection of errors only. There is no way to determine which bit in a character is incorrect if an error is detected. There is even the possibility that the received parity bit itself is bad. The only way to assure that the correct information gets to the receiver is to request retransmission of any message that contains a parity error. Applications that involve a large amount of data would require a considerable loss of time to retransmit a message that had contained an error. Instead, these systems use an error-*correction* process rather than one that only *detects* errors. With error correction, detected errors are corrected at the receiver instead of forcing a retransmission of the data. The trade-offs are that an error-correction method requires additional circuitry and an error code character to be added to the message.

An error-correction process used with asynchronous data streams, called **LONGITUDINAL REDUNDANCY CHECK (LRC),** makes use of the parity process. The message is sent with the normal parity bit included with each character. These parity bits are called **VERTICAL REDUNDANCY CHECK (VRC)** bits. As before, any parity bit that does not compare correctly with one generated at the receiver indicates an incorrect character. The problem is to determine which bit in the character is not correct and then to invert it to make it correct. Longitudinal redundancy check is used to create a cross-grid matrix pattern to pinpoint a bad bit. As each character is loaded into the universal asynchronous receiver transmitter (UART) circuitry in parallel from the terminal, that character is added to the character that preceded it. All carries are ignored. The sum of all the characters is transmitted as an LRC error code character at the end of the message. At the receiver, normal VRC parities are generated and compared with the received parities. The received data are summed in a similar manner to the transmitted data, creating a second LRC character. The finished sum is compared with the LRC error character received from the sending station. Any error will be detected if any parity bit is not correct and the two LRC characters do not match. A bad parity bit coupled with a bad LRC error character is used to determine which bit of which word is bad. This can be illustrated best by example.

Example 3.5   What is the LRC error character for the message Help!? How can the LRC error character be used to determine which bit in Hehp! in example 3.3 is bad?

Solution: The LRC character is formed by summing the character bits and discarding the framing and VRC parity bits. Carries produced by the addition are ignored:

```
0001001 H
1010011 e
0011011 l
0000111 p
1000010 !
0000100 LRC
```

The LRC error character is framed with start and stop bits, and a parity bit is generated for it. It is appended to the end of the message and sent to the receiver. (00000100111)

The receiver, after shifting in each character and checking for parity errors, detects the error in the middle character as shown in example 3.3. The receiver then computes the LRC for the received message and compares the two characters. The bit position that does not compare is the one that is incorrect in the bad character.

Received LRC:

```
0001001 H
1010011 e
0001011 h
0000111 p
1000010 !
0010100 receiver-generated LRC
0000100 received LRC with message
0010000 comparison result—a 1 indicates the bad bit position
```

The detection of a bad parity bit in the "1" character designated the location of an error in that character. As a result of the comparison of the LRCs, bit 2 of that character is detected as incorrect and would be inverted to yield the corrected character.

Another way to illustrate the detection of the bad bit is by using a matrix diagram as shown in Figure 3.3. This is the matrix for the preceding example word Help! The character codes are arranged so that they are stacked vertically. Each individual parity bit is shown to the right of each character. These are the vertical redundancy bits. Longitudinal redundancy bits are shown at the bottom of the matrix. Note that the longitudinal parity bits form the LRC character, which includes its own parity (VRC) bit. In Figure 3.3b the matrix is shown with the error from example 3.5. The comparison of the LRC bits from the transmitter and

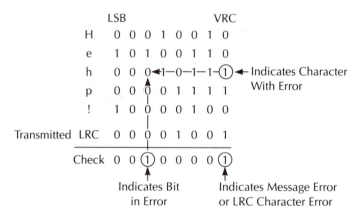

```
                                          Parity
                    LSB                    (VRC)
              H    0  0  0  1  0  0  1     0
              e    1  0  1  0  0  1  1     0
              l    0  0  1  1  0  1  1     0
              p    0  0  0  0  1  1  1     1
              !    1  0  0  0  0  1  0     0
                  ─────────────────────────
              LRC  0  0  0  0  1  0  0     1
```

a) Correct LRC Sent


```
                    LSB                    VRC
              H    0  0  0  1  0  0  1     0
              e    1  0  1  0  0  1  1     0
              h    0  0  0←1─0─1─1─(1)  ◄─ Indicates Character
              p    0  0  0  0  1  1  1  1     With Error
              !    1  0  0  0  0  1  0  0
Transmitted  LRC   0  0  0  0  1  0  0  1
                  ────────────────────────
             Check 0  0 (1) 0  0  0  0 (1)
                        ▲              ▲
                  Indicates Bit   Indicates Message Error
                     in Error      or LRC Character Error
```

b)   Receive Message with Error

Figure 3.3 LRC Matrix Example

receiver is accomplished by adding the LRC character received to the entire message (including transmitted LRC). This works because adding binary numbers and comparing them is the same process of using exclusive ORs as shown earlier. Note the circled parity (VRC) and LRC bits that point to the bad bit.

Error-correction methods described in this chapter are classified as **FOR-WARD ERROR CORRECTION (FEC)** because errors are corrected as the message is received. There is no requirement for retransmitting the messages as long as the errors are infrequent. A burst type of interference destroying several bits cannot be corrected by these methods. Burst errors when using LRC and VRC are indicated by more than one VRC parity error occurrence. This also assumes that the burst corrupted more than one character, and there is the possibility that it did not. For instance, a single character can experience a burst error affecting 6 of its bits. The VRC parity may or may not indicate an error. There would definitely be a change in the LRC character, but the receiver cannot determine where the errors are.

In this case a catastrophic error condition exists, requiring retransmission of the faulty message.

● 3.4 ERROR DETECTION FOR SYNCHRONOUS TRANSMISSIONS

Synchronous data are transmitted at higher data rates in as an efficient manner as possible. Start and stop framing bits and parity bits are omitted from the data stream. It should be noted that in some synchronous data systems parity is occasionally employed for error detection. Most high-speed synchronous transmissions, however, do not follow that practice. The reason is that most errors in high-speed transmissions occur in bursts. That is, some external interference or other effect on the line causes several bits to be corrupted at once. Single-bit errors occur less frequently. Because of this, error-detection methods have evolved to detect single or multiple errors within a data stream.

An additional one or two error characters are sent in addition to message data. The computation of these error characters is carried on quickly to support the higher data rates of transmission. One of the most common and widely used error-detection methods for synchronous data transmissions is **CYCLIC REDUNDANCY CHECK (CRC),** developed by IBM. On paper this method uses a pseudobinary division process to obtain the value of the CRC characters added to the message. The circuitry used to produce the actual CRC character is included in the transmit and receive shift register circuits. Both of these are described—the first to aid in understanding the process and the second to illustrate how it is done electronically.

IBM uses CRC-16 as the specific application of the CRC method. This process uses a constant "divisor" that appears in binary as:

1000 1000 0001 0000 1

The process, by hand, is begun by adding 16 zeros (1 less bit than the number of bits in the "divisor") after the LSB of the message to be sent. The message is then written down as it would be transmitted, with the extra bits sent first. On paper it is oriented with the **MOST SIGNIFICANT BIT (MSB)** on the left. Leading zeros are ignored. The "divisor" is exclusive ORed with the 16 most significant bits of the message. Enough bits from the message are appended to the result of this to form another 16 bits of data headed by a 1. The exclusive ORing is repeated along with the rest of the process until all bits in the message are exhausted. The final exclusive OR result is the CRC characters. Enough leading zeros are appended to the CRC character to form a total of 16 bits.

Example 3.6   Compute the CRC-4 character for the following message using a modified "divisor" constant of 10011:

1100 0110 1011 01   (the LSB is on the right)

Solution: The modified CRC-4 is used for illustration purposes since an example using CRC-16 looks cumbersome on paper and is difficult to follow. To use the shortened CRC-4 divisor, four zeros are first appended to the message. There is one less zero than the total number of bits in the "divisor," similar in manner to the CRC-16 described above. The extended message becomes

$$1100011010110000$$

Next, set up the problem:

$$10011 \overline{)\, 1100011010110000}$$

Exclusive OR the first five bits with the "divisor":

$$
\begin{array}{r}
10011 \overline{)\, 1100011010110000} \\
\underline{10011} \\
1011
\end{array}
$$

Bring down the next bit from the message and repeat the process until all the bits in the appended message are used:

$$
\begin{array}{r}
10011 \overline{)\, 1100011010110000} \\
110011 \\
\underline{10111} \\
10011 \\
\underline{10010} \\
10011 \\
\underline{11011} \\
10011 \\
\underline{10000} \\
10011 \\
\underline{11100} \\
10011 \\
\underline{11110} \\
10011 \\
\underline{11010} \\
10011 \\
\underline{1001} = \text{CRC}
\end{array}
$$

The original message followed by the CRC-16 bytes is sent to a receiving station. At the receiving station the data are shifted in and a CRC is computed using the entire message plus the CRC bytes. If the same constant "divisor" is used, the result of the receiver's CRC computation is all zeros when there are no errors. In some cases different CRC "divisors" are used at the transmitting and receiving stations. For this case the resultant CRC at the receiver has some given value if no errors occur.

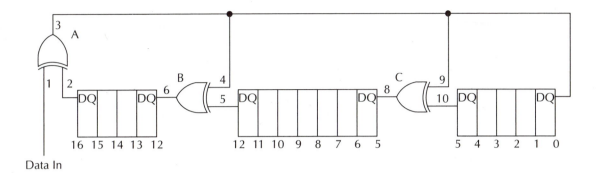

Figure 3.4 CRC-16 Block Diagram

A block diagram of the circuit that implements CRC-16 is illustrated in Figure 3.4. Exclusive OR gates are placed in the transmit and receive shift registers at the bit positions indicated by 1s in the "divisor." To facilitate the drawing of the CRC-16 or any other CRC circuit block diagram, a form of "quadratic" equation is used to represent the "divisor." This equation is developed using the powers of 2 for the bit positions in the divisor that contain a logic 1 state. For CRC-16 they are

$$
\begin{array}{cccc}
16 & 12 & 5 & 0
\end{array}
$$
$$10001000000100001$$

which is written

$$G(X) = X^{16} + X^{12} + X^5 + 1$$

The flip-flops in the register are numbered by similar bit position numbers. XOR gates are inserted at the indicated bit positions for b12 and b5. An additional XOR combines the output of b15 and the serial data stream (this represents nonexisting b16). The output of this XOR supplies a second input to the gates at b12 and b5 as well as the data input to b0. Once more, a simpler example illustrates the process.

Example 3.7   Draw the block diagram for the CRC-4 circuit of example 3.6.

Solution: First determine the bit positions of the 1s in the divisor:

$$
\begin{array}{ccc}
4 & 1 & 0
\end{array}
$$
$$1 0 0 1 1 = X^4 + X^1 + 1$$

Next lay out the data flip-flops and XOR gates, one between bits b0 and b1 and one combining the output of b3 with the incoming data stream:

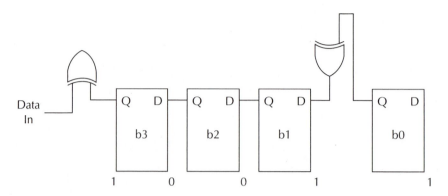

To complete the block diagram, draw a line from the output of the XOR gate at b3 to the input of the other XOR gate and the ''D'' input of b0 flip-flop:

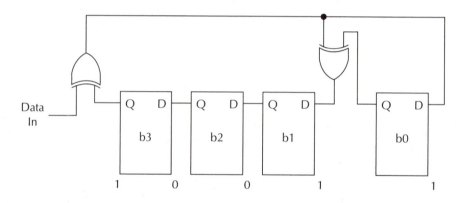

The final contents of these CRC registers are the CRC characters, which are shifted out following the transmitted message. In the mathematical example, done by hand, the CRC bits replace the zeros appended to the message to perform the calculation. At the receiving site, the received CRC-16 bytes are sent through the circuit following the last message bytes. If no errors occurred, then the CRC-16 receiving register will hold all zeros.

It was noted earlier that parity and LRC methods are unreliable if more than one error occurs within a character or message. One major advantage of the CRC method is its ability to detect multiple errors within any length of message. While it is mathematically possible for different multiple occurrences of errors that would produce the same CRC result, the odds against that happening are astronomical.

Thus multiple errors, as well as single errors, are expected to create CRC results that differ from that produced by an error-free message.

● 3.5 CHECKSUM ERROR DETECTION

An error-detection process called **CHECKSUM** generates an error character by adding all the bytes of a message together. Any carryover resulting from the addition is ignored. The process is repeated at the receiver. The two checksum bytes are compared, and if they match, an errorless message is assumed to be received. A mismatch indicates an error in the message. This method, like CRC, is also capable of detecting the presence of multiple errors. The major advantage of checksum is that it is simple to implement—each character being transmitted is exclusive ORed with an accumulated total of all previous characters. The final accumulated total is the checksum character sent with the message. Once again, at the receiver exclusive ORing all the characters and the transmitted checksum should produce a result of zero. Any other result indicates an error has occurred. In comparison with CRC, checksum is more likely to experience similar values for different messages. While the odds that enough errors would cause a duplicate checksum to the one originally sent are reasonably high, that possibility is still far more likely than a similar occurrence using the CRC method.

Example 3.8

What is the checksum character for the synchronous EBCDIC message Help!?

Solution: The EBCDIC codes for Help! are C8, 85, 93, 97, and 5A. Their sum, ignoring carryover, is the checksum and equals D1.

Synchronous data, generally run at higher speed than asynchronous, are more susceptible to various forms of maladies that result in errors. Still, most land-based communications are content with relying on checksum or CRC to detect errors. CRC correction requires retransmission of faulty messages. Now imagine a satellite rushing around Jupiter or Saturn, transmitting visual data as binary streams of information. The time it takes those messages to arrive at an Earth station from the satellite is measured in hours. During this time the satellite has adjusted its orbit and is soaring across new territory and sending additional data. Correcting errors in these messages cannot be done by retransmission. A request for that retransmission takes as long to get to the satellite as the original message took to get to Earth. Then consider the time it would take to resend the message. First, it would have to be found. During the time it took to send the original message and get the request to resend it, the satellite has gathered billions of additional data. The memory needed to hold data long enough to assure they were sent properly is staggering, to say the least. Instead, an error-correction method such as the Hamming code is used so that errors can be corrected as they are detected.

• 3.6 HAMMING CODE FOR ERROR CORRECTION

HAMMING CODES provide another method for error correction. This method is the most complex from the standpoint of creating and interpreting the error bits. Error bits, called Hamming bits, are inserted into the message at random locations. It is believed that the randomness of their locations reduces the odds that these Hamming bits themselves would be in error. This is based on a mathematical assumption that because there are so many more message bits compared with Hamming bits, there is a greater chance for a message bit to be in error than for a Hamming bit to be wrong. Determining the placement and binary value of the Hamming bits can be implemented using hardware, but it is often more practical to implement them using software. The number of bits in a message (M) are counted and used to solve the following equation to determine the number of Hamming bits (H) to be used:

$$2^H \geqq M + H + 1 \tag{3.1}$$

Once the number of Hamming bits is determined, the actual placement of the bits into the message is performed. It is important to note that despite the random nature of the Hamming bit placements, the exact same placements must be known and used by both the transmitter and receiver. This is necessary so that the receiver can remove the Hamming bits from the message and compare them with a similar set of bits generated at the receiver. Once the Hamming bits are inserted into their positions, the numerical values of the bit positions of the logic 1 bits in the original message are listed. The equivalent binary numbers of these values are added in the same manner as used in previous error methods by discarding all carry results. The sum produced is used as the states of the Hamming bits in the message. The numerical difference between the Hamming values transmitted and that produced at the receiver indicates the bit position that contains a bad bit, which is then inverted to correct it. As with the CRC-16, an example best illustrates the process.

Example 3.9   What is the Hamming code for the message in example 3.3, and how is it used to correct the error in the middle character?

Solution: When sending the message Help! using synchronous data, start and stop bits as well as parity are not used. Each ASCII character contains 7 bits, for a total of 35 bits. The number of Hamming bits is computed using equation 3.1:

$$2^H \geqq 35 + H + 1$$

H = 6 is the smallest value that satisfies the equation:

$$64 > 35 + 6 + 1 > 42$$

For simplicity we will insert the Hamming bits, less randomly, at every other bit position, starting with the least significant bit:

H   e   l   P   !

1001000110010111011001 11000001H0H0H0H0H1H

To begin the process of determining the states of each of the Hamming bits, list the numerical value of each bit position whose state is a 1. Start with the least significant bit as 1 and increase the count by 1 for each succeeding bit position. H bits must be included in the counting (but not the listing) process. In our example, the first bit position with a 1 in it is position 2. The following are all the bit positions containing a 1: 2, 12, 18, 19, 20, 23, 24, 26, 27, 28, 30, 33, 34, 38, and 41.

The next step is to list the numbers vertically along with the binary equivalent of each one. The value of the Hamming bits (H) is created by adding each binary bit column in the list, ignoring any carry condition:

2	0 0 0 0 1 0
12	0 0 1 1 0 0
18	0 1 0 0 1 0
19	0 1 0 0 1 1
20	0 1 0 1 0 0
23	0 1 0 1 1 1
24	0 1 1 0 0 0
26	0 1 1 0 1 0
27	0 1 1 0 1 1
28	0 1 1 1 0 0
30	0 1 1 1 1 0
33	1 0 0 0 0 1
34	1 0 0 0 1 0
38	1 0 0 1 1 0
41	1 0 1 0 0 1
H =	0 1 1 0 1 1

These H values are substituted for the ''H'' bits in the message in the order shown. The receiver repeats the process, again ignoring the Hamming bit positions in the list. If everything is all right, a comparison of the H values sent and those generated by the receiver produces zero. Using the error in the middle bit of example 3.3 as illustrated earlier (a 68 was received in place of the 6C), bit 23 arrived at the receiver low instead of high. When the receiver does the Hamming process, number

23 is omitted from the list. This causes the Hamming bits to have a value of 0 0 1 1 0 0 instead of 0 1 1 0 1 1. When the H bits computed at the receiver are compared with the transmitted H bits the result is as follows:

H sent: 0 1 1 0 1 1 as shown above
H computed at receiver: 0 0 1 1 0 0
comparison results: 0 1 0 1 1 1 = 23 decimal

To correct the message the receiver would invert bit 23.

There is no "black magic" mystery to why the Hamming code works. The originally transmitted codes are formulated by adding binary bits together, ignoring carries. The process of this addition is nothing more than exclusive ORing these bits together. A similar process occurs at the receiver. If a bit has changed state between being sent and received, it either will not be included in the process or, in the case of changing from a 0 to a 1, will be added to the process. By exclusive ORing the two Hamming codes, the process is reversed. The errant bit appears as the difference between the transmitted and received Hamming bit values.

The limitations on the success of the Hamming code in correcting errors is twofold. First, like parity and longitudinal redundancy checking, it works only if a single bit is bad. Multiple errors will actually cause the receiver to invert another bit, thereby resulting in an additional error in the message. The second difficulty with the Hamming method results from an error in one of the H bits. At the onset it was mentioned that locating the H bits in the message randomly is done to increase the odds against a Hamming bit error. However, this does not eliminate that possibility entirely. If a Hamming bit is bad, it will cause the receiver to invert a good message bit that it thinks is bad because of the difference in Hamming bit values. The proof of this is left as an exercise for the student. Using the previous example, change one of the H bit values but leave the message bits correct. Now compare the good Hamming bit values with the bad set you created by changing the H bit. Which message bit will the receiver now make incorrect?

● 3.7 SIGNETICS ERROR DETECTION/CORRECTION IC

An example of an **INTEGRATED CIRCUIT (IC)** created for the purpose of error detection and correction is the Signetics 74F630. It belongs to the **FAST COMPLEMENTARY METAL OXIDE SEMICONDUCTOR (FCMOS)** logic family and is capable of detecting and correcting single-bit errors in a parallel transfer of 16 data bits. Additionally, the 74F630 is capable of correctly detecting dual-bit errors that may occur in 16 bits of data. It cannot correct those dual-bit errors, nor can it correctly identify more than two errors in the 16-bit data word. The chip (Figure 3.5) is composed of a 16-bit data latch, a 6-bit check bit latch, error detection and correction logic circuitry, and function selector logic.

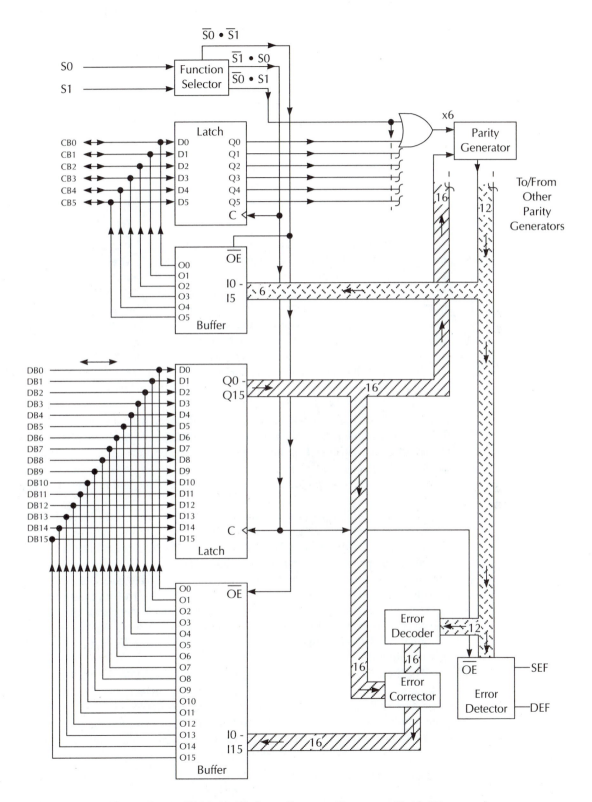

Figure 3.5 74F630 16-Bit Error Detector/Corrector Block Diagram

Table 3.2 74F630 Select Inputs

S_1	S_0	Function
$\overline{CS}$	R/$\overline{W}$	Interface to Memory
0	0	Read Data and Check Bits
0	1	Write Data and Check Bits
1	0	Disable Parity Generator
1	1	No Operation

The chip can be addressed by selecting any one of four combinations on the select inputs, S1, used as an active low chip select ($\overline{CS}$) and S0, used as a read/write (R/$\overline{W}$) input pin. Table 3.2 summarizes the chip functions based on the states of S1 and S0. When both of these pins are high, the 74F630 is not being accessed. Whatever information held in the data and check bit latches (Figure 3.5) remains in those registers. Sixteen-bit data are written into the data latches when S0 is brought low and S1 remains high. The data bits are presented a tree of 16 parity generators (one set pictured in Figure 3.5), where they create a pseudo–Hamming code (recall that parity generators and the Hamming code process use essentially the same type of methodology—exclusive OR addition of the bits ignoring carries), which is stored into the check bit latch.

When the data are ready to be sent, they are first read from the 74F630 (S1 and S0 both low), which places the 16 data and 6 check bits onto the bidirectional lines connected to the latches. This information can then be transferred in parallel to another 74F630 at the receiving end, or converted into a serial stream of data, moved to the receiving end, converted back to parallel data, and fed to the second 74F630.

At the receive 74F630, the data and check bits are written into the latches and the parity generators create a new pseudo–Hamming code from the 16 data bits. This is compared with the check bits that were sent with the data bits, and the results are fed to the error detector and corrector logic. This logic is capable of detecting one or two errors in the data *and* check bits. The result of the detection is set onto the single-error flag (SEF) or dual-error flag (DEF). Single errors in the 16-bit data word are corrected by inverting the bad bit. Single errors in the check bit word are detected but not corrected. The upper section in Table 3.3 shows the SEF and DEF indications of detected errors. Single errors cause SEF to be set high, and double errors cause both flags to be set high.

The error syndrome table in Table 3.3 shows the condition of the check bits when an error has occurred with respect to the bit that is bad. Note that for every single data bit error, three check bits are low and three are high. Single check bit errors result in a single check bit (the bad one) in a low state and the remaining 5 check bits high. With this logic the circuitry in the detector and corrector blocks can pinpoint the bad bit. The middle section of Table 3.3 is another way of looking at the error syndrome; the Xs indicate which check bits are low whenever a particular data bit is bad. Compare the Xs in column 1 (for bit 0) with the error syndrome table for b0 error. Dual errors in the data bits cause 4 check bits to be set low, while dual check bit errors cause 2 check bits to be bad.

Gross errors—all 16 data bits high or all low—are also detectable and cause the SEF and DEF flags to be set. Any other error conditions can cause the logic of the chip to produce results that generate more than 4 low check bits or that are similar to detecting no error, a single error, or double errors. These results are erroneous, which places the functional limitation onto the chip. It can be used only with systems that experience occasional single- or double-bit errors but not burst-type errors. Applications for this chip would include systems with short transfer distances whose transfers may be parallel rather than serial.

● 3.8 ERROR-TESTING EQUIPMENT

As the digital and data communications field has erupted, so has the proliferation of equipment used to test and troubleshoot communications systems. Two types of equipment used to check for error rate occurrences in communications systems are the **BIT ERROR RATE TESTER (BERT)** and the **ERROR-FREE-SECOND (EFS) TEST BOX.** Bit errors are measures of the occurrence of an incorrect bit in a stream of data. They are classified as one bit error in so many bits transmitted. For example, 1 bit in 10^6 bits is 1 error bit in 1 million bits transmitted. Frequently, this specification is shortened to a bit error rate of 10^{-6} since the one bit error is understood. Another measure of bit error rate is by percentage and is calculated by using the following formula:

$$\text{bit error rate} = \frac{\text{number of bad bits}}{\text{total number of bits sent}} \times 100\% \qquad (3.2)$$

Bit error rate testers are available that can test a data link line for a number of different types of error occurrences. Chief among them is the bit error rate, but also included are parity and framing error testing. The tester can be used singularly by tapping into the line between the terminal and the modem as shown in Figure 3.6. In this placement the tester can be used to monitor the line or to inject test data sequences into the line. Generally, the receiving end is terminated in a **LOOP-BACK** arrangement at some point. Loop-backs take the received data and return them to the sending station. This is accomplished by connecting the transmit and receive data lines together at the originating line (point A on Figure 3.6) to test the local sending loop; the remote end of the telephone line (point B) to test the telephone line connections, or point C, which also includes testing the secondary's modem.

A known pattern of data is generated by the bit error tester and sent down the line. One common and familiar pattern is "Quick brown fox jumps over the lazy

Table 3.3 Error Detection Correction FAST 74F630, 74F631 Signetics

FUNCTION TABLE

Total Number of Errors		Error Flags		Data Correction
16-BIT DATA	6-BIT CHECK WORD	SEF	DEF	
0	0	L	L	Not Applicable
1	0	H	L	Correction
0	1	H	L	Correction
1	1	H	H	Interrupt
2	0	H	H	Interrupt
0	2	H	H	Interrupt

H = High voltage level, L = Low voltage level

Check Word Bit	16-Bit Data Word															
	0	1	2	3	4	5	6	7	8	9	10	11	12	13	14	15
CB0	X	X		X	X				X	X	X			X		
CB1	X		X	X		X	X		X			X			X	
CB2		X	X		X	X		X		X			X			X
CB3	X	X	X				X	X			X	X	X			
CB4				X	X	X	X	X						X	X	X
CB5									X	X	X	X	X	X	X	X

NOTE:
1. The six check bits are parity bits derived from the matrix of data bits as indicated by "X" for each bit.

Table continues next page

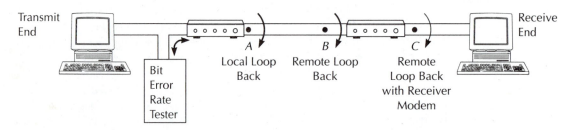

Figure 3.6 Line Test Using a Bit Error Rate Tester

Table 3.3 — Continued

ERROR SYNDROME TABLE

Error Location	Error Flags					
	CB0	CB1	CB2	CB3	CB4	CB5
DB0	L	L	H	L	H	H
DB1	L	H	L	L	H	H
DB2	H	L	L	L	H	H
DB3	L	L	H	H	L	H
DB4	L	H	L	H	L	H
DB5	H	L	L	H	L	H
DB6	H	L	H	L	L	H
DB7	H	H	L	L	L	H
DB8	L	L	H	H	H	L
DB9	L	H	L	H	H	L
DB10	L	H	H	L	H	L
DB11	H	L	H	L	H	L
DB12	H	H	L	L	H	L
DB13	L	H	H	H	L	L
DB14	H	L	H	H	L	L
DB15	H	H	L	H	L	L
CB0	L	H	H	H	H	H
CB1	H	L	H	H	H	H
CB2	H	H	L	H	H	H
CB3	H	H	H	L	H	H
CB4	H	H	H	H	L	H
CB5	H	H	H	H	H	L
No Error	H	H	H	H	H	H

H = High voltage level   L = Low voltage level

dog's back,'' which is used because it contains all of the letters of the alphabet. Other patterns include alternating 1s and 0s and single characters. When the data reach the loop-back, they are echoed back to the sender. The bit error tester monitors the returned data and checks for errors. A counter in the tester keeps track of how many errors occurred to determine the bit error rate. Most bit error testers are capable of operating at a wide range of data rates and can be used to test asynchronous and synchronous data systems.

Another measure of bit errors, called **ERROR-FREE SECONDS,** is used as a measure of error occurrences for digital data transmissions from 2.4 kbps to 2.5 Mbps. Instead of measuring bit errors occurring in a specified number of transmitted bits, error-free seconds is a measure of the number of seconds of transmission time that contain at least one error. An error-free second percentage is computed using the following relationship:

$$\text{error-free seconds} = 100\% - \frac{\text{seconds with an error}}{\text{total seconds of transmission}} \times 100\% \quad (3.3)$$

Error-free-second testers are used in the same manner as bit error rate testers and serve a similar purpose for digital data systems.

Example 3.10

Contrast the specifications, bit error rate, and error-free seconds for a system that experienced five errors in 25 Mbytes of data transmitted in 20 seconds. Two errors occurred in the 7th second of transmission, one error in the 12th second, and the last two errors in the last second of transmission.

Solution: The bit error rate is found by dividing the total number of transmitted bits by the total number of errors. This will yield the total number of bits sent on an average for each error occurrence: Bit error rate = 25 Mbytes $\times$ 8 bits/byte divided by five errors, or one error for every 5 million bits. This translates into a bit error rate of 5×10^6. The bit error percentage is computed using equation 3.2:

$$\text{bit error rate} = (5 \times 100\%)/(25 \text{ M} \times 8) = 2.5 \times 10^{-6}\%$$

The total number of seconds in which an error occurred is 3 (7th, 12th, and 20th seconds). Equation 3.3 is employed to compute the error-free second percentage:

$$\text{error-free seconds} = 100 - \frac{3 \text{ error seconds}}{20 \text{ total seconds}} \times 100\% = 85\%$$

SUMMARY

In summation, error-detection and error-correction methods do a highly reliable job of detecting and correcting errors. Parity and LRC, used with low-speed asynchronous transmissions, find heavy application in the home computer field. Connections, via modems, between personal computers and public services such as

DIGITAL & DATA COMMUNICATIONS

Compuserve rely on parity for error detection. This application does not require instant error correction and is generally satisfied to retransmit messages that were detected as having errors. Table 3.4, which summarizes the error-detection and error-correction methods described in this chapter, gives the student a quick comparison reference showing data type and use.

GLOSSARY

Checksum—Error-detection process that uses the sum of all data bytes.

Cyclic redundancy checking (CRC)—Error-detection process that uses a quasi-division process.

Error-free seconds—Amount of transmission time in which there are no errors in the data.

Hamming code—Error-correction code based on the number of ''1's'' in a message.

Longitudinal redundancy check (LRC)—Error-correction process using parity.

Loop-back—Circuit which returns a received signal without amplifying or modifying it.

Parity—Error-detection process based on the number of ''1's'' in a character.

Table 3.4 Error methods summary

Error Method	Data Type	Detection Correction	No. of Errors Detectable
Parity	Asynchronous	Detection	one per character
LRC	Asynchronous	Correction	one per message
Checksum	Either	Detection	unlimited
Hamming code	Synchronous	Correction	one per message
CRC	Synchronous	Detection	unlimited

Error Method	Overhead
Parity	LRC character added to message
LRC	LRC character added to message
Checksum	Checksum character added to message
Hamming code	H bits added to message
CRC	CRC bytes added to message

QUESTIONS

3.1. What error-detection method is generally used with asynchronous data transmissions? What process is frequently used for error correction using asynchronous data?

3.2. What is the prime limitation of the longitudinal redundancy error-correction method?

3.3. What is ARQ, and how is it used?

3.4. What is meant by burst errors? What significance do burst errors have on the error-detection or error-correction method used?

3.5. What is meant by forward error correction? Contrast this with automatic request for retransmission.

3.6. Select which error-detection method you would use for each of the conditions below. Support your choice:
   a. Direct memory transfer of data from floppy disk to computer RAM memory
   b. Bulletin board messages via the phone line to a personal home computer
   c. Messages between workstations on a production line
   d. Communications between bank automatic teller machines and central computer
   e. Computerized accounting network

3.7. Describe the reason that automatic request for retransmission is not used for messages sent from the *Voyager* space probe.

3.8. Which of the scenarios below would the 74F630 be suited for? Explain.
   a. Transferring a block of data from user RAM to video RAM
   b. Messages sent from a bank automatic teller machine to the central office
   c. Sending inventory from a main branch to a local distributor
   d. Transferring keyboard data to user RAM
   e. Sending a file to a local laser printer

3.9. Is the 74F630 design to be used with asynchronous or synchronous data type? Explain your choice.

PROBLEMS

3.1. What is the hexadecimal value of the LRC character for the following message:

Our Last Date.

Use ASCII character code with odd parity for the normal parity bits.

3.2. Show how an error that causes the word "Last" in problem 3.1 to be received as "Tatt" is detected and corrected using VRC and LRC methods.

3.3. What is the CRC-4 character, in hexadecimal, for the message

This 1

Use the EBCDIC code and the abbreviated divisor constant:

10011

3.4. Draw the schematic diagram for the following CRC-6 "divisor":

3.5. What is the checksum character, in hexadecimal, for the following synchronous message using the ASCII code?

Come to our aid now!

3.6. How many Hamming bits are used with the Hamming code for the message in problem 3.5?

3.7. What is the value of the Hamming bits, placed starting with the least significant bit position and every third place after that for the ASCII synchronous data message May 5?

Hamming bit placement sample:

. . . . H   b3   b2   H   b1   b0   H

3.8. The hexadecimal number 1234 is written into a 74630. The resultant 16-bit data and 6-bit check word are read and transferred to another 74F630 at the receive end. What is the data word and check bit word that is read from the receive 74F630 if the last hexadecimal digit has an error that changed it from a 4 to a 5? What are the states of the receive 74F630s SEF and DEF lines?

3.9. Four errors are detected in a synchronous transmission of 125 Mbytes of data. What is the bit error rate and the bit error rate percentage?

3.10. A transmission takes ¼ minute to complete. In that time, four errors are detected, three in the 9th second and one in the 12th second. What is the error-free-second percentage for this transmission?

DESIGN PROBLEMS

3.1. Design, construct, and verify the operation of a parity generator circuit. The circuit allows selection of even or odd parity and indicates when a parity error occurs. Another option the circuit is to operate with is a choice of either 7 or 8 data bits. A separate input is available to enter the parity bit from the received data to be compared with the one being generated. This input is selected to be active (for use in a receiver) or inactive (for use in transmitters).

3.2. Design, construct, and verify the operation of a CRC-16 generating circuit. The divisor to be used is the one discussed in the text. The requirements for the circuit are as follows:

a. While the CRC-16 bytes are being formed, the original data are shifted through unchanged.

b. After the last data bit is shifted out, the CRC-16 bytes are shifted out from the circuit.

c. Before data are shifted through and after the CRC-16 is shifted out, the data line is to be in an idle line 1 condition.

Essentially, the design of the CRC circuit follows the process shown in the text for CRC-4. However, this design project is extended to include the logic circuitry in conjunction with the CRC-16 circuit to meet the requirements above.

Answers to Odd-Numbered Questions

3.1. Parity.   LRC

3.3. Automatic request for retransmission—upon detecting an error, the receive station automatically requests that the message containing the error be retransmitted.

3.5. Forward error correction is any method that corrects errors as the data are received. Unlike ARQ, there is no need to retransmit the message.

3.7. The distance between the *Voyager* space probe and receiving stations on Earth is vast. By the time a message is sent by *Voyager* and received on Earth, the probe has moved quite a distance. Also, the probe sends a continual stream of data. Asking the *Voyager* to retransmit a message disrupts the steady flow of information. The probe's computer would have to search back through its temporary data bank to find the message to be retransmitted. All in all, not practical.

3.9. Synchronous data. There is no provision for checking parity bits usually associated with asynchronous data. Even though the transfers in and out of the 74F630 are done in parallel, this device would be used with synchronous data if serial data are to be used.

Solutions to Odd-Numbered Problems

3.1. 78H   3.3. 0110 = 6H   3.5. 6CH
3.7. H = 111100   3.9. 250×10^6 $0.4 \times 10^{-6}\%$

Solutions to Design Problems

3.1. and 3.2. See Suggested Circuits in Figures 6.7 and 6.8.

Figure 3.7 Solution to
Design Problem 3.1

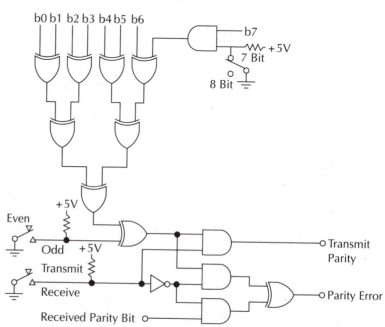

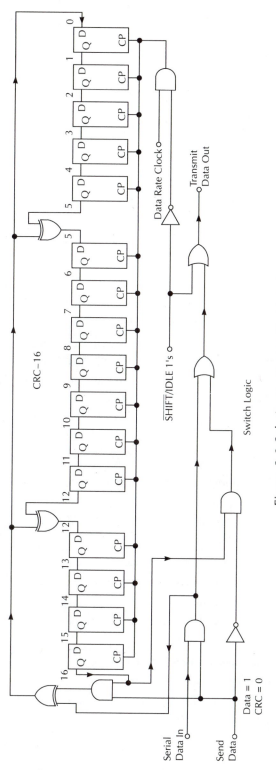

Figure 3.8 Solution to Design Problem 3.

Four • Data Link Protocols

OBJECTIVES

After studying this chapter, the student will be familiar with data link protocol specifications. Included in this chapter is material on the following:

1. Asynchronous and bisynchronous character protocols
2. Synchronous data link control, explained in depth as a synchronous bit-oriented protocol and as a prelude to networking protocols used in later chapters
3. XMODEM communications protocol

ACRONYMS

ACK—Positive **ACK**nowledge

ADM—**A**synchronous **D**isconnect **M**ode

ARM—**A**synchronous **R**esponse **M**ode

ASCII—**A**merican **S**tandard **C**ode for **I**nformation **I**nterchange

BCC—**B**lock **C**heck **C**haracter

BISYNC—**BISYNC**hronous protocol

CAN—**CAN**cel command

CRC—**C**yclic **R**edundancy **C**heck

DC—**D**evice **C**ontrol character

DISC—**DISC**onnect

DLE—**D**ata **L**ink **E**scape character

DM—**D**isconnect **M**ode command

EBCDIC—**E**xtended **B**inary **C**oded **D**ecimal **I**nterchange **C**ode

ENQ—**ENQ**uiry character

EOT—**End Of T**ransmission character

ESC—**ESC**ape character

ETB—**E**nd of **T**ransmission **B**lock character

ETX—**E**nd of **TeX**t character

FCS—**F**rame **C**heck **S**equence

FRMR—**FRaMe R**eject

GS—**G**roup **S**eparator

HDLC—**H**igh-level **D**ata **L**ink **C**ontrol protocol

ITB—**E**nd of **I**ntermediate **T**ransmission **B**lock

NAK—**N**egative **AcK**nowledge character

NL—**N**ew **L**ine

NR—Frame **N**umber expected to be **R**eceived next

NS—Frame **N**umber being **S**ent

PC—**P**ersonal **C**omputer

RD—**R**equest **D**isconnect as a SDLC/HDLC frame. **R**eceive **D**ata in reference to communications hardware.

RIM—**R**equest **I**nitialization **M**ode

SDA—**S**tation **D**evice **A**ddress

SDLC—**S**ynchronous **D**ata **L**ink **C**ontrol

SIM—**S**et **I**nitialization **M**ode

SNRM—**S**et **N**ormal **R**esponse **M**ode

SOH—**S**tart **O**f **H**eading

SPA—**S**tation **P**olling **A**ddress

SSA—**S**tation **S**election **A**ddress

STX—**S**tart of **TeX**t

SYN—**SYN**chronization character

UA—**U**nnumbered **A**cknowledge frame

UI—**U**nnumbered **I**nformation frame

● **4.1 INTRODUCTION**

It has been shown in chapter 1 how character codes are used to define characters used for digital data transfers. Alphanumeric and graphics characters formulate the actual messages sent and received. Data link characters are used to establish the beginning and end of messages, set attributes applied to the message, acknowledgments of transmissions, error responses, and station identification. A set of rules that establish which of these characters to use and how to respond to them is a **PROTOCOL**. Four widely used protocols that serve as a good representation are the character-oriented **ASYNCHRONOUS** and **BISYNCHRONOUS (BI-SYNC)** protocols and bit-oriented **SYNCHRONOUS DATA LINK CONTROL (SDLC)** and **HIGH-LEVEL DATA LINK CONTROL (HDLC)** protocols.

● 4.2 PROTOCOLS IN GENERAL

Besides establishing how a station reacts to data link and other characters and which responses are given for certain conditions, protocols also determine the method of identifying the secondary station that a primary wishes to communicate with. Recall from chapter 1 that in a multipoint system, a single primary controls the communications interface between it and many secondaries. After the electronics system link is operational, the communications link is established by the primary in one of two ways. The primary can **POLL** a specific secondary to inquire if that secondary has any messages to send to the primary, or the primary can send a **SELECTION** to a secondary to inquire if it is ready to receive messages (**TRAFFIC**). Responses to polls and selections depend on the state of the communications and peripheral links at the secondary station. The secondary could be in any one of three modes: SEND, RECEIVE, or LOCAL. In all three cases the secondary is required to respond to the poll or selection. The most common responses to a poll are as follows:

1. The secondary is in the send mode and transmits an actual message.
2. The secondary acknowledges the poll but is in the receive mode; it informs the primary that it has no messages to send but is capable of receiving traffic.
3. The secondary is "off-line" (local mode) for maintenance or local use, in which case it sends back an acknowledgment informing the primary of this condition.

The responses to a selection are similar, but have a slight variation:

1. The secondary is in the receive mode and sends a positive acknowledgment that it is ready to receive the primary's traffic.
2. The secondary is in the send mode and is not ready to receive messages but has something to send.
3. The secondary is in the local mode and is not ready to send or receive traffic.

Each protocol includes a method to handle polls and selections and the secondary's specific responses. In addition to a specific format of data for these functions, secondary addresses are included to determine which station is being polled or selected. Polling addresses are unique for each secondary station in the link, which means that the primary can poll only one station at a time.

Selection addresses fall into three forms. Each secondary station has a *UNIQUE SELECTION ADDRESS,* allowing the primary to select it individually. Groups of secondaries, not including every secondary in the system, are assigned **GROUP ADDRESS.** These allow a number of stations to be selected to receive the same message simultaneously. The last selection address type is called a **BROADCAST ADDRESS.** Every secondary station is selected when this address is used. The primary can send the same message to every secondary by first selecting them using the broadcast address.

Besides establishing the formats for polls and selections and secondary addresses, protocols also designate message formats and responses to data link characters. Recall from chapter 1 that character codes such as ASCII contain three types of characters—alphanumeric, graphics, and data link. The first two types dealt with printed characters and where and how they appeared on the screen or printed page. Data link characters were used to establish the software communications link. These include functions such as **START OF TEXT (STX)** and **END OF TRANSMISSION (EOT).** Responses to these characters differ for each protocol but overall they perform the same necessary functions. Bit-oriented protocols, on the other hand, recognize specific groups of bits or single bits to perform the data link functions.

● 4.3 SAMPLE FORMS AND INTERPRETATIONS OF CHARACTER-ORIENTED PROTOCOLS

It is not the intention of this text to dwell on protocols but instead to illustrate enough about them to understand their use in applications later on. The asynchronous and BISYNC protocols are used here to illustrate the formats and requirements for character-oriented protocols.

The asynchronous protocol has no requirement to establish bit and character synchronization between the sender and receiver. This task is performed through the nature of asynchronous data. Recall that start and stop bits are used to frame asynchronous data characters. The line is held at a logic 1 (mark) condition, often referred to as idle line 1s, as long as the link is established and data are not being sent. The first character transmitted is detected by the receiver as a change from the idle line 1 to a 0 or space (start bit) condition. This establishes both bit and character synchronization. Bit synchronization is established as beginning with the detection of the start bit and lasting for a time period dependent on the data rate (one bit time period being the reciprocal of the data rate). Character synchronization is established because the least significant bit of the character follows the start bit. Since there is no form of "preamble" message required to establish synchronization, actual messages are sent starting with the first character transmitted. A poll in the asynchronous protocol is recognized by the reception of a sequence of three characters:

E D
O C R
T 3

Each character is recognized by reading its name vertically. They are written this way so that each character occupies only one space. Spaces between characters are used for clarity and are not included in the sequence. Each character is recognized in the character code set (ASCII, EBCDIC, etc.) used. The end of transmission (EOT) character is used as an alerting character. All secondaries will react to this character in anticipation of receiving a poll or selection. Device control character 3 (DC3) further tells these stations that this message is a poll. Finally, the third character (R in this case) is a secondary's polling address. The secondary polled is required to respond in one of three ways described earlier. If the secondary being polled is in the send mode, it sends its messages in response to the poll. A secondary that is in the receive mode responds with the character sequence:

A
\ C
K

to indicate this condition in response to the primary's poll. Lastly, if the secondary is in the local mode, the sequence for the response is:

\ \

BISYNC is also a character-oriented protocol, but it is used for synchronous data. Synchronous data have no framing bits and rely on the receiver to regenerate a synchronizing clock from the incoming data stream. Once the receiver has recovered the clock, it must then establish bit and character synchronization. It does this by recognizing specific data character patterns sent by the transmitting station in the preamble to a message. This preamble and a BISYNC poll requires many characters and appears as:

P P S S E P P S S S S S E P P
A A Y Y O A A Y Y P P D D N A A
D D N N T D D N N A A A A Q D D

The two leading PAD characters are alternating 1s and 0s, which are ideal for clock recovery and bit synchronization by a receiver. The first pair of SYN characters aids the receiver to establish character synchronization by supplying a recognizable sequence of 1s and 0s to lock in on. The end of transmission (EOT) character is used here, as it is in the asynchronous protocol, to alert secondaries to a poll or selection. The second PADs are groups of four 1s each, which supply a short time delay to allow the secondaries to respond to the EOT character. A pair of SYN characters reestablishes character synchronization in case it gets lost during the time delay. These are followed by a **STATION POLLING ADDRESS (SPA)** and **STATION DEVICE ADDRESS (SDA),** the latter for communication with a specific device at a station. An SDA equivalent to the character '' is recognized as a general poll of all devices at the station. The address characters are followed by an **ENQUIRY (ENQ)** character and two more delay PAD characters. Usually, synchronous data do not use parity for error checking to reduce the number of additional bits to be sent with each character. Normally, a **BLOCK CHECK CHARACTER (BCC)** is generated using CRC-16. For polls and selections, this character is not generated. Instead, error detection is performed by duplicating significant characters. There are two SYN characters and two polling address and device characters. Each pair contains identical characters. Any errors in these characters prevent the secondaries from recognizing the poll, and there is no response. The primary will retransmit the poll if it fails to get a response from a secondary within a specified time period.

A secondary, in the send mode, responds to a BISYNC poll by sending a message. The sequence used for responding when the secondary is in the receive or local mode and there are no problems at the secondary station is called a **HANDSHAKE** and is:

S S E P
Y Y O A
N N T D

If the secondary is experiencing some difficulty with any part of its system and needs to inform the primary of this, it will respond to the poll with a sense and status message. This message informs the primary that the secondary cannot communicate with the primary. There are codes within the message that define most of the common reasons for the breakdown in communication.

Selection in asynchronous protocol is performed by sending an EOT followed by two selection address characters. These characters identify the secondary or secondaries selected. Selection in BISYNC is done by using a selection address in place of the polling address in the polling example shown above. This requires the definition of a unique set of polling addresses and a unique set of selection addresses by the BISYNC protocol. Device addresses are coded using the same characters as polling addresses, but since their position in the poll or selection differs from polling address and selection address positions, mistakes are not made.

There are specific secondary responses to selections that are similar to the ones used for polling. Additionally, there are responses designated to indicate the result of checking the **CYCLIC REDUNDANCY CHECK** (CRC-16) block check character used for error detection.

Actual message formats for these protocols include data link, graphics, and alphanumeric characters. Both protocols use similar formats, but the use of the characters differs significantly. How they affect the message as it appears on a terminal screen or printer page is also significantly different.

For the asynchronous protocol the message begins with a start of text (STX) or **START OF HEADING (SOH)** character. These characters, when sensed by stations that are not being polled or selected, cause these stations to become "blind" to the message—that is, the station cannot receive it. The polled or selected station does read and interpret the message. Either starting character is followed by text, which may be any combination of alphanumeric and graphics characters. Various applications respond to SOH characters in different ways. Some protect heading data from being altered at the receiving site. Some highlight the heading area, while others do nothing different in response to SOH or STX. A heading is terminated by an STX character, which also denotes the remaining text as data rather than heading. STX text is terminated by an EOT (end of transmission) character, which also "unblinds" the other secondaries on the line. Graphics characters, besides line feed and carriage return, are designated by a preceding **ESCAPE (ESC)** character. For instance, an ESC 3 highlights text that follows it, while an ESC 1 sets a tab at the present cursor location. Numerous others are used and are defined by the asynchronous protocol specification.

An example of an asynchronous message is as follows:

```
S E   E                                           E    E
T S R S 3 D A T A S C O M M U N I C A T I O N S S 4 N O
X C   C           P                               C    L T
```

As mentioned above, STX indicates the beginning of the message and causes any receiver not previously polled or selected to be "blinded" or effectively taken "off-line." ESC R sequence causes the cursor on a CRT to be placed in the "home" position and also clears the screen of anything on it. ESC 3 causes the following data to be highlighted. DATA COMMUNICATIONS is the actual text, followed by an ESC 4, which removes the highlight attribute. NL stands for new line, which is a combination of carriage return and line feed. Finally, the message is ended by an EOT character.

Example 4.1   In the message just shown, pick out the alphanumeric, graphics, and data link characters.

Solution: Alphanumeric characters are those that are printable, which are DATA COMMUNICATIONS. Graphics characters dictate how and where the message appears on the screen. They include all of the ESC n sequences. Finally, data link characters are used to establish the software linkage and include STX and EOT.

What about message efficiency? Earlier it was shown that the inclusion of start and stop bits along with parity reduces the efficiency of transmission. In its broadest meaning, transmission efficiency takes into consideration anything sent that is not

PAD 55	Preamble	Opening Data Link	— Text —	Closing Data Link	BCC	PAD FF

Figure 4.1 BISYNC Message Format

purely information. For the asynchronous message above, start and stop bits and parity are still noninformation bits, but now all data link and graphics characters must also be included as noninformation. They also reduce the efficiency of the message transmission. There are a total of 30 characters transmitted in the example message. Using one start, two stop, and one parity bit and ASCII characters, there are 30 characters $\times$ 11 bits per character, or 330 bits in the total message. The text bits in DATA COMMUNICATIONS, including the space character between the two words, total 19 characters $\times$ 7 bits per character, or 133 message bits. Using equation 1.1, it can be shown that the efficiency of sending this message using asynchronous data is:

$$133/330 \times 100\% = 40.3\%$$

A BISYNC message format (Figure 4.1) starts with a PAD and a Preamble. The opening data link character is an STX or SOH character similar to the asynchronous protocol. This is followed by a erase-write command character preceded by an ESC character. This sequence determines which portion of a message is being sent — specifically, whether that portion is unmodified or modified (altered) text or has some other designated attribute. This is useful when data entered into a fixed form are being sent from one station to another. Since the body of the form is fixed (and therefore not modifiable), both stations already have that portion of the message. The only section that is new is the data entered into the form's blanks. Because only modified data are sent, the message can be much shorter than when sending both the body of the form and the entries into it.

The opening sequence, which is illustrated in example 4.2 following this discussion, is followed by a write control character, which initializes the receiver terminal to the correct size and form of the message. The beginning location of the message, which is a memory buffer address associated with a specific terminal screen or printer location, follows the command sequence. Actual text is next, which includes an assorted set of characters as before. Graphics definitions called **ATTRIBUTES,** which define how the text looks, are defined by characters designated by a preceding **GROUP SEPARATOR (GS)** character. An **END OF TEXT (ETX), END OF TRANSMISSION BLOCK (ETB),** or **END OF INTERMEDIATE TRANSMISSION BLOCK (ITB)** followed by block check character (BCC) and trailing time delay PADs which end the message block. ETX also designates this block as the last block of the message. ETBs and ITBs indicate that additional blocks of the message are due to follow. ETX and ETB characters require the receiver to respond to the condition of the block check character (which is CRC-16 error characters) after they are received. ITBs are used to designate the end of the block but do not require an immediate response from the secondary. Instead, the secondary will respond to the condition of all BCCs when it receives the ETX terminating the message. ETBs and ITBs have the advantage that if an error occurs, only that block of the message needs to be retransmitted. ITBs have the advantage that a message can be broken into blocks as with ETB, but there is no time lost by the receiver responding to each block individually. Instead, all BCCs are acknowledged after receipt of ETX when ITBs are used. However, the receiving station can inform the sender of which block is bad as part of the response.

Example 4.2   How does the message **DATA** COMMUNICATIONS appear using the protocol? It is to be displayed on an 80-character screen and is modified data.

Solution: The actual characters are taken from the BISYNC protocol and are shown here for illustration purposes. Assume the opening dialogue (poll and responses, etc.) have already occurred. The message appears as:

```
P S S E P S S E     D     G              G
A Y Y O A Y Y T S 5 0 C S S I D A T A S S S C O M M
D N N T D N N X C   1                 P     P

                      D E B B P
U N I C A T I O N S C T C C A
                      3 X C C D
```

The meaning of each character as defined by the BISYNC protocol is:

First PAD—55 or AA to establish clock recovery and bit synchronization.

SYN SYN—Establish character synchronization.

EOT—Alerting character.

Second PAD—FF to allow time to respond to EOT.

SYN SYN—Reestablish character synchronization.

STX—Start of text indication.

ESC 5 (erase/write character)—Clears the screen and homes the cursor

0 (write control character)—Sets 80-character-per-line mode.

DC1 SS—signifies that the message will be stored in video memory starting at the location associated with the home position on the screen.

GS I—The group separator indicates that an attribute character (I) follows. Attribute I causes the field of the message to follow to be intensified and protected from being altered.

DATA (data followed by a space)—Actual text.

GS SP (group separator and attribute character space)—Removes intensify and protect from the next field of the message.

COMMUNICATIONS—Actual text.

DC3—Inserts the cursor at the current video memory location (placing it following the S in communications in the message).

ETX—Indicates the end of the message. Block check characters are expected next.

BCC—Two block check characters to hold the results of CRC-16 error detection are next.

PAD—Time delay PAD (FF hex) allows the receiver to check the BCCs for any indication of an error.

What about the efficiency of this message? With synchronous data there are no framing or parity bits within each character, so all 7 bits of each ASCII character are data bits. There are still only 19 information characters in DATA COMMUNI-CATIONS, but there are a total of 42 characters in the message. Since all bits are character bits, the efficiency of a synchronous message is the total information characters divided by total message characters:

$$19/42 \times 100\% = 45.24\%$$

This is not much more efficient than the asynchronous version, but keep in mind that this example is a short message, which is not typical for synchronous data transmissions. The difference becomes much greater as message length grows. Actual data transmissions will include hundreds to thousands of characters. As an example, compare the efficiency of an asynchronous and synchronous 256-character message. Assume the same number (23) of graphics characters (ESC and GS/DC characters) are used in both messages. You will soon note that the efficiency of synchronous data transmissions will far outdistance that of asynchronous. The reason is simple enough. Each character added to an asynchronous transmission is 63.6% efficient by itself (7 ASCII bits divided by 11 total, using parity and 2 stop along with a start bit). Each additional synchronous character is 100% efficient (all data bits) by itself.

$$\text{efficiency} = \frac{256}{279} \times 100\% = 91.8\% \text{ for BISYNC}$$

The asynchronous message retains its 63.6% efficiency.

One additional overhead character is used in the BISYNC protocol to facilitate a function called **TRANSPARENCY.** There are occasions when data messages may contain numerical values or other characters that have the same codes as the data linking characters. An example application is the sending of an inventory to the accounting department of a huge store. It is very possible that the count of any number of items might appear as link control characters. To avoid this, a data link escape character (DLE) is used as a delimiter character that precedes the start of text

(STX) character of the message. All characters within the message that are intended to be data link characters are preceded by the DLE character. The receiver then interprets only those characters following DLE as data link characters. All others are treated as information or graphics characters.

This description of the character data link protocols is a stripped-down version for the purpose of illustrating the concept of character interpretation in establishing the actual message. There are a number of detailed functions and code tables, responses, and requirements that these character-oriented protocols have. Furthermore, information and further details about these character-oriented protocols can be obtained from the individual protocol specifications themselves.

● 4.4 SYNCHRONOUS DATA LINK CONTROL (SDLC)

SDLC and the enhanced version high-level SDLC or HDLC are bit-oriented protocols that were developed to allow smaller sections of messages to be transmitted at a given time. This lets a primary station interleave messages between secondaries by sending each a portion of its message at one time. As the communication continues through a network, each secondary is linked to the primary over and over until all data are sent. The secondaries then reassemble their messages from the smaller sections they have received from each link up. To do this it is necessary to maintain the sequence of the message sections so that a secondary can correctly reassemble its information. These message sections are sent as SDLC or HDLC messages called **FRAMES.**

Being bit-oriented means that the data linking processes are established by the particular grouping of bits rather than by characters such as STX or SYN. SDLC was conceived by IBM to be used solely with the EBCDIC character code. Additionally, IBM selected b0 in a group of 8 bits (byte) to designate the most significant bit and the b7 as the least significant bit. This is the inverse of the standard notation most students are familiar with. For example, converting the hexadecimal numbers 36 to binary by IBM appears as:

b0 b1 b2 b3 b4 b5 b6 b7

36 = 0 0 1 1 0 1 1 0

The basic format for SDLC and HDLC messages is shown in Figure 4.2. Each message transmission is framed by a starting and ending flag, which is recognized as six 1s between two 0s, or 01111110, or 7E in hexadecimal form. Do not forget that SDLC and HDLC are bit-oriented protocols, so while it is easier to refer to six 1s surrounded by 0s as 7E, it is still 8 bits that are being referred to. If this sequence of six 1s and two 0s appears *anytime or anywhere* in an SDLC/HDLC message, the SDLC/HDLC protocol recognizes it as a flag, intended or not. The reason for this is that this protocol is *bit-oriented*. How is this a problem? Suppose that in the body of text the EBCDIC character G is followed by a U. The EBCDIC code for G is

7E Flag	Secondary Address	Control Field	Text Field	Frame Check Sequence	7E Flag

Figure 4.2 SDLC/HDLC Frame

11000111 and for U is 11100100. Together, the bit sequence becomes 1100011111100100. Notice that a 7E sequence is created in the middle of the data sequence. SDLC/HDLC reacts to this combination as a flag and would end the current message at this point. Since a flag sequence does not follow this sequence, it will assume that this ending flag is also the beginning flag of the next frame! To avoid this problem, a method called **ZERO STUFFING** or **ZERO INSERTION** is employed. A 0 is inserted following five consecutive 1s anywhere in a message except the flag areas. The receiver removes any 0s it detects following five consecutive 1s before interpreting the message contents.

Example 4.3

For the EBCDIC message sequence for GUTTER, where are 0s inserted to avoid mistaken flags?

Solution: The binary sequence for the message GUTTER is:

110001111110010011100011111000111100010111011001

Zeros are inserted following five consecutive 1s. The message sent becomes (with inserted 0s highlighted to aid in showing their location):

1100011111010010011100011111000011110001011011001

Notice that a 0 is inserted following the second group of five 1s even though they would not be interpreted as a flag, which requires six consecutive 1s.

You can use as many consecutive 7E sequences as desired before beginning a message. The SDLC protocol will not interpret a beginning flag until the bits following that flag no longer look like a flag. With consecutive flags, an ending 0 of one flag can be used as the beginning 0 of the next flag. So these are both two consecutive flags:

0111111001111110   and   011111101111110

As stated earlier, the ending flag of one frame can be used as the beginning flag of the next frame.

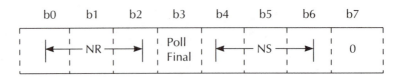

b0	b1	b2	b3	b4	b5	b6	b7
← NR →			Poll Final	← NS →			0

Figure 4.3 Information Frame Control Field

The next field in the message is the secondary address field. All messages between a primary and a specified secondary contain that secondary's address in this field. This identifies the station that the primary is communicating with. In SDLC this field is 8 bits wide, allowing for 256 different addresses. In HDLC this field is unlimited. Each address must be in multiples of 8 bits (bytes) and can be any total length. A 1 in the most significant bit of the address byte designates that byte as the last byte of an address. All other bytes in the address have a 0 in the most significant bit position. Effectively, there is no limit to the number of secondaries used with the HDLC protocol.

Following the address field is the control field, which defines what type of message is being sent. There are three different message frames used in SDLC and HDLC—information, supervisory, and unnumbered frames. Figure 4.3 shows the control field for an information frame, which is identified by b7 of the control field being in a logic low state. Information frames sent by a primary can be either a poll (b3= 1), which requires a response from the secondary, or not a poll (b3= 0), which does not require a secondary to respond. Generally, the last frame sent by a primary is a poll frame requiring the secondary to reply to the message sent. Messages sent from the secondary can be a final frame (b3 = 1) or not a final frame (b3 = 0). Again, generally, the last frame sent by a secondary is a final frame. The primary is not required to respond to a final frame. It is used to inform the primary that the secondary is sending its last frame.

The remaining bits in the control field of an information frame are interpreted in groups of three for SDLC. One group contains the frame number being sent (**NS**) and is composed of b4, b5, and b6. Bits b0, b1, and b2 yield the frame number expected to be received next (NR) from the replying station. These frame numbers are used to keep track of the number of successful frames sent and received. An example best illustrates how they work.

Example 4.4   Starting with a reset condition, how does the frame numbering sequence operate to keep track of frames?

Solution: For now, a reset condition sets all the number bits to 0. Initially, the primary sends a single polling frame. The control frame in binary (LSB on the right) is 00010000. NS and NR are both 0, indicating that this is the first frame sent and that the primary expects frame number 0 as the first frame in reply. A high in b3 makes this a poll frame, and a low in b7 designates this as an information frame. The secondary replies by sending three information frames:

00100000

00100010

00110100

Realize that each of these control fields is part of a complete message and is extracted for illustration purposes. All of the control fields have an NR of 001, telling the primary that the secondary expects frame number 1 as the next frame to be sent by the primary. Bit b3 is low for each of the first two fields and high for the last control field. The last field is the final frame sent by the secondary in this sequence. The NS indicate that frames 0, 1, and 2 have been sent.

Suppose the primary receives the first two frames without any problem, but something goes wrong with the third frame. The primary wants this frame retransmitted. The primary also has five frames to send. To accomplish this the primary sets its NR to 010. This tells the secondary that the primary received frames 0 and 1 all right but did not receive frame 2. The secondary, when it finally replies, starts by sending frame 2 and any additional frames it has ready to send. The primary's control fields for the next five frames are:

01000010

01000100

01000110

01001000

01011010

Now suppose this system really gets fouled and the receiver does not get frame number 011 correct from the primary. The secondary has two frames to send besides the one to be retransmitted. It will send an NR of 011 with the three frames it sends. The primary will respond by transmitting frames 3, 4, and 5. There is no way for a sending station to detect whether frames following a bad frame were good or bad, so it must retransmit all frames starting from the bad frame until the last frame is sent. The sequence, first from the secondary and then from the primary, is as follows:

sent by the secondary	sent by the primary
01100100	10100110
01100110	10101000
01111000	10111010

The secondary is required to reply to the poll to complete the dialogue.

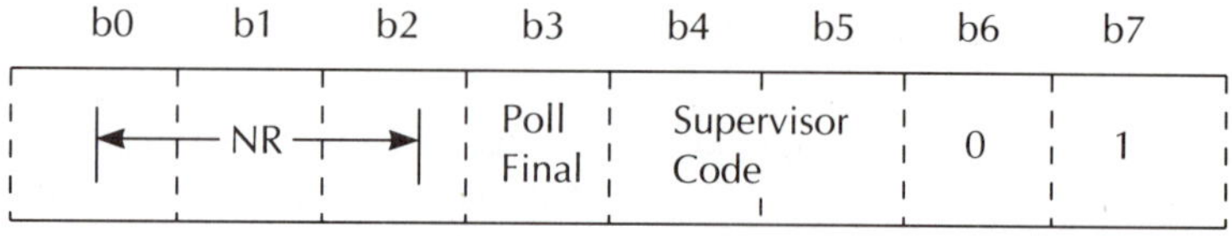

b0	b1	b2	b3	b4	b5	b6	b7
←— NR —→			Poll Final	Supervisor Code		0	1

Supervisor Code: b4	b5	Meaning
0	0	Ready to receive
0	1	Not ready to receive
1	0	Reject
1	1	Frame reject*

*HDLC only

Figure 4.4 Supervisor Frame Control Field

A maximum of seven consecutive frames can be sent by a station without a response from the receiving station. When frame number 7 (111) is followed by the next frame, the NS and NR wrap around to 000 and resume counting. This is the reason for the limiting of the maximum consecutive frames to seven. A request to retransmit frame 2 (010) in response to a sequence of more than seven frames in a row would be confusing. Which frame 2 is to be retransmitted, the first one or the second one?

In the HDLC control field, both NS and NR are expanded to 7 bits, increasing the maximum consecutive frames to 127. A logic 0 state of the LSB of the control frame still indicates an information field, and the bit between the NS and NR numbers still indicates polls and final frames.

SDLC and HDLC supervisory frames (Figure 4.4) are recognized by b7 = 1 and b6 = 0. There is no NS number. Instead, a code is used to define the type of supervisory frame being sent. Supervisory frames do not include text within the message. All supervisory information is conveyed by bits b4 and b5. NR (bits b0, b1, and b2) and poll/final function of bit b3 remain as before.

Supervisory fields are most commonly sent by a station if it has no information to send in reply to a sending station but it is desirable to maintain the dialogue. For example, a primary may have twenty frames of information to send and the secondary only two. After the secondary sends its two frames, it uses supervisory "ready to receive" formats in response to primary polls to maintain the dialogue while the primary completes sending its information.

The "not ready to receive" function informs the primary that the secondary is off-line and incapable of receiving future messages. The "reject code" informs the primary that the frames sent starting with the one indicated by the NR will have to be retransmitted. HDLC includes an additional supervisory code for frame reject. This is used by HDLC to indicate a specific bad frame. In the earlier scenario, if a bad frame is detected, the receiving station indicated this by sending the appropriate frame number in the NR. The sending station was then required to retransmit all frames beginning with the bad frame. With HDLC, if only one frame is bad, that frame number can be indicated using the "frame reject" supervisory message. The sending station is then required to retransmit only the faulty frame.

The third frame type, the unnumbered frame, is recognized by bits b6 and b7 both being high. Bit b3 still signifies polls and final frames as before. The remaining bits in the control field define the type of unnumbered frame being sent. NR and NS are not used with unnumbered frames. These frames are used by the primary to send commands to the secondary and by the secondary to send status messages or to request an action from the primary. Table 4.1 details the control field codes and their related unnumbered frames. Hexadecimal equivalents of the control field are used with bit b3 high (poll and final frame condition) in one column and low in the other. The next columns give the unnumbered frame acronym the originator of the frame (primary or secondary station) and a brief description of the frame. Essentially, request frames are initiated by the secondaries and command frames by the primary station. The reason for this lies at the very core of a multipoint system—the primary always has control of the system and is the only station that can issue commands.

If a secondary wishes to go to a local mode, it must first request the primary to disconnect it from the line by sending a **REQUEST DISCONNECT (RD)** frame to the primary. The primary then issues a **DISCONNECT (DISC)** command, which allows the secondary to go off-line. The secondary can be placed back on-line when the primary sends a **SET NORMAL RESPONSE MODE (SNRM)** frame.

Table 4.1 Unnumbered Frames

Binary Code	Hex Equivalent P/F: 0	1	Frame Name	Sender	Description
000*0011	03	13	UI	Both	Information
000*0111	07	17	SIM	Primary	Set initialization
000*0111	07	17	RIM	Secondary	Request initialization
000*1111	0F	1F	DM	Secondary	Disconnect mode
010*0011	43	53	RD	Secondary	Request disconnect
010*0011	43	53	DISC	Primary	Disconnect command
011*0011	63	73	UA	Both	Acknowledge
100*0011	83	93	SNRM	Primary	Set normal mode
100*0111	87	97	FRMR	Secondary	Frame reject

*P/F bit.

7E Flag	Secondary Address Field	FRMR Code 87 or 97	Rejected Frame Control Field	0-NS-0-NR-b8 b9 b10 b11	FCS	7E Flag

Figure 4.5 Frame Reject Format

NR and NS numbers are reset to 0 with the SNRM command. When a primary polls a secondary that is off-line, the station is still required to respond to the poll. In the local or disconnected mode, it does this by sending a **DISCONNECT MODE (DM)** status frame. The secondary can also request to be placed back on-line by sending a **REQUEST INITIALIZATION MODE (RIM)** in response to a primary's poll. This is sent in place of the DM status message. A primary would then respond with an SNRM, which reinitializes the secondary's NR and NS numbers and places that secondary back on-line.

When communication is first established with a secondary, the primary issues a **SET INITIALIZATION MODE (SIM)** command. This command's function is to initialize the secondaries' NR and NS numbers to 0 so that frames can be correctly tracked. In this manner the primary controls the start of the dialogue with a secondary station.

The remaining frames include an **UNNUMBERED INFORMATION (UI)** frame, **UNNUMBERED ACKNOWLEDGE (UA)** frame, and a **FRAME REJECT (FRMR)** frame. UI is used whenever text outside of normal traffic is to be sent between the primary and secondary—for instance, details about problems one station is having. UA frames are used by the secondary or primary as a means of acknowledging an unnumbered transmission.

Frame reject is used by the secondary to inform the primary as to why the secondary is rejecting the frame sent by the primary. The format for a frame reject frame is shown in Figure 4.5. The flag, secondary address, control field (containing the FRMR code), frame check sequence, and ending flag constitute the standard SDLC/HDLC frame format. In the text field, instead of character code information, a specific bit format is employed to identify the problem. Sixteen bits are used to fit the format shown in Figure 4.5. The NR and NS are those frame numbers at the point at which the frame was rejected. A high state in one of the bits 8, 9, 10, or 11 specifies the reject cause. These causes are problems other than error detection requiring simple retransmission. Those are satisfied using the NR/NS system detailed earlier.

Bit b8 denotes an invalid frame received by the secondary. This is used if a station sends an unnumbered code in the control field that is not a code included in the protocol. The received station cannot interpret the frame, so it sends an unnumbered reject with bit b8 set high. Bit b9 is used when text is sent with a field that does not allow text. If text is included with a supervisor frame, this frame would be rejected and bit b9 would be set. Bit b10 indicates when too many consecutive frames are sent without a poll frame (more than 7 for SDLC and 127 for HDLC).

Finally, bit b11 indicates when an NR does not agree with the transmitted NS. For example, three frames are sent by the primary, with NS of the frames being bits b3, b4, and b5. The secondary responds with a ready to receive supervisor frame setting the NR to 6. The primary starts its next sequence by sending a frame with an NS of 7 instead of an NS equal to 6 as it was expected to do. The secondary is confused because it expected the primary to send frame number 6 first. The secondary has no idea why the primary did not follow the correct sequence. In response to the poll frame, the secondary sends a frame reject. In it, the secondary includes the frame number (6) that the primary was expected to send and sets bit b11 of the text field high, informing the primary that it transmitted out of sequence.

The next field in SDLC frames that allow text is the text field, which follows the control field. Text is generally sent in groups of bytes using EBCDIC character code for SDLC and any character code for HDLC. Error detection is accomplished using CRC-16. The error characters are computed on the data following the starting 7E flag up to and including the last character of text. The CRC-16 bytes, called a **FRAME CHECK SEQUENCE (FCS),** are placed following the text and preceding the ending 7E flag. The ending flag completes an SDLC/HDLC frame sequence.

Example 4.5   How does the SDLC message Help! appear transmitted as the final frame number 5 from secondary 3E? Assume the primary already sent three frames.

Solution: Start by determining the contents of the control field. This message is an information frame, so bit b7 is low. NS is 101 to indicate frame 5 is being sent. NR is set to 011 since frame number 3 is expected from the primary. Frame 3 is the fourth frame, since the first frame is 000. Bit b3 is high, indicating that this is a final frame. Combining this information into the control field produces 01111010, or 7A in hexadecimal.

To clarify where each piece of information is placed, hexadecimal bytes and the fields of the frame are separated by a space. This is *not* the way it is transmitted. Actual frames do not have the spaces; they are shown here in this form for clarity only. The message is:

7E 3E 7A C8 85 93 97 5A FD C2 7E

The first 7E is the starting frame flag. This is followed by the secondary's address, 3E. 7A is the control field derived above. The next five hexadecimal bytes are the EBCDIC code for Help! FD C2 is the actual CRC-16 bytes computed for the data starting with 3E and ending with 5A. The FCS is followed by the 7E ending flag.

● 4.5 ADDITIONAL HDLC DIFFERENCES

As with SDLC, the secondary using HDLC is required to respond if the primary sends a poll frame. However, in the **ASYNCHRONOUS RESPONSE MODE (ARM),** specified in the HDLC protocol, the secondary can also respond to nonpoll frames. The final (f) bit no longer indicates a final frame from the secondary. It now indicates when the secondary is responding to a poll frame. This bit is low for secondary responses to nonpoll frames.

With the **ASYNCHRONOUS DISCONNECT MODE (ADM)** of HDLC, the secondary can initiate a disconnect (DISC) and go ''off-line'' at any time. It no longer requires a command to be sent from the primary. The secondary is still required to issue a disconnect status (DM) in response to a primary's poll frame if the secondary is off-line (disconnected).

Data link protocols are used to establish the communications link between one or more stations. They set prescribed rules and procedures for originating calls and for station responses. Message formats and sequences and error-detection methods are also included within the protocol specifications. The four most common types of protocols are the character-oriented asynchronous and BISYNC protocols and the bit-oriented SDLC and HDLC protocols, summarized in Table 4.2.

● 4.6 XMODEM COMMUNICATIONS PROTOCOL

BULLETIN BOARDS are services that are made available to computer users for the sharing of information, programs, and electronic mail. In order to access one of the bulletin boards from a home computer, a user is required to have a computer, a modem, and some form of communications software. In addition, many bulletin board services require a fee and sign-up procedure before a user can access them. Among many services found on these boards are various computer user's clubs that list Public Domain software programs that anyone may use. Public domain

Table 4.2 Protocol Summary

Protocol	Data Type	Data Link Form	Transmission Size	Overhead (nonmessage/nongraphics)
Asynchronous	Asynchronous	Character	Message	Parity/framing bits/data link characters
BISYNC	Synchronous	Character	Block	Preamble/data link characters/CRC (BCC)
SDLC/HDLC	Synchronous	Bit	Frame	Flags/control field/CRC (FCS)

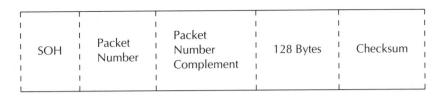

| SOH | Packet Number | Packet Number Complement | 128 Bytes | Checksum |

Figure 4.6 XMODEM Frame Format

generally means that there is no charge for the use of the software and no copyright restrictions. A similar type of program called **SHAREWARE** may require a small fee by its creator. Generally the fee is used to assure that the user gets updates to the shareware software. All this leads to the need for communications software to enable a user to ''down-load'' or copy these programs from the boards.

One such program in the public domain is called **XMODEM.** XMODEM, written by Ward Christensen, has been around for quite a while. It is an asynchronous file transfer communications protocol program. Data are transferred on the local public telephone system at data rates of 1200 bps or less. It uses ASCII codes embedded in 8-bit codes—that is, the 7-bit ASCII code is expanded to 8 bits by adding a leading 0. A form of ASCII called extended ASCII, which includes many printable characters not specified in the original ASCII code, is equally usable. The limitations of data rate and distance result from an increase in error rate at higher data rates or degraded telephone line signals that occur on long-distance lines.

XMODEM transfers data in blocks of 128 bytes of data surrounded by framing information (Figure 4.6). These blocks or packets begin with an SOH character followed by a packet number and its complement. The data field and a checksum error-detection character round out the XMODEM frame. XMODEM uses a stop-and-wait process of half-duplex transmission. A frame is sent, and then the transmitter must stop and wait for an acknowledgment from the receiver that the frame was received. The packet numbers tell the receiver if the correct frame has been received. They follow sequentially, and if one is missing, the receiver can detect it by an incorrect packet number.

ACKNOWLEDGE (ACK) characters are used as positive acknowledgments and to terminate a file transfer. **NEGATIVE ACKNOWLEDGE (NAK)** characters reflect a bad checksum result and are also used to initiate a file transfer. The checksum used by XMODEM is created by first adding the data bytes together, ignoring all final carry bits. The resulting value is divided by the hexadecimal value FF. The remainder of that division is the checksum byte. The receiver will repeat the process and compare checksums. If they do not match, the receiver will send an NAK and the sender will be obliged to retransmit the bad frame.

The first character sent to begin a file transfer is an NAK character. The receiver returns an ACK telling the sender that it is ready to receive messages. After each frame sent, the receiver acknowledges it with an ACK or NAK depending on the checksum results. The last frame must end with an EOT character. The receiver is then required to send a final ACK to end the file transfer. If a catastrophic problem arises, transmission can be aborted by sending a **CANCEL (CAN)** command. One indication of a major problem is the failure to receive a good packet after retransmitting it ten times. After the tenth time, the receiver acknowledges that something must be inoperable in the system by sending a CAN causing the file transfer to be aborted.

XMODEM is easy to implement using higher-level programming languages such as PASCAL or C. It requires a 256-byte buffer in the computer memory to handle the transferred data. Since it is a public domain program, the original cost is negligible. Numerous computer and telecommunications companies include XMODEM in their software packages. Other equally available communications protocol packages exist that overcome the data rate, distance, and character-size limitations of XMODEM. One such program that is popular today is **KERMIT.**

KERMIT was developed by Columbia University and is essentially similar to XMODEM. Its main difference from XMODEM is that it is designed to run at full duplex and at higher data rates. While XMODEM operates between **PERSONAL COMPUTERS (PCs)** over the phone lines, KERMIT is used to communicate between PCs, minicomputers, and mainframes. There are many versions of KERMIT, and the user is advised to consult the specific documentation accompanying the copy of KERMIT in use.

SUMMARY

Data link protocols were developed to establish a standard set of rules to govern the establishment and maintenance of a communications link. For character-oriented protocols, numerous data link characters such as STX and ESC are used to define when messages begin and how the text is to be handled. Methods for error detection are also specified within the protocol. Bit-oriented protocols rely on specific groups of bits for the definition of message beginnings and ends, the definition of each message unit (frame), and a method for flow control (frame number sequencing).

A number of public domain communications software packages provide a basic protocol such as KERMIT and XMODEM for everyday use by business and home computer enthusiasts.

GLOSSARY

Attributes—Those accents of a message that bring attention to a part of the message. They include boldface, underline, blinking, etc.

Asynchronous protocol—Character-oriented asynchronous data protocol.

BISYNC—Character-oriented synchronous data protocol.

Broadcast address—In a multipoint system, the address used by the primary to send a message to all of the secondaries in the system.

Bulletin boards—On-line message centers that are sometimes general and sometimes oriented to a specific topic.

Contention—Two or more stations vying for use of a line.

Frames—Message unit in the SDLC/HDLC protocol.

Group address—In a multipoint system, an address used to designate a number of secondaries but not all secondaries.

Handshake—Message used for secondary responses when secondary has no message to send.

HDLC and **SDLC**—High-level and standard synchronous data link control bit-oriented protocols.

KERMIT—Communications protocol program.

Poll—Message sent by a primary station asking if a secondary station has traffic to send.

Protocol—Set of rules for establishing data links and message interpretation.

Selection—Message sent by a primary asking a secondary station if it is ready to receive traffic.

Shareware—Public domain software that has a licensing fee. This fee is usually low and covers the cost of creating and updating the shareware programs.

Traffic—Data messages.

Transparency—Sending and receiving messages without mistaking data as data link control characters.

XMODEM—Popular communications protocol used by home PC and small systems users.

Zero stuffing or **insertion**—Addition of zero level bits to avoid mistaking a series of 1 data bits as a SDLC flag.

QUESTIONS

4.1. Define protocol. Why are data link protocols necessary?

4.2. What is the difference between a poll and a selection? When is each used?

4.3. Of the four data link protocols discussed in this chapter, asynchronous, BISYNC, SDLC, and HDLC, which one(s)
  a. Is used with asynchronous data
  b. Uses CRC-16 exclusively for error detection
  c. Uses EBCDIC exclusively as a character code
  d. Can send messages in blocks
  e. Can send messages in frames
  f. Is bit-oriented
  g. Is character-oriented
  h. Makes no distinction between a poll and a selection
  i. Uses frame numbers to indicate retransmission needs

4.4. Under which circumstances must a secondary respond to a primary's message?

4.5. What responses in the asynchronous and BISYNC protocols are used for a positive acknowledgment? A negative acknowledgment?

4.6. What is the purpose of the following BISYNC characters?
  a. STX　　b. DC1　　c. EOT　　d. GS　　e. SYN

4.7. What is a BISYNC handshake message used for? What is the sequence of characters in the handshake message?

4.8. When is a 7E flag sequence recognized as a flag in an SDLC message? What is used to prevent such a sequence from being mistaken for a flag?

4.9. What types of message frames can be sent with SDLC?

4.10. Which SDLC frame type never includes a data field?

4.11. What are the six fields of an SDLC information frame?

4.12. Which address is used in the SDLC address field?

4.13. What sequence of bits is used by the SDLC protocol to recognize a message abort?

4.14. What is the maximum number of stations that can be used with an SDLC protocol?

4.15. Which station in a multipoint environment sends unnumbered command messages? Unnumbered request messages? Unnumbered status messages?

4.16. What is the effect of sending an unnumbered set initialization mode command? Which station does it affect?

4.17. How do the fields in an HDLC message differ from an SDLC message?

4.18. How many consecutive SDLC messages can be sent without an acknowledgment from a secondary station? How many HDLC messages?

4.19. Compare the number of stations that can be used with HDLC and with SDLC.

4.20. Which supervisor frame is included in the HDLC protocol and not the SDLC protocol? What is that frame used for?

4.21. How do the receivers in a multipoint environment know when an HDLC address is completed?

4.22. What are the types of rejects that can be specified in an unnumbered frame reject message?

4.23. What is the XMODEM protocol used for?

4.24. Describe the fields of an XMODEM frame.

4.25. How many bytes of data are sent with each XMODEM frame?

4.26. How does the XMODEM checksum differ from the checksum method described in chapter 3?

PROBLEMS

4.1. An SDLC frame is sent from the primary to the secondary with a control field of 10110010 (least significant bit to the right).
   a. What type of frame is it?
   b. Is it a poll frame?
   c. What is the frame number being sent?
   d. What is the frame number of the expected response frame?

4.2. Three frames are sent by the primary, with the last frame's control field being 11010110. The secondary responds with a first frame control field of 01001100.
   a. Is this response correct?
   b. What must the primary do in response to this frame?

4.3. Stuff zeros in this message wherever appropriate. The message is framed correctly with one start and one ending flag. The least significant bit is at the left and is the first bit transmitted.

0111111000111010101110011111110000111100111111100001111110

4.4. Compare the efficiency of sending a 512-byte message using asynchronous, BISYNC, and SDLC protocols. Assume EBCDIC character code is used for each message. For the character-oriented protocols, 30 bytes are used for graphic characters (including ESC, NL, DC, GS, and attribute characters). BISYNC sends the message in three even blocks, and SDLC uses three frames to complete the transmission. CRC-16 is used for error checking for the synchronous protocols. Asynchronous data use 1 start, 1 stop, and even parity.

4.5. Develop the HDLC message for a single frame with the following requirements:
   a. Information frame to station address 3CD4F9 (in hex).
   b. Frame number 39 (decimal) is being sent, and frame number 78 (decimal) is expected as the next frame from the secondary.
   c. This is the final frame in the sequence.
   d. EBCDIC characters are used for the data field.
   e. The message is: HDLC is a more versatile Protocol.
   f. CRC-16 is used for error detection.

Answers to Odd-Numbered Questions

4.1. A set of rules for successful data link communications. They are required to establish and maintain control over a communications link

4.3. a. Asynchronous
   b. BISYNC and SDLC
   c. BISYNC and SDLC
   d. BISYNC
   e. SDLC and HDLC
   f. SDLC and HDLC
   g. Asynchronous and BISYNC
   h. SDLC and HDLC
   i. SDLC and HDLC

4.5. Asynchronous positive acknowledge = \ACK negative \\ or ** BISYNC positive acknowledge = ACK0 (DLE0) and ACK1 (DLE1), WACK (DLE;) negative = NAK

4.7. When secondary has no message but must respond to primary, it sends a handshake message:

SSEP
YYOA
NNTD

4.9. Information, supervisory, unnumbered

4.11. Start flag, secondary address, control, text, FCS, end flag

4.13. Seven to fourteen consecutive 1s

4.15. Primary sends unnumbered commands. Secondary sends unnumbered requests and status messages.

4.17. Larger addresses, longer NS and NR numbers, additional supervisory code, larger data fields allowed

4.19. SDLC's address field is limited to 1 byte, which limits the total number of addresses to 256 including the broadcast address. HDLC's address field is unlimited, with the specification that the last byte of the address contain a least significant bit state of 1. This allows for an unlimited number of stations to be addressed using HDLC.

4.21. b0 of last address field byte is high

4.23. XMODEM is used for low-speed data communications. It is very popular with home PC and bulletin board users

4.25. 128

Solutions to Odd-Numbered Problems

4.1. a. Information
      b. Yes
      c. 1 (001)
      d. 5 (101)

4.3. 01111110001110101011001111101100001110011111011000011111110

4.5.

7E	3C	D4	F9	9D	4E	C8	C4	D3	C3	40	89	A2	40	81
flag	address		contr.	H	D	L	C			sp	i	s	sp	a

40	94	96	99	83	40	A5	85	99	A2	81	A3	89	93	85
sp	m	o	r	e	sp	v	e	r	s	a	t	i	l	e

40	D7	99	96	A3	96	83	96	93	48	xx xx	7E
sp	P	r	o	t	o	c	o	l	.	CRC	flag

Five • Low-Speed Data Communications

OBJECTIVES

The objective of this chapter is to acquaint the student with the elements of a low-speed data communications system. The chapter begins with a general discussion of the components of a frequency shift keying communications link. This is followed by sections on:

1. Universal asynchronous receiver/transmitter (UART)
2. Motorola's ACIA as a specific UART example
3. Intel's 8251 USART
4. RS232C or other serial data interface standards
5. RS232C drivers and receiver buffers
6. FSK modems
7. Intel's 2970 FSK modem
8. Data access arrangement

ACRONYMS

ACIA—Asynchronous Communications Interface Adapter
CCITT—International Consultative Committee for Telegraphy and Telephony
CD—Carrier Detect
CPU—Central Processing Unit
CTS—Clear To Send

DAA—**D**ata **A**ccess **A**rrangement

DCD—**D**ata **C**arrier **D**etect

DCE—**D**ata **C**ommunications **E**quipment

DSR—**D**ata **S**et **R**eady

DTE—**D**ata **T**erminal **E**quipment

DTR—**D**ata **T**erminal **R**eady

FSK—**F**requency **S**hift **K**eying

IRQ—**I**nterrupt **R**e**Q**uest

MODEM—**MOD**ulator **DEM**odulator

PC—**P**ersonal **C**omputer

PLL—**P**hase **L**ock **L**oop

RLSD—**R**eceived **L**ine **S**ignal **D**etect

RTS—**R**equest **T**o **S**end

SIO—**S**erial **I**nput/**O**utput interface

SYN—**S**ynchronization character

SYNDET—**SYN**chronization character **DET**ect

UART—**U**niversal **A**synchronous **R**eceiver/**T**ransmitter

USART—**U**niversal **S**ynchronous/**A**synchronous **R**eceiver/**T**ransmitter

VCO—**V**oltage **C**ontrolled **O**scillator

● 5.1 INTRODUCTION

Most home data communications applications involve the use of **FRE-QUENCY SHIFT KEYING (FSK)** modems connected to **PERSONAL COMPUTERS (PCs).** This system of data transfer links was also the prime method used in the early days of data communications. Today, operating with FSK links is considered low-speed data communications utilizing data rates between 300 and 1200 bps. At the time of this writing, FSK modems operating at 2400 bps are being introduced onto the market. A low-speed system consists of the data terminal, which includes a **UNIVERSAL ASYNCHRONOUS RECEIVER/TRANSMITTER (UART)** and a **MODULATOR/DEMODULATOR (MODEM)** at each communication site. Also required for a number of applications is a **DATA ACCESS ARRANGEMENT (DAA)** to allow modem data lines to be interfaced onto the telephone system.

The basic units of a low-speed data communications system are illustrated in Figure 5.1. Regardless of the original source of information, it is eventually presented to the communications network as a parallel binary word, usually of either 8 bits (byte), 16 bits (referred to as a word or half-word, depending on the system developers), or 32 bits (long word or word, depending on the system used). The deciding factor on the size of the parallel data (and, hence, the actual devices used in the data communications system) is the size of the data bus employed by the **DATA TERMINAL EQUIPMENT (DTE)** that delivers the data. The functional block which receives the parallel data from the computer terminal that is to be sent eventually to the secondary station is the UART.

The UART accepts the parallel data information and converts it into a serial data stream. Asynchronous start and stop framing bits and a parity bit, if used for error checking, are added by the UART to the data stream. On the receiving side, the UART detects the presence of a start bit and reads in the serial stream. The data are converted back to parallel form and checked for errors. The parallel data are presented out on the data lines from the UART to the DTE data bus.

Once a communication link has been physically established, the transmitting side of each station places a mark or logic 1 level on the data lines. As long as the data link remains active without data present on the line, the line remains in an idle line 1 or marking state. The beginning of a digital character is detected by the reception of a start bit, which is a change from the mark condition to logic 0 or space state. Character information bits follow the start bit, then the parity bit, and finally the stop bits. The number of stop bits actually determines the minimum bit time a line must return to the mark condition between characters. Each character is synched with the UARTs receiving circuits by the detection of the start bit. Figure 5.2 shows a complete serial data stream for an ASCII character Q as it is sent using asynchronous data format.

To utilize telephone lines, which were designed to carry audio signals, the serial data stream leaving the UART is converted from digital form to analog form at the transmitter and returned to digital at the receiving side. Two audio tones need to be generated, one for the mark logic 1 level and one for the space logic 0 level. The rate at which the digital information changes in bits per second (bps) causes the frequency to change from mark to space values at the same rate. The symbols for an FSK modem are the frequencies of the signals being transmitted. Each symbol represents 1 data bit.

The task of converting digital levels into analog signals is performed by the modem or data set, which is one type of **DATA COMMUNICATIONS EQUIPMENT (DCE)**. The modem has additional circuitry to detect the presence of the receive tone on the telephone line. A **HANDSHAKE** control signal called **CARRIER DETECT (CD)** is used by the modem to signal the terminal that it has detected the received signal tone. Handshaking is a process by which control signals are used to establish the validity of a communications link.

132

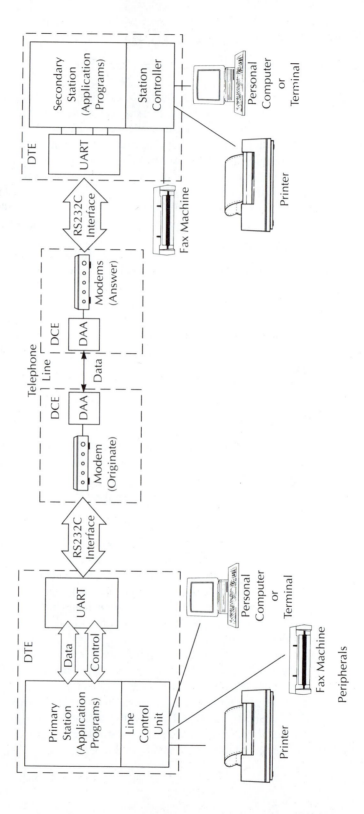

Figure 5.1 Basic Units of a Low-Speed Data Link

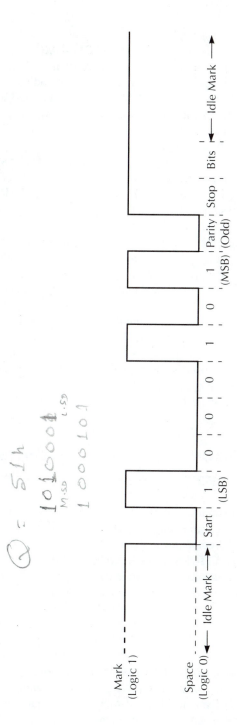

Figure 5.2 Asynchronous ASCII Q

Local handshaking between the modem and data terminal equipment is accomplished by the use of two signals, **REQUEST TO SEND (RTS),** sent by the DTE to initiate a data path through the modem, and **CLEAR TO SEND (CTS),** returned by the modem once it is ready to send and receive data. A delay between RTS and CTS called an RTS/CTS delay may be used to allow time for modem circuits to come on-line and stabilize. Any required internal synchronization by the modem also takes place during this time. Some systems require the establishment of the RTS/CTS sequence before any form of tones is sent out on the telephone lines, while others are configured to begin the RTS/CTS sequence after the DTE receives an active carrier detect (CD), also known as **DATA CARRIER DETECT (DCD),** and **RECEIVED LINE SIGNAL DETECT (RLSD).** The CD signal is generated in response to the detection, by the modem, of the presence of a mark or space signal on the telephone lines.

Numerous manufacturers produce computer terminals and modems. Some modems are made to be interfaced with a specific terminal, such as the internal modem cards that are installed in IBM PCs and compatibles. Most other modems are made to "stand alone" and interface with any computer terminal. In order to do this, both the terminal and the modem must be equipped with standard interface capabilities. The job of providing this interface is outlined in the RS232C standard briefly introduced in chapter 1. This standard outlines the electrical and physical characteristics of the connectors and cabling used to interface between DTEs and DCEs for data transmissions below 20 kbps at distances under 50 feet.

This, then, is the essence of an FSK data communications network. It is time now to explore the separate functional areas in greater detail.

● 5.3 UNIVERSAL ASYNCHRONOUS RECEIVER TRANSMITTER

There are quite a number of UART devices on the market today. Most of them are manufactured on a single IC chip, which is incorporated onto a circuit board installed in the computer terminal itself. One such IC, the **ASYNCHRONOUS COMMUNICATIONS INTERFACE ADAPTER (ACIA)** produced by Motorola, is used in this text to illustrate the application of UART functions. The ACIA is selected since it has the basic elements of most UARTs. Figure 5.3 is a block diagram and pin assignment for the MC6850 ACIA.

The ACIA consists of three main sections: transmit, receive, and control unit. It was designed to be used with Motorola's 6800 microprocessor-based computing systems. However, it is sufficiently universal for application with any **CENTRAL PROCESSING UNIT (CPU).** Eight bidirectional data lines (D0–D7) interface between the ACIA and a computer's data bus to allow parallel byte-size data to be transferred between the CPU and ACIA. Three chip selects (CS0, CS1, and $\overline{CS2}$), along with the register select (RS) input lead, allow the CPU to directly access the registers within the ACIA. The chip selects are asserted as a result of address

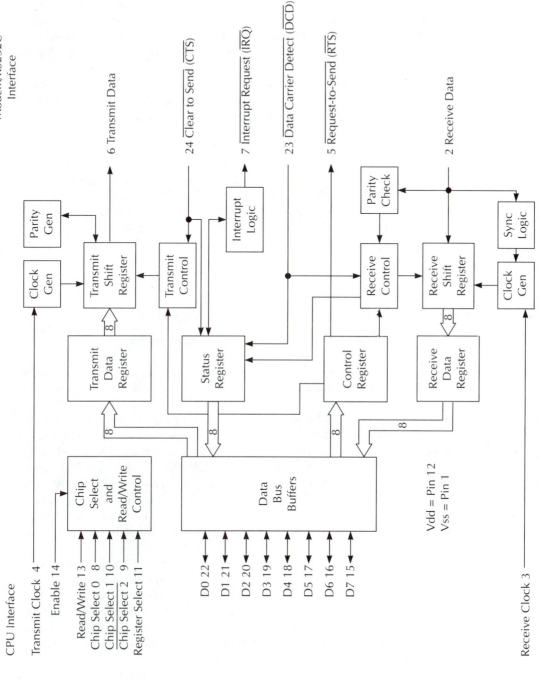

Modem/RS232C Interface

CPU Interface

Transmit Clock 4

Enable 14

Read/Write 13
Chip Select 0   8
Chip Select 1 10
Chip Select 2   9
Register Select 11

D0 22
D1 21
D2 20
D3 19
D4 18
D5 17
D6 16
D7 15

Receive Clock 3

6 Transmit Data

24 Clear to Send (CTS)

7 Interrupt Request (IRQ)

23 Data Carrier Detect (DCD)

5 Request-to-Send (RTS)

2 Receive Data

Parity Gen

Clock Gen

Transmit Shift Register

Transmit Data Register

Transmit Control

Interrupt Logic

Status Register

Control Register

Receive Control

Parity Check

Receive Shift Register

Sync Logic

Clock Gen

Receive Data Register

Data Bus Buffers

Chip Select and Read/Write Control

Vdd = Pin 12
Vss = Pin 1

Figure 5.3 ACIA Block Diagram (Courtesy of Motorola, Inc.)

decoding circuitry within the computer. The register select is usually driven by the lowest address line. Because of this, the ACIA is easily included in a computer's memory map area. No provision is made to distinguish between memory or input/output addressing to access the ACIA. This flexibility permits the ACIA to be included as an I/O port as well as part of a computer's memory map.

The enable (E) input is connected to the computer's clock signal to provide timing for ACIA registers. The remaining CPU interface signal is the R/$\overline{\text{W}}$ (read/write) input line that determines whether data are being read from or written into the ACIA.

On the modem interface side, transmit (TX) and receive (RX) data lines allow serial data transfers between the ACIA and the modem. The transmit and receive clocks (TX Clk and RX Clk) are used to establish the transmission data and receive sampling rates. $\overline{\text{CTS}}$, $\overline{\text{RTS}}$, and $\overline{\text{DCD}}$ are the clear to send, request to send, and data carrier detect handshake signals discussed in the previous section. Before the ACIA can be used to send and receive data, it needs to be programmed by the computer's operating software to configure it to transfer data in a given format and at a specific data rate.

● **5.4 DATA CHARACTER OPTIONS PERFORMED BY UARTs**

To be universal, a UART, and the ACIA is no exception, has to be able to recognize different data character formats and operate at different data rates. Since system data bus sizes are usually fixed, a byte-size data bus forces 8 bits of digital data to be sent to the ACIA by the central processing unit. However, not all 8 bits are necessarily used because some character codes use fewer than 8 bits to define a character. The ASCII code is an example of such a code since it requires only 7 bits to represent a single character. Others such as EBCDIC use all 8 bits. Older codes such as the Baudot code used a mere 5 digital bits per character. The ACIA is flexible enough to manage 7- or 8-bit codes but not any others. Zilog, the makers of the popular Z80 8-bit processor, makes a UART called the **SERIAL INPUT/OUTPUT INTERFACE (SIO),** which allows selectability of 5-, 6-, 7-, or 8-bit data sizes.

The minimum mark time between characters is set by the number of stop bits used to frame the end of the asynchronous character. The number of these bits is different for one system compared with another. They have been set at 1, 1.5 or 2 stop bit periods as required by the hardware or the software protocol in use. Most applications today have reduced the selection to 1 or 2 stop bits, allowing systems that used to use 1.5 to operate using 2 minimum stop bit periods. The other framing bit, the start bit, is always fixed at one bit time.

Error checking for asynchronous data is commonly performed by the use of parity as explained in chapter 2. UARTs provide selection options of even or odd parity or no parity at all. In the first two cases, the appropriate parity bit state for each character is generated and added to the character's data stream. When selecting no parity, some UARTs add an extra bit that is always in a low state. Others do not add a bit if no parity is selected. The ACIA falls into the latter category.

The last option performed by a UART is a combination of hardware signaling and software programming. Parallel data are converted to serial and shifted out at a specific data rate. This data rate is determined by the frequency of the TX Clk input and an internal divide ratio set in the ACIA. The clock frequency is divided by a ratio factor that generates the shift timing for the serial data stream. Today, the de facto standards for data rates used by FSK systems are 100, 300, 1200, and 2400 bps. The 100-bps rate has now all but disappeared from use, while the 2400 bps rate is emerging as the highest standard rate used for FSK communications on telephone lines.

Example 5.1   A serial data stream is to be transmitted at a rate of 1200 bps. What is the relationship between the transmit clock (TX Clk) and data rate if a ratio of 16 is used?

Solution: The internal circuitry of the UART divides the TX Clk by the ratio factor to produce a shift pulse. This shift pulse determines the rate at which the serial data bits are transmitted. For this example, a TX Clk of 19.2 kHz is required to generate a data stream at 1200 bps using a ratio of 16.

The ACIA has selectable ratios of 1, 16, and 64. Other UARTs, such as Zilog's serial input/output interface (SIO), have greater selectability in the choice of a value for this ratio. However those values used by the ACIA are the most common. The increased flexibility in the SIO facilitates a wider range of transmit clock frequencies for a given data rate. The use of the divide ratio and its importance have more significance when analyzing the receive side of the UART, which is done in section 5.6.

● 5.5 ACIA CONTROL UNIT

The options discussed in section 5.4 are selected by programming the appropriate state of the bits in the ACIA control register (Figure 5.4). Once the choices are made, a control word representing the desired contents of the control register is sent to the ACIA. The control register is accessed by the central processing unit by making the register select (RS) and the read/write (R/$\overline{\text{W}}$) lines low.

7	6	5	4	3	2	1	0
Receive Interrupt	RTS	Transmit Interrupt	Word Size	Stop and Parity		Counter Divide	
0 = Dis	0 = On	0 = Dis	0 = 7 bit	0 = 2 Stop	0 = Even	0 0 = ÷ 1	
		1 = En		1 = 1 Stop	1 = Odd	0 1 = ÷ 16	
1 = En	1 = Xmt (Transmit Interrupt Enable)	0 = RTS Off	1 = 8 bit	0 = No parity	0 = 2 Stop	1 0 = ÷ 64	
		1 = RTS on Transmit Break			1 = 1 Stop	1 1 = Reset	
				1 = 1 Stop	0 = Even		
					1 = Odd		

Figure 5.4 ACIA Control Register

The two least significant bits, b0 and b1, of the control register serve two purposes: (1) when they are both high, they cause the ACIA to reset, and (2) other combinations select the divide ratio between the transmit clock and data rate on the transmit data line. This same ratio is used to set the sampling rate for the ACIA receive side. A reset clears any pending interrupt requests (discussed later) and resets status register flags except clear to send (CTS) and data carrier detect (DCD) to their inactive state. Bit b1 is set low and b0 high to select the divide-by-16 ratio. When this is done, the data rate is one-sixteenth of the transmit clock rate.

The opposite states of bits b1 and b0 select a divide ratio of 64, which would require a clock frequency of 76.8 kHz to produce the 1200-bps data rate of example 5.1. If bits b0 and b1 are both set low, the divide ratio selected is 1, which means the data rate and clock frequency would have the same value. This selection is used when the ACIA is adapted to pass synchronous rather than asynchronous data.

Bits b2, b3, and b4 are used to select the character format to be used. Bit b4 allows a choice between 8- or 7-bit character length. When bit b3 is high, the number of stop bits is set to 1. When this bit is low, most of the time the number of stop bits is 2. The one exception occurs when bit b4 is high and bit b3 is low. For this case, bit b2 is used to select between 1 or 2 stop bits as shown in Figure 5.4. Additionally, for this set of conditions, parity is not being used. For other combinations of bits b4 and b3, b2 is used to select between even (b2 = 0) and odd (b2 = 1) parity.

Bits b6 and b5 determine whether transmit interrupt requests are enabled or disabled. These bits also affect the state of the request to send ($\overline{\text{RTS}}$) handshake output of the ACIA. Details on ACIA interrupts are discussed in section 5.7 and on the RS232C handshake signals in section 5.9. Lastly, bit b7 enables or disables receive interrupt requests.

Example 5.2   What hexadecimal data are sent to the ACIA control register to configure it to handle ASCII asynchronous data with odd parity and 2 stop bits? Disable both interrupts and assert $\overline{\text{RTS}}$. Use a divide ratio of 16.

Solution: ASCII uses a 7-bit character code that requires bit b4 to be low. Bits b3 and b2 are set to 0 and 1, respectively, to pick odd parity and 2 stop bits. Bits b5 and b6 are low to disable transmit interrupts and assert $\overline{\text{RTS}}$. Receive interrupts are disabled by making bit b7 low. Lastly, bit b1 is set low and bit b0 high to select a divide ratio of 16. In binary the control register looks like the following:

b7	b6	b5	b4	b3	b2	b1	b0
0	1	0	1	1	0	1	0

which is 05 as a hexadecimal value.

● 5.6 CLOCK DIVIDE RATIO

The reason for a divide ratio between the input clock frequency and the data rate is to assure successful data reception by the UART. The input receive data line is constantly monitored for a character's start bit. Once a low-level state is detected, to verify that this condition is an actual start bit (as opposed to a noise pulse or dropout condition), the receiver monitors the receive data line for half a bit time period. For a divide ratio of 16 (sixteen clock cycles per data bit time), this takes eight clock cycles to occur. If the level remains low for this time period, the low condition is accepted as a start bit. Shift pulses are generated every sixteen clock cycles following the verify of the start bit. This shifts the data bits into the UART receiver at the center of each bit given ideal circumstances (the transmit clock is exactly sixteen times the incoming data rate). The shifting continues until the last stop bit is pulled into the receiver's shift register. The UART circuits then resume monitoring the receive data line for the next start bit, repeating the process for each character received. Because of this, the synchronization of each character occurs with each start bit. Figure 5.5 illustrates the timing relationship between the data stream and the shift pulses created by this process. An ASCII Q discussed earlier is used for this illustration.

Because each data bit is shifted in at its center, the transmit clock of the transmitter and receive clock of the receiver do not have to be *exactly* the same. It would be fairly difficult for these clocks, one at the **ORIGINATING STATION** and one at the **ANSWERING STATION,** to be exactly the same frequency. They are as close as possible but rarely exactly the same frequency. This difference, called **CLOCK SKEW,** allows the receive clock to be significantly different from

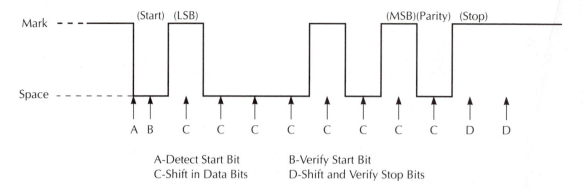

A-Detect Start Bit   B-Verify Start Bit
C-Shift in Data Bits   D-Shift and Verify Stop Bits

Figure 5.5 Receive Timing/Data Relationship for ASCII Q

the transmit clock without loss of any data information. The reason for this is that the shift process is reinitialized with every start bit. As long as the receive clock frequency allows the sampling and correct shifting in of every bit of a single character word, successful data reception occurs. The following discussion illustrates how much skew can exist between the receive and transmit clock frequencies and still allow successful data to be shifted into the receive side of the UART. The example uses an ASCII character length of 7 bits with parity and 2 stop bits (11 total bits) at a rate of 300 bps. The ideal receive clock, using a divide ratio of 16, is 4,800 Hz.

At 300 bps the time of a single bit is 1/300, or 3.33 ms. The total time from the start bit verify to the middle of the last stop bit is 10 × 3.33 ms, or 33.3 ms. The first half of the start bit and the last half of the second stop bit are beyond the shift area, as illustrated in Figure 5.6. To assure that a complete character is entered into the receive shift register, the last stop bit must be included. The last shift pulse must fall sometime during the last stop bit's time period. It can occur anywhere during that bit period (beginning, middle, or end). This allows the total time from the start bit verify to the last shift pulse to be between 9.5 bit periods (to the beginning of the last stop bit) and 10.5 bit periods (to the end of the last stop bit). Computing the total time from start verify to last shift pulse for both of these cases yields a range from 9.5 × 3.33 ms = 31.667 ms to 10.5 × 3.33 ms = 35 ms.

Sixteen clock cycles of receive clock occur between each shift pulse, for a total of 160 clock cycles between start verify and the last shift pulse. Dividing each of the total times by 160 gives the time period of a single clock cycle for both extremes. Taking the reciprocal of each cycle time period yields the receive clock frequency for each extreme. As long as the receive clock frequency is in that range, data will be successfully received by the UART. This all presumes that the transmit clock was exactly 4,800 Hz.

$$31.667 \text{ ms}/160 = 198 \ \mu\text{s} \qquad 35 \text{ ms}/160 = 218.8 \ \mu\text{s}$$
$$1/198 \ \mu\text{s} = 5,050 \text{ Hz} \qquad 1/218.8 \ \mu\text{s} = 4,570 \text{ Hz}$$

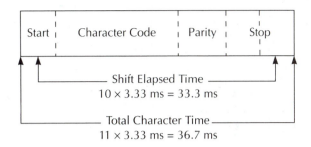

Figure 5.6 ASCII Character Time Period at 300 bps

The percentage of error or skew between the transmit clock of the sending station and the receive clock of the receiving station for this example is:

$$(5,050 - 4,800)/4,800 \times 100\% = 5.2\%$$

This means that there can be an error of $+/-5.2\%$ between these clock frequencies and data will still be successfully transferred between the stations.

Dividing the clock frequencies by 64 instead of 16 divides the sampling rate into smaller increments. This increases the accuracy in the detection and shifting in of the characters. It also provides for larger variations between the receive and transmit clock frequencies (though the percentage change remains the same). It should also be noted that it requires a much higher clock frequency (by a factor of 4) to generate data at a given rate. For 300 bps used in the preceding illustration, TX Clk and RX Clk frequencies of 19.2 kHz are required. This is not necessarily an advantage or disadvantage.

Interestingly enough, the originating station can use a divide ratio of 16 to send data and sample the received data, while the answering station can use a divide ratio of 64. Both stations require the character formats to be the same to be able to communicate. The only differences are the required transmit and receive clock frequencies. For this scheme to work, the answering station clocks would have to be four times the originating station clocks.

• 5.7 DATA LINK USING THE ACIA

Once the control register has been set (example 5.2) to configure ACIA, an active request to send ($\overline{RTS}$) is sent to the modem. The ACIA is ready to begin the sequence of operations to transfer data between two stations. It is assumed that both stations have similarly configured ACIAs and that both modems are on and have returned an active clear to send ($\overline{CTS}$) to their respective ACIAs.

b7	b6	b5	b4	b3	b2	b1	b0
$\overline{\text{IRQ}}$	PE	OVRN	FE	$\overline{\text{CTS}}$	$\overline{\text{DCD}}$	TDRE	RDRF

RDRF   — Receive data register full
TDRE   — Transmit data register empty
DCD   — Data carrier detect
CTS   — Clear to send
FE   — Framing error
OVRN — Overrun
PE   — Parity error
IRQ   — Interrupt request

Figure 5.7 ACIA Status Register

The process begins with the originating station placing a telephone call to the answering station. The answering station comes on-line and sends a mark signal back to the originating station. When the telephone company connects the call between the stations, a mark signal produced at the originating station is also sent out on the telephone lines. In order to send and receive both mark signals simultaneously on a single line, the frequencies of these mark signals are different. Filters in the modem receivers select the correct mark signal intended for that receiver. The received mark signals are acknowledged by each modem by asserting the data carrier detect ($\overline{\text{DCD}}$) line.

At the originating station, the application program directs the central processing unit to read the ACIA status register (Figure 5.7) and check the status of the $\overline{\text{CTS}}$ and $\overline{\text{DCD}}$ bits. The $\overline{\text{CTS}}$ informs the central processing unit that its own modem is ready, and the $\overline{\text{DCD}}$ status indicates when the answering station is on-line. Finally, the transmit data register empty (TDRE) bit is examined to assure the central processing unit that there are no current valid data in the transmit buffer register. If these three status bits are active ($\overline{\text{CTS}}$ and $\overline{\text{DCD}}$ low, and TDRE high), the program enters the data phase. This application will follow a full-duplex process in which data are sent and received at the same time. It is necessary to understand or at least accept that a microprocessor-type central processing unit (CPU) can execute many instructions in the time it takes to serially send or receive a data word (even at 1200 or 2400 bps).

After verifying that the TDRE bit is high, the CPU fetches and sends a message character to the ACIA. The parallel binary code for the character is stored in the ACIA's transmit buffer register, which causes the TDRE bit in the status register to go low. The ACIA adds framing and parity bits and automatically transfers the entire character to the transmit shift register. At this time the TDRE bit is returned high to indicate that the transmit buffer register is again available for data from the CPU. The data are then shifted out through the transmit data (TX) lines to the modem, which converts it to mark and space tones that are placed on the telephone lines. Because there are two registers placed between the data inputs to the ACIA and the serial data out, this device is referred to as a **DOUBLE-BUFFERED** device. Being double-buffered simply means that the ACIA has the ability to hold two data words at a single time (transmit buffer and transmit shift register data).

While the transmit process is in progress, the CPU again reads the status register. This time, the application program directs the CPU to monitor the receive data register full (RDRF) bit. Signals representing data sent by the answering station are converted into a digital data stream by the receive side of the modem and passed to the ACIA on the receive data (Rx) line. After the receive logic of the ACIA detects and verifies the presence of a start bit, the incoming data are shifted into the receive shift register. Once the appropriate number of bits are entered into the receive shift register, the complete character is transferred to the receive buffer register and the RDRF bit in the status register is set high. At this time the parity bit is checked to determine whether an error exists. A parity error causes a corresponding status bit to be set to indicate the presence of the error. Stop bits are also examined to verify that they are logic 1s. If either stop bit shows a low, the framing error bit in the status register is set.

The receiving side, like the transmitting side, is double-buffered. Once the receiver's data buffer register has been loaded from the shift register, monitoring circuits begin to look for the next start bit. It is now up to the CPU to read the receiver's buffer register and take the character into the computer via the data bus. Double-buffering in the ACIA gives the CPU a full word time to read the received character. If the CPU fails to do this before the next received character has been completely shifted in and transferred to the receiver's buffer register, a receiver overrun occurs. Overrun indicates that the current data in the receiver's buffer have overrun the last data that the CPU failed to read. This condition causes the corresponding bit in the status register to be set. After reading the received data, the CPU checks the status register for any parity (PE), framing (FE), or overrun (OE) errors and returns to the transmit operation to send the next character.

The sequence is repeated until the messages sent by both stations have been transferred and acknowledged. The applications programs, called *DATA LINK PROTOCOLS* (chapter 4), determine how the beginning and end of a message are recognized. This description illustrated a full-duplex operation accomplished through polling of the status register. The state of the bits in the status register guided the CPU toward which portion of its software programs to run. They included a transmit, receive, and error-response sequence. Details of these programs are embedded in various protocols.

All forms of data transfers can utilize the interrupt function in the ACIA to signal the CPU to perform a certain sequence. The establishment of the communications link takes place as it did before with the completion of the call and the assertion of both stations' data carrier detect ($\overline{DCD}$) lines. The originating station begins the process by sending the first data word. All subsequent transfers occur in response to interrupt requests sent by the ACIA to the CPU. This system allows the CPU to return to other types of processing between sending characters or between receiving them. The reason this works is the rapid speed at which CPUs operate (clock speeds for microprocessor CPUs range from 1 to 50 MHz) in comparison to the rate at which data are sent (using a maximum of 2400 bps for low-speed applications).

The data sent to the answering station enter the ACIA as before. When the receive data register full (RDRF) bit in the status register goes high, a receive interrupt occurs, causing the **INTERRUPT REQUEST (IRQ)** pin of the ACIA to go low. This signal is sent to the CPU, requesting it to interrupt its current processing and divert to its communications programs. The CPU reads the status register to determine whether the interrupt request is due to a receive data register full, error, or transmit data register empty. For this example sequence, since the answering station was not sending data, only the RDRF bit is checked. Sensing an active RDRF, the CPU at the answering station reads the data in and checks the parity and framing status bits for an error indication. If they are both low, the data are presumed correct and the CPU returns to its original processing.

Back at the originating station, after the CPU sent the first data word, the transmit data register empty (TDRE) bit in its ACIA status register goes low. As soon as the data are transferred to the transmitter's shift register, TDRE goes high and generates an interrupt request on the IRQ line. The CPU responds to this interrupt by fetching the next data word and sending it to the ACIA. As these data are stored in the transmitter's buffer register, the TDRE bit returns low and remains that way until the first word is completely sent out of the ACIA. The second word is transferred to the transmitter's shift register and TDRE goes high, causing the process to repeat until the message is fully sent. After each time the CPU sends a word of data, it resumes its original processing until it detects the next interrupt request.

The IRQ is made high (inactive) each time data are stored in the transmitter's buffer register or read from the receiver's buffer register. Error flags in the status register are reset by each read of the status register. This makes them available for use by the receiver logic in time for the next received data character.

● 5.8 TESTING THE SERIAL DATA STREAM

To verify that transfer of serial data is correct, a regular oscilloscope can be used, but it is difficult to keep track of the data as they are sent 1 bit at a time. A special type of test equipment called a **DATASCOPE** is available to facilitate the reading of serial data. It displays and holds a serial stream of data after the data have been applied to the datascope's vertical input. The first bit sent to it would be the start bit. This is followed, in time, by the least significant bit of the character word, the remaining character bits, parity, and stop bits. Figure 5.8 illustrates how a transmitted character would appear on the datascope.

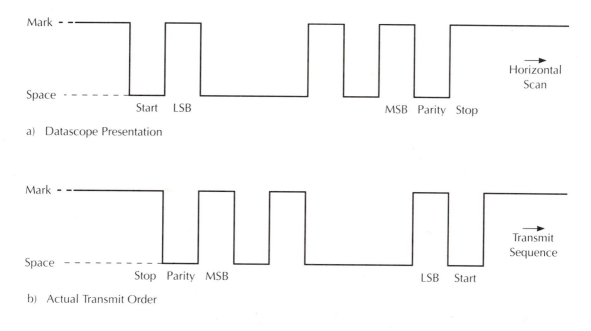

Mark

Space

Start LSB   MSB Parity Stop

Horizontal
Scan

a) Datascope Presentation

Mark

Space

Stop Parity MSB   LSB Start

Transmit
Sequence

b) Actual Transmit Order

Figure 5.8 Comparison of Transmit Data and Datascope Display
Using ASCII Q

● 5.9 INTEL'S 8251 USART

Another major manufacturer of integrated circuits, Intel Corporation, produces
a UART version known as the 8251A, which is also capable of handling
synchronous data transmissions. This **UNIVERSAL SYNCHRONOUS/ASYN-
CHRONOUS RECEIVER/TRANSMITTER (USART)** is designed to be used
with microprocessor-based computer systems. An 8-bit data bus (D0–D7) of the
USART (Figure 5.9) provides an information path to pass control, status, and
message information between the central processing unit (CPU) of the system and
the USART. The control information includes mode and control words used to
configure and direct the operation of the USART.

Additional interface signals, shown on the left side of Figure 5.9, are used by
the system to access the USART and control the flow of message data between the
CPU and the USART. The chip select ($\overline{CS}$) and control/data select ($C/\overline{D}$) inputs are
derived from decoding the system address selected to access the device. $\overline{CS}$, the
chip select, is necessary whenever the CPU writes to or reads from any of the
internal registers of the 8251A. $C/\overline{D}$, or control/data select, determines whether the
CPU is communicating with the chip's control or message data registers. This pin
is usually tied to the lowest address line (A0) to assure that addresses associated

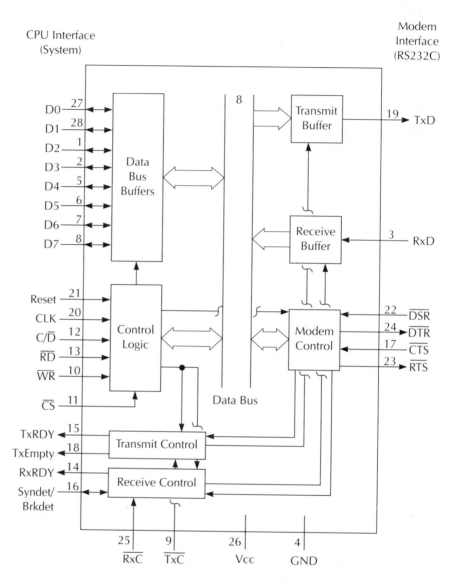

Figure 5.9 8251 Functional Diagram and Pin Outs

with the USART are sequential. For example, an address such as 5000H would be electrically decoded to assert the $\overline{\text{CS}}$ and select a data register (A0 = 0), while the next consecutive address (5001H) would also assert $\overline{\text{CS}}$ but would be used to access a control register (A0 = 1). The relationships between $\overline{\text{CS}}$, C/$\overline{\text{D}}$, read ($\overline{\text{RD}}$), write ($\overline{\text{WR}}$), and the internal registers of the USART are summarized in Table 5.1. When C/$\overline{\text{D}}$ is low, the transmiter data register is written into and the receiver data register is read by the CPU. When C/$\overline{\text{D}}$ is high, a read by the CPU gathers status register information, while a write sends control information to the USART.

Table 5.1 8251 Register Access

C/$\overline{D}$	$\overline{RD}$	$\overline{WR}$	$\overline{CS}$	Access
x	x	x	1	None
x	1	1	0	None
0	0	1	0	Receive Data Reg
0	1	0	0	Transmit Data Reg
1	0	1	0	Status
1	1	0	0	Control

Read ($\overline{RD}$) and write ($\overline{WR}$) inputs determine whether a selected register in the USART is being read ($\overline{RD} = 0$) or written ($\overline{WR} = 0$) into. A clock (CLK) input supplies the clocking signal for internal synchronization of the logic and registers of the USART. This clocking signal is separate from the data clocks used to set bit rate operation for transmit and receive data. The clocks that perform that task are the transmit clock ($\overline{TxC}$) and the receive clock ($\overline{RxC}$) shown at the bottom of Figure 5.9. $\overline{TxC}$ is used to set the transmit bit rate in the same manner as it does in the ACIA. $\overline{RxC}$ is used to set the sample rate for the receive side. Both of these clocks relate to the bit rate by a selectable factor of 1, 16, or 64, again, much the same as the ACIA.

The reset input is used to set the initial conditions for the USART and prevent erroneous data from being sent or received until the 8251A is configured (programmed) for correct operation. The effect of reset is discussed with each functional section of the USART. The remaining four interface signals shown on the bottom left portion of the figure are handshake signals that assist in the transfer of message information between the CPU and the USART and are discussed in relation to that data transfer in the following sections.

On the right side of Figure 5.9 are the modem interface pins. These control pins are interfaced through an RS232C or other standard interface to interconnect a modem to the USART. The modem then supplies the interface to the transmission medium used to convey messages between two stations. Two lines from the USART contain the actual message information in serial data format. They are transmit data line (TxD) and receive data line (RxD), which are internally connected to separate transmit and receive data buffers Two signals are provided as indications between the USART and modem that both are ready to handle data transmissions. Data terminal ready ($\overline{DTR}$) is an output to the modem that says to it that the USART is ready, while data set ready ($\overline{DSR}$) is the input from the modem conveying its ready state to the USART. Once the USART is configured and the ready signals set and detected, in preparation to begin sending and receiving data, the USART asserts the ready to send ($\overline{RTS}$) line and waits for a clear to send ($\overline{CTS}$) response from the modem. Once an active $\overline{CTS}$ is detected by the USART, message data transmission and reception can begin.

After following a reset of the 8251A, the first word sent to it is interpreted by the 8251A as a mode word (Figure 5.10). This word is used to configure the type of message data that the USART will be sending and receiving. The asynchronous data format (Figure 5.10a) contains selection for the number of stop bits (1, 1.5, or 2), the number of data bits (5, 6, 7, or 8), parity selection (odd, even, or none), and clock/data rate ratio. Besides the selection of a clock/data rate ratio, an additional option is made in the lower 2 bits of the asynchronous format. If these bits are both low, then the format is no longer for asynchronous data but is instead for synchronous data as shown in Figure 5.10b.

In the synchronous mode format, selection of parity and data information length is the same as for asynchronous. Since the data are to be synchronous, there

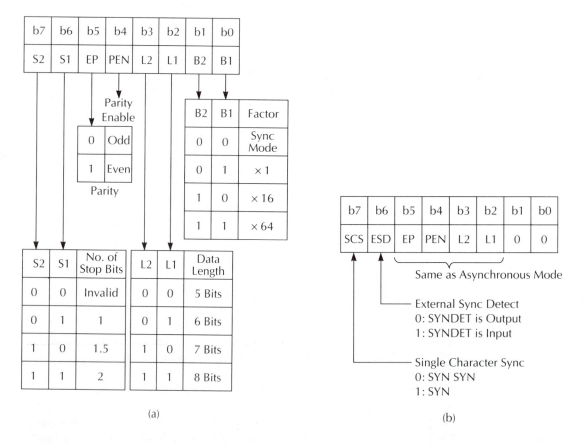

Figure 5.10 8251 Mode Format (a) Asynchronous Mode (b) Synchronous Mode

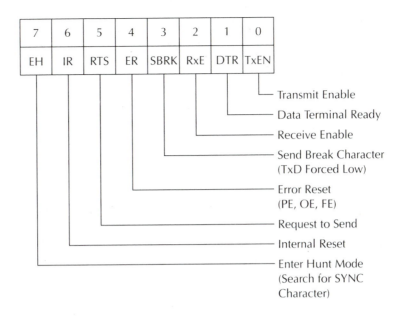

7	6	5	4	3	2	1	0
EH	IR	RTS	ER	SBRK	RxE	DTR	TxEN

Transmit Enable

Data Terminal Ready

Receive Enable

Send Break Character
(TxD Forced Low)

Error Reset
(PE, OE, FE)

Request to Send

Internal Reset

Enter Hunt Mode
(Search for SYNC
Character)

Figure 5.11 8251 Command Instruction Format

are no framing bits. Instead, there are options involving synchronous characters (SYN) and how those SYN characters are detected. Setting the most significant bit of the synchronous mode format high selects a single synchronous character required for synchronization. A low picks dual synchronous bits for systems that require SYN pairs. The next bit in the format determines whether or not synchronization detection should be internal or external. In the case of an internal synchronization, when the USART detects the correct SYN character sequence, it sets the SYNDET (synchronization character detect) line high. This says to the CPU that the USART is synchronized and knows when data characters begin and end. For external synchronization, the SYNDET line is forced high by circuitry external to the USART. This circuitry must be capable of detecting synchronization patterns and sending the appropriate indication to the USART via the SYNDET line.

Mode format information is always sent immediately following a reset. If the synchronous format has been sent, the SYN character code must be sent to the USART so that it can be recognized by the transmit and receive logic. If double SYN has been selected, two SYN characters must be written into the USART. Additional data sent to the control access of the USART following the asynchronous mode word or the SYN character words are interpreted as command words (Figure 5.11).

The command word is used to begin or stop a communication sequence or to internally reset the USART. The transmit and receive functions are independently enabled or disabled by TxEn and RxE bits in the command word. Asserting the DTR pin is accomplished by setting bit 1 of the command format high. If a break

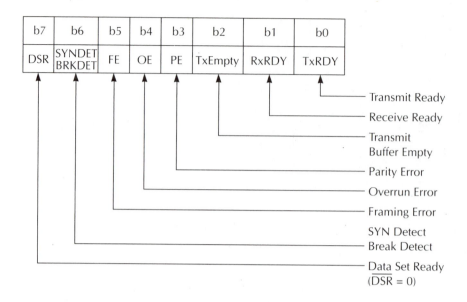

b7	b6	b5	b4	b3	b2	b1	b0
DSR	SYNDET BRKDET	FE	OE	PE	TxEmpty	RxRDY	TxRDY

Transmit Ready

Receive Ready

Transmit Buffer Empty

Parity Error

Overrun Error

Framing Error

SYN Detect

Break Detect

Data Set Ready ($\overline{DSR}$ = 0)

Figure 5.12 8251 Status Register

in transmission (holding transmit data line low a minimum of two characters) is required, it can be accomplished by sending a command word with bit 3 set high. Errors detected in the receive data cause selected bits in a status register to be set. Bit 4 of the command word is used to reset these status bits after the CPU has made use of them. The request to send ($\overline{RTS}$) output pin is made active by setting the RTS bit 5 of the command register high. The USART can be reset by asserting the RESET pin or by setting b6 of the command register high. One effect of either reset is to prepare the 8251A to receive a new mode word, since the first data word sent to the 8251A with C/$\overline{D}$ high is interpreted by the chip as a mode word.

The most significant bit of the command word places the 8251A into the hunt mode if it has been selected to handle synchronous data. In this mode the 8251A searches (or hunts) for a match between the receive data line and the SYN character previously stored in the USART. For single SYN a match indicates that the USART is synchronized to the incoming data. If the SYNDET has been programmed as an output (internal synchronization), the SYNDET line will go high when a match is found. For double SYN mode, the SYNDET line will be asserted in coincidence with the middle of the last bit of the second SYN character, assuming both characters match between the receive data line and the stored SYN characters.

Reading the USART with the C/$\overline{D}$ line high allows the CPU to read the status register (Figure 5.12). This register contains bits that reflect the state of the $\overline{DSR}$ and SYNDET lines. The $\overline{DSR}$ line is sent by the modem to indicate that it is ready to transmit and receive data. In the asynchronous mode, the SYNDET line is used to detect whether a break in transmission has occurred. In the synchronous mode, the SYNDET pin indicates when a match has been found for the SYN character. Three error bits in the status register are parity, framing, and overrun error. They perform the same error-detection function for the 8251A as they did for the ACIA. They are not modified or changed during a read of the status register, nor are they cleared by a reset function. The error bits can only be cleared by issuing a command word with the error reset (ER) bit set.

The transmit and receive ready (TxRDY and RxRDY) and transmit empty (TxEmpty) status bits match the state of the output control pins of the 8251A. This allows these functions to operate in an interrupt (hardware) or polled (software) environment. Their functions are discussed in the following sections dealing with the operation of the 8251A.

● 5.11 ASYNCHRONOUS OPERATION OF THE 8251A

After a reset has been applied to the 8251A, the transmit data line is placed into a high (marking) state. The transmit and receive enable bits in the command logic are forced low, along with data terminal ready (DTR) (causing their associated control pins to be set inactive). A mode word is sent to the USART to establish the type of data the system intends to transfer.

Example 5.3   Set up an asynchronous operation to follow the text discussion. This example will use ASCII characters (7-bit data words), odd parity, and 1 stop bit. The ratio factor is 16.

Solution: The asynchronous mode format shown in Figure 5.10a takes these states:

b7	b6	b5	b4	b3	b2	b1	b0
0	1	0	1	1	0	1	0

which is 5AH in hexadecimal notation. Bits b7 and b6 select 1 stop bit time, bit b5, odd parity which is enabled in bit b4. Bits b3 and b2 select 7 data bits, and bits b1 and b0 set a ratio of 16.

Since the mode selected is asynchronous, the next word sent to the USART with control/data select (C/D̄) set high is a command word. It will be necessary to set the error reset (ER) bit initially to clear any possible false-error condition in the status register. It will also be desirable to enable the transmit and receive logic and inform the modem that the USART is ready to operate.

Example 5.4   What is the command word that allows the system to begin transmission of data immediately following that word?

Solution: Besides enabling the transmit and receive logic and sending a data terminal ready ($\overline{\text{DTR}}$), the request to send ($\overline{\text{RTS}}$) signal will have to be asserted to allow transmissions to begin. The command word, following the format in Figure 5.11, is:

b7	b6	b5	b4	b3	b2	b1	b0
0	0	1	0	0	1	1	1

Hunt mode is disabled since this is an asynchronous application. A reset—both internal and error—if desired will have to have been sent preceding the enabling control word. A break in transmission is not being sent. Highs in the remaining bits enable the transmit and receive logic and assert the DTR line. When the modem detects the active $\overline{\text{DTR}}$ and $\overline{\text{RTS}}$, it asserts the clear to send ($\overline{\text{CTS}}$) line, and the USART is prepared to send and receive data.

A high on the transmit ready (TxRDY) line may be used as an interrupting signal to the CPU. This line is high when the USART transmit logic is enabled and data have not yet been sent to the transmit buffer register. Figure 5.13 shows the timing involved in sending several words of data through the transmit side of the USART.

Sending a character data word requires writing data to the data register of the USART. $\overline{\text{CS}}$ and $\overline{\text{WR}}$ are asserted whenever information is sent to the 8251A. $\text{C}/\overline{\text{D}}$ is forced low by decoding an address with line A0 low. When these three signals go low, they force transmit ready line low. This indicates that the transmit buffer will not be ready to receive another character until the present one is sent to it. These data are presently latched into the data bus buffer. The write cycle is completed by $\overline{\text{WR}}$ returning high. Several operations occur on the trailing edge of the $\overline{\text{WR}}$ signal. The data in the data buffer are transferred to the transmit data buffer, forcing TxEmpty signal to go low; start and stop framing bits and parity are added to the character; and the completed word starts to shift out serially on the TxD line. Also on the trailing edge of the $\overline{\text{WR}}$ signal, TxRDY is set back high, signaling the CPU that it can now send the next character's data word. The word is sent as before, but the previous word is not yet fully sent by the USART. As such, the TxRDY line is held low, even though the $\overline{\text{WR}}$ has returned high. It will stay low until the last stop bit of the current word is serially transmitted on the transmit data (TxD) line. At that time the next character's data are transferred to the transmitter buffer register and the TxRDY line returns to a high state. Notice that the TxEmpty line remains low as long as characters are sent continuously. When the last character is sent to

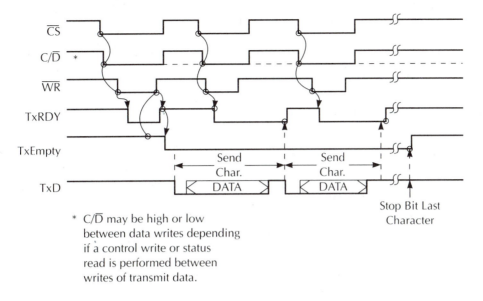

Figure 5.13 Transmit Control Signal Relationships (Asynchronous Data)

* C/$\overline{\text{D}}$ may be high or low
between data writes depending
if a control write or status
read is performed between
writes of transmit data.

the USART, it also will be transferred to the transmitter's buffer and serially sent out on the transmit data line. TxEmpty is set high since no additional data are transferred into the transmitter's buffer register.

On the receive side, the operation is very similar to the ACIA's. Reception of data begins when a start bit is detected when the receive data (RxD) line goes low. This starts a counter in the receive logic, which counts one-half the ratio factor number of clock cycles. During this time, one-half of the start bit has been shifted into the USART. The level is checked again to assure that it is still low and that this low state is a valid start bit. The remaining data bits are shifted in at their center by a shift pulse generated by a counter set to count the ratio factor number of clock pulses.

The USART checks for the first stop bit and to see that the receive data line remains high for the minimum stop time. Failure to do this causes the framing error bit in the status register to be set. The state of the parity bit is checked for parity error, which also sets a status bit if it occurs. Once the complete character is shifted in and regardless of errors, the receive register ready (RxRDY) line is set high, informing the CPU that the receive buffer contains a data word for it to read. The CPU must read that data before the next character is fully shifted in. If the CPU fails to do that, the overrun bit in the status register is set and the new word is transferred to the buffer, overwriting (and losing) the previous word.

The status register also contains TxRDY, RxRDY, and TxEmpty status bits so that the sequence can be done using polling (reading and reacting to status bit states) instead of hardware interruptions.

There are many similarities between asynchronous and synchronous data. A mode word must be sent following a reset to configure the system to handle the synchronous data.

Example 5.5

This time data are synchronous EBCDIC without parity. Internal synchronization is recognized as the presence of a double SYN character (32H). What is the sequence of data sent to the USART preceding the first command word?

Solution: The mode word reflects the 8-bit EBCDIC character size without parity:

b7	b6	b5	b4	b3	b2	b1	b0
0	0	0	0	1	1	0	0

Bit b7 is set to select double SYN characters, and bit b6 picks SYNDET as an output (internal synchronization). Since parity is not used, both bits b5 and b4 are set low. Bits b3 and b2 set 8-bit character length, and bits b1 and b0, being low, place the USART into the synchronous data mode. The hexadecimal equivalent of the mode word is 0CH, which is sent to the USART by the CPU following a reset. Being in the synchronous mode with double SYN selected, two data words of 32H are sent following the mode word. This sets the USART to look for the double EBCDIC SYN characters preceding actual message character transmissions. Also, if the CPU fails to send a character word during transmission, the USART will insert SYN characters to maintain synchronization.

The first command word sent following the mode and SYN character data accomplishes the same functions as the asynchronous command with one addition. The most significant bit of the command word is set high to place the USART into the hunt mode. This mode remains until the USART detects the prescribed number of SYN characters. Upon successful detection, the USART asserts the SYNDET line, indicating that it is synchronized to the data character length.

If external synchronization is used, external circuits are used to detect synchronization and indicate success by asserting SYNDET, which causes the USART to exit the hunt mode. The SYNDET input can be removed one clock cycle following exiting of hunt mode. The USART then will maintain synchronization as long as data continue to flow on the receive data line.

Several interface standards have been formulated to interconnect any one of the many modems also known as **DATA SETS** or other data communications equipment (DCE) on the market to any of the many data terminal equipment (DTE) systems made today. One of the most popular and widely used standards is EIA's RS232C. This standard designates the physical connector type used, the electrical characteristics, and functions of the signals applied to the lines of the interconnect cabling.

Figure 5.14 is a picture of a typical twenty-five-pin connector that is used to meet the RS232C standard. Table 5.2 lists the functions associated with each pin of the connector. Refer to this table during the following discussion.

On the transmit data (TX data) line, BA, the RS232C standard specifications require that the output voltage level of a driving amplifier for mark data lies between -5 and $-15V$ and for space logic data lies between $+5$ and $+15V$. Terminating circuits on the receive data (RX data) line, BB, recognize a mark level as a voltage of at least $-3V$ and a space level of at least $+3V$ providing for a 2V noise immunity between sending and receiving data. Thus the data lines operate using negative logic since a logic 1 is more negative than a logic 0.

For all control lines, the voltage levels have the same values as the data lines, but the logic type is positive. An "ON" condition produces positive voltages, while an "OFF" condition is indicated by negative voltage values. The main control signals used to establish a data link are as follows:

1. RTS (request to send)—Sent by the DTE to the modem, this signal begins the process by which a data link is established. The modem, after a short delay, responds by asserting its CTS (clear to send). The delay allows the modem to respond to the RTS input. Time is allotted for tone generators to stabilize and for

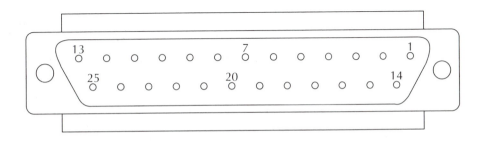

Figure 5.14 DB 25 Connector for RS232C

Table 5.2 RS232C Pin Functions

Pin #	EIA Symbol	Common Abbrev.	Function	Originate From
1	AA	GND	Ground	DTE/DCE
2	BA	TxData	Transmit data	DTE
3	BB	RxData	Receive data	DCE
4	CA	RTS	Request to send	DTE
5	CB	CTS	Clear to send	DCE
6	CC	DSR	Data set ready	DCE
7	AB	COM	Signal return	DTE/DCE
8	CF	RLSD	Rcv line signal detect	DCE
9–11			Not used	
12	SCF	SRLSD	Secondary RLSD	DCE
13	SCB	SCTS	Secondary CTS	DCE
14	SBA	STxData	Secondary TxData	DTE
15	DB	TxSigC	Transmit Signal Timing	DCE
16	SBB	SRxData	Secondary RxData	DCE
17	DD	RxSig	Receive signal timing	DCE
18			Not used	
19	SCA	SRTS	Secondary RTS	DTE
20	CD	DTR	Data terminal ready	DTE
21	CG	SQ	Signal quality	DCE
22	CE	RI	Ring indicator	DCE
23	CH/CI	DSRS	Data signal rate select	DTE/DCE
24	DA	TxSigT	Transmit signal timing	DTE
25			Not used	

any synchronization required in the use of synchronous modems when establishing a synchronous data communications link. RTS and CTS act to establish the local data link between the DTE and the modem. After these signals are set active, a mark tone is sent on to the telephone line by the modem.

2. CD (carrier detect), DCD (data carrier detect), or RLSD (receive line signal detect)—When a modem detects a mark tone on the telephone line, it responds

by asserting the CD signal. The DTE detects an active CD and interprets it to mean that a remote station is connected on-line and the data link is completed.

3. DTR (data terminal ready) from the DTE and **DSR** (data set ready) from the modem (data set)—These lines are used to inform the two main units (DTE and modem) in the data link that they are electrically ready to establish a data link. In some systems these signals are connected directly to positive voltage sources. In this case they only indicate that the RS232C cable is firmly connected between the two units. Other systems generate an active DTR when the DTE is up and running (actually in a ready state). Modems in this type of system assert DSR shortly after the modem is powered and operating, again, essentially saying that it is "ready."

4. Common signal return—This line establishes a common signal reference point between the modem and the DTE.

5. RI (ring indicator)—This line is asserted by modems capable of detecting an incoming telephone ring signal. These modems are equipped to automatically answer the call. A DTE connected to an auto answering modem responds to the RI signal by asserting its RTS line, beginning the process required to establish the data link.

The remaining lines are used for secondary channel signaling and for timing control for modems requiring data stream synchronization. A signal quality detector line SQ is available for DCEs that have the electrical means for error detection incorporated into them.

Secondary channel data rates are much lower than the data channel. Their primary use is for control signal information. Frequently, they are configured to send acknowledgment or control data in the opposite direction from the data channel. When they are configured this way, the secondary channel is referred to as a **BACK CHANNEL.**

A maximum of $+/-25V$ can be applied to any RS232C line without damage to any DTE or modem circuits. Other limitations and specifications imposed by the RS232C standard include:

1. Cable impedance (330 ohms minimum for driver lines and 7,000 ohms maximum for terminating load circuits).

2. A maximum line capacitance of 2,500 pf on any terminating line.

3. A maximum cable length between DTE and modem of 50 feet is recommended but not specifically denoted in the RS232C specification. Longer cables are usable as long as the line capacitance does not exceed the 2,500-pf specification.

Numerous integrated circuit manufacturers produce buffer interfaces to facilitate the translation of common digital logic levels (0 = 0V and 1 = 5V for TTL, as an example) into RS232C levels. One such manufacturer, National Semiconductor, produces an RS232C line driver (Figure 5.15) and line receiver (Figure 5.16). The driver, a DS1488, converts TTL levels to one of two pairs of

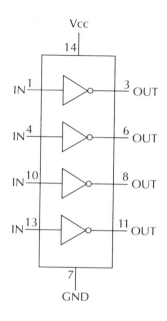

Vin (min)	Vcc	Vout
Logic 0 3.0V	5.0V	0.4V max
Logic 1 −3.0V	5.0V	2.8V min

Figure 5.15 RS232C Line Receiver—National Semiconductor DS14C89A

RS232C levels ($+/-9V$ or $+/-13.2V$). The actual output levels are dependent on the supply voltage applied to the chip; these are shown on the chart in Figure 5.15. The provision for selection of different RS232C levels allows this chip to be incorporated in a wide variety of RS232C interface applications.

The line receiver, a DS14C89A, accepts voltages as low as $+/-3V$ from the RS232C lines and coverts them into TTL logic levels. Note in the charts for both interface chips that there is a logic change as well as a voltage change. That is, TTL logic 1 (5V) is translated to a negative voltage and logic 0, a positive voltage. This is in keeping with the mark and space RS232C logic ($-$voltage for a mark or logic 1 and a $+$voltage used for a space level). To use these interface chips to translate control signals (like RTS and CTS), an inverter is required on the TTL side so that

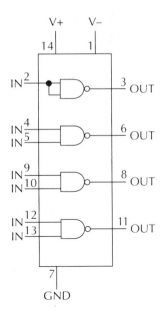

Both Inputs	V+/V–	Vout	Iout
Logic 0 = 0V	9.0V	7.0V	10 ma
Logic 1 = 5V	9.0V	– 6.8V	10 ma
Logic 0 = 0V	13.2V	10.5	10 ma
Logic 1 = 5V	13.2V	–10.5	10 ma

Figure 5.16 RS232C Line Driver—National Semiconductor DS1488

the logic levels of on and off will correspond to the correct RS232C levels (on, a positive voltage, and off, a negative voltage).

An international standards organization based in France and known as the **INTERNATIONAL CONSULTATIVE COMMITTEE FOR TELEGRAPHY AND TELEPHONY (CCITT)** has issued a specification very similar to RS232C. This specification, V.24, and its relationship to RS232C are shown in Table 5.3. RS232C has experienced wide acceptance and usage for systems with maximum data rates of 20 kbps, the maximum rate specified for this standard, over relatively short distances. Requirements for an interface for systems using higher data rates and/or desiring more distant interconnections led Electronics Industries Association (EIA) authors to develop the RS449 standard.

Table 5.3 Comparison of RS232C and V.24

RS232C	Pin	CCITT V.24 Designation	Signal Name (V.24)
GND	1	101	Shield (ground)
TxData	2	103	Transmit data
RxData	3	104	Receive data
RTS	4	105	Request to send
CTS	5	106	Clear to send
DSR	6	107	Data set ready
COM	7	102	Signal common return
RLSD	8	109	Receive line signal detect
None*	9	none	+12V V.24 only
None	10	none	−12V V.24 only
None	11	110	Signal quality indicator
Pins	12, 13, 14	none	Not used by V.24
TxSig	15	141	Transmit clock (DCE)
SRxData	16	none	Not used by V.24
RxSigC	17	115	Receive clock
None	18	142	Local loopback
SRTS	19	none	Not used by V.24
DTR	20	108.2	Data terminal ready
SQ	21	140	Remote digital
RI	22	125	Ring indicator
DSRS	23	111	Data rate select
TxSigT	24	113	External transmit clock
None	25	142	Test indicator

*RS232C pins designated with "none" are not used.

● 5.14 TESTING RS232C INTERFACES

A device called a **BREAKOUT BOX** (Figure 5.17) is used to monitor and test devices that use the RS232C interface. Light-emitting diodes (LEDs) are used to constantly monitor the lines. Some boxes are equipped with switches and test access

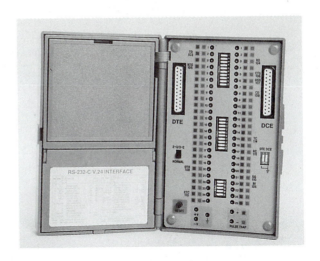

Figure 5.17 RS232C Breakout Box—Navtel Datacheck (DC)7
Photo Courtesy of GN Navtel Limited

points that allow users to make or break each line and/or inject test signals onto each line. The access points also allow other monitoring equipment, such as oscilloscopes, datascopes, and power meters to be connected to the RS232C data and control lines. These breakout boxes are of great assistance in determining what is occurring electrically between terminal equipment and modems.

At times it is desirable to connect two terminals over a short distance without the need for a modem. Essentially, since these terminals are in close proximity with each other, there is no need to use telephone lines to interconnect them. In this case a "null modem" (Figure 5.18) can be used for the connection. A null modem is actually an RS232C cable that has a few wires crossed. Transmit and receive data between both ends are crisscrossed so that the transmit data of one terminal are fed into the other on the receive data line. Data terminal ready (DTR) and data set ready (DSR) lines are also cross-coupled so that one terminal's DTR output becomes an indication on the other's DSR input. Lastly, the request to send (RTS) output of each terminal is fed back to its own clear to send (CTS) and to the other terminal's carrier detect (CD) input. A common line between both terminals supplies a common ground reference for electrical voltages and signals. By using the null modem, two terminals can establish a data link without the need for a modem. An active RTS from one terminal signals its own CTS and the other terminal's CD. Similar action is established at the other terminal, and the data link is set. Data can now flow in serial digital form between the two terminals.

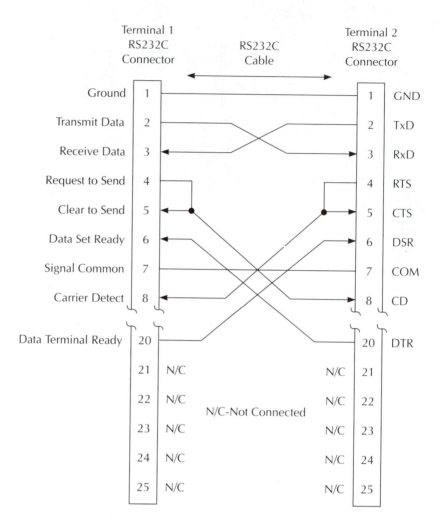

Figure 5.18 Null Modem

● 5.15 RS449 INTERFACE STANDARD

The RS449 standard contains most of the signal lines and electrical specifications outlined in the RS232C. Table 5.4 lists the RS449 functions by pin number and includes notations for equivalent RS232C signal lines. Note that the EIA selected more appropriate designations for the control signal lines. Abbreviations closer to the line functions are used.

RS449 specifies two connectors (Figure 5.19), one with thirty-seven pins for the main data channel and one containing nine pins for secondary channel signaling. The increased number of pins are used to accommodate the added functions detailed

Table 5.4 RS449 Pin Assignment

37-Pin Connector (main channel):

PIN NO.	RS232	CIRCUIT	NAME	DIRECTION
1			Shield	Common
2	CI	SI	Signal rate indicator	From DCE
3			Not used	
4, 22	TD	SD	Send data	From DTE
5, 23	DB	ST	Send timing	From DCE
6, 24	RD	RD	Receive data	From DCE
7, 25	RTS	RS	Request to send	From DTE
8, 26	DD	RT	Receive timing	From DCE
9, 27	CTS	CS	Clear to send	From DCE
10		LL	Local loop-back	From DTE
11, 29	DSR	DM	Data mode	From DCE
12, 30	DTR	TR	Terminal ready	From DTE
13, 31	RLSD	RR	Receiver ready	From DCE
14		RL	Remote loop-back	From DTE
15	RI	IC	Incoming call	From DCE
16	CH	SF SR	Select frequency Signaling rate selector	From DTE
17, 35	DA	TT	Terminal timing	From DTE
19	SG	SG	Signal ground	Common
20		RC	Receive common	From DTE
23	CG	SQ	Signal quality	From DCE
28		IS	Terminal in service	From DTE
32		SS	Select standby	From DTE
34		NS	New signal	From DTE
36		SB	Standby indicator	From DCE
37		SC	Send common	From DCE

9-Pin Connector (S—secondary channel):

PIN NO.	RS232	CIRCUIT	NAME	DIRECTION
1			Shield	Common
2	SCF	SRR	Receiver ready	From DCE
3	SBA	SSD	Send data	From DTE
4	SBB	SRD	Receive data	From DCE
5	SG	SG	Signal ground	Common
6		RC	Receive common	From DCE
7	SCA	SRS	Request to send	From DTE
8	SCB	SCS	Clear to send	From DCE
9		SC	Send common	From DCE

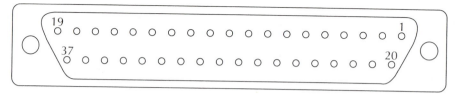

a) 37 Pin Connector

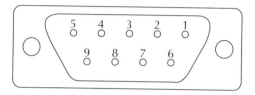

b) 9 Pin Connector

Figure 5.19 RS449 Connectors

in the standard. A number of pins share the same function to assist in reducing line capacitance. This allows for data transfer rates of up to 2 Mbps over cables that can reach a maximum of 200 feet in length.

New functions include control indications for local and remote **LOOP-BACK** test functions. Local loop-back (LL) indicates the system is in the local loop-back test mode. Data placed on the transmit data line are looped back through the modem onto the receive data line to the DTE. This mode is used to test the modem functions. Since data are not placed onto the medium, the test is restricted to the local equipment at a station. In contrast, the remote loop-back test mode (indicated by an active RL line) checks the telephone system connections and lines as well as the modems at both stations. The data sent by a station are looped back through the remote modem onto the telephone lines back to the sending station. The remote DTE is not included in this test mode. A third control line, test mode (TM), signals the DTE that the modem is in one or the other test mode. A similar type of control signal, the data mode (DM), indicates that the modem is in a normal data transfer mode. This last control line replaces the DSR function of the RS232C standard.

The modem can be transferred to a standby channel by using the select standby (SS) signal. The modem returns a status indication that it is connected to the standby channel by asserting the standby indicator (SB) line. If the DTE desires to prepare the DCE to receive a new line signal, it asserts the new signal line (NS). This causes the level on receive data line (RD) to go to a mark state and disables the carrier detect (CD) line. Once the modem receives the new signal, it places it on the receive data line and again asserts the carrier detect line. The send common (SC) and receive common (RC) lines indicate which direction data are flowing (out on transmit data line or in on receive data line).

The secondary channel lines are included in the nine-pin connector/cable along with a few duplicated functions found in the thirty-seven-pin connector. These leads are summarized in Table 5.3 with their equivalent RS232C functions.

● 5.16 RS422 AND RS423 INTERFACE STANDARDS

The RS232C and RS449 specifications detail basic requirements for interfacing DTE and DCE equipment. Two additional standards specify electrical and mechanical requirements for interconnecting any two circuits used for transferring serial data. They are the RS422 for balanced circuits and the RS423 for unbalanced circuits. Balanced networks, as shown in Figure 5.20, generate two outputs that are the opposite of one another. The voltage (or power) of each output (V1 and V2) can

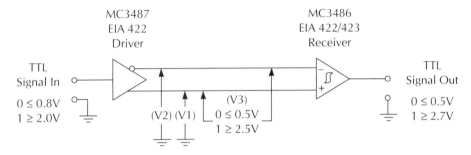

a) Balanced Network

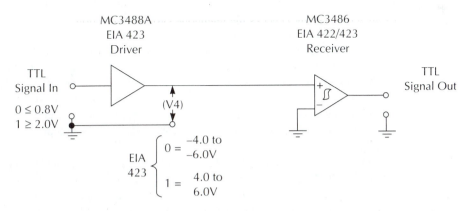

b) Unbalanced Network

Figure 5.20 Balanced and Unbalanced Interfaces

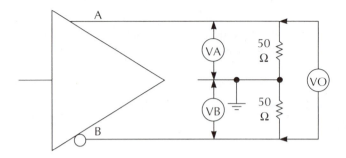

Figure 5.21 RS422 Generator Output

be measured in relation to a common reference point (signal ground, for example). However, the full output is generally measured differentially between both outputs (V3).

The output of an unbalanced circuit is taken from a single output with reference to a common return such as ground (V4 in Figure 5.20). Balanced networks are used to assist in the elimination of noise experienced on the lines between the generator and receiver units. Unlike the desired outputs, noise on one output is approximately identical to the noise from the other output. Since they nearly agree, the potential difference between the noise levels is negligible. This reduction in noise power, called common-mode rejection, results in less noise presented to the receiver from the generator.

The RS422 and RS423 specifications set requirements for balanced and unbalanced voltage digital interface circuits. The details for each are similar, with differences in numbers applicable to the specific type of circuit (balanced or unbalanced) the specification addresses. Because of this, this section will illustrate the type of information referred to in the RS422 specification. Exact information for the RS423 standard is left to the individual requiring its use.

Figure 5.21 shows the output loading circuits for an RS422 balanced generator. It shall be referenced in discussion involving RS422 electrical specifications. RS422 specifies that a generator is to be able to produce a voltage between 2 and 6V for each logic level while sourcing a 100-ohm differential impedance. In Figure 5.21 this is represented by 50-ohm load resistors between each output and signal ground. The differential output (V0) is sensed from output A to output B across the combined 100-ohm resistance.

The generator is limited to 6V from either output to ground as well as from output to output. Essentially, this says to the designer of the receiving end that, at no time, are the inputs to the receiver to expect more than a 6V potential. The polarity of the voltage generated determines the logic level. RS422 does not specify which logic level matches which polarity. It does, however, state that the magnitude of the voltages for each polarity is not to differ by more than 0.4V. That is, if a logic 1 is +5V, a logic 0 must fall into the range of −4.6V to −5.4V. While the

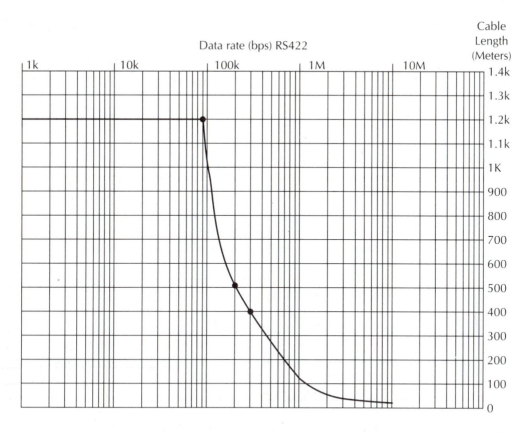

Figure 5.22 Data Rate versus Cable Length RS422

specification does not assign polarity to logic level, it does state that a logic 1 is a mark and is to be the same level as a control line off condition. Logic 0, a space, is to be the same polarity as a control line on signal.

The input impedance of the balanced receiver is greater than or equal to 4 kohms. Internally, the receiver is to be able to respond to a change of $+/-0.2$V as a logic switch. That is, the receiver must be able to detect a 200-millivolt signal and determine the correct logic level. Once the logic level is detected, the receiver must indicate that level as long as the common mode input voltage remains between $+/-7$V. Receivers are built to withstand a maximum differential input of 12V.

Another area specified by RS422 is the length of the cable between the generator and receiver. As with the RS232C standard, RS422 specifies the capacitance per foot as the limiting factor to a cable's length rather than attempting to specify actual lengths. The capacitance of a cable is specified as 52.5 pf/meter (or 16 pf/foot) terminated into 100 ohms. In addition, for data rates under 90 kbps, the maximum length of a cable is 1,200 meters (approximately 4,000 feet). Figure 5.22 is a graph of the usable data rates and maximum cable lengths expected when using the RS422 standard.

The rationale for presenting information for the EIA standards here is to provide a feel for the type of information contained within them. Realize that there are many additional details within each standard and that the specifications discussed here are presented so that the student can perceive how their use can facilitate the standardizing of the interconnection of various devices.

● 5.17 FSK MODEMS

The last functional block in a basic low-speed data link is the modem, data set, or DCE. It is the job of the modem to convert the serial digital data stream to a series of analog sine waves or tones. One tone represents a logic 1 (mark), and another is used for a logic 0 (space). On the receive end these tones are converted back to digital levels by the receiver's modem. Figure 5.23 illustrates the functional blocks of a frequency shift key (FSK) modem. On the transmit side serial data are supplied by the DTE on the TX data input and fed to a **VOLTAGE CONTROLLED OSCILLATOR (VCO).** The VCO acts as a frequency modulator by converting DC voltages into different analog sine waves. These sine waves have different frequencies for each level of DC voltage applied. Hence, a tone at one frequency is produced for a logic 1's DC voltage, and a second tone at a different frequency is produced for the space logic DC voltage level.

The request to send (RTS) input from the terminal (or UART) initiates the modem's RTS/CTS delay, which varies with each modem type and requirements. Once the delay is completed, the modem generates a clear to send (CTS) signal back to the terminal, completing the local data link handshaking. An active CTS enables the voltage controlled oscillator, which, sensing idle line 1s from the UART, immediately outputs a mark tone on the telephone lines.

Data terminal ready (DTR) and data set ready (DSR) inputs to the modem and DTE, respectively, can be as simple as a steady logic level 1. They are used for signaling the terminal and modem that they are interconnected (usually through a RS232C cable). These logic inputs must be sensed before the modem or terminal will react to the RTS/CTS sequence. See section 5.9 for details on DTR and DSR.

The mark signal received on the input to the receive side of the modem passes through the mark filter to the carrier, data carrier, or receive line signal detect (CD, DCD, or RLSD) and demodulator units. An active full-wave rectifier, in the carrier detect section, converts the tone into a DC voltage that is sent out of the modem on the CD line to the DTE to complete the remote handshake necessary to establish the data link. The remote station may already be "on-line," having gone through the same RTS/CTS sequence performed by the initiating station. It is also possible that the remote site will establish its local link through its RTS/CTS sequence in response to an active CD. The choice is selected by the application software that controls the system and the type of modem used. In either case, once the CD lines

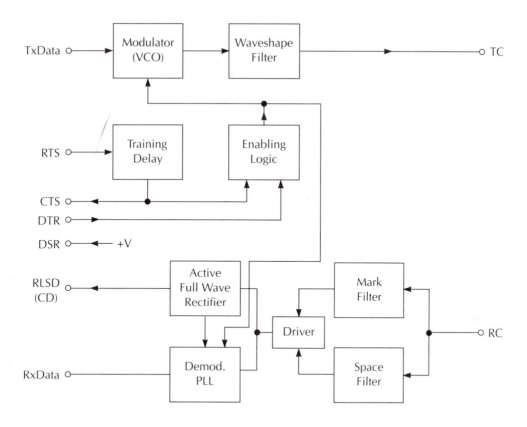

Figure 5.23 Modem Functional Blocks

at both stations are active, a data link is established and actual data can now be sent back and forth through the modems.

The mark signals from the mark filter and space signals (which are passed through a space filter) are detected by a **PHASE LOCK LOOP (PLL)** circuit in the demodulator section. Three operational blocks form a PLL. They are a phase detector, a buffering amplifier, and a VCO (Figure 5.24). For a FSK demodulator, the VCO is designed to produce a square wave signal at a frequency (fc) that lies between the mark and space frequencies. This frequency is never sent between modems since it is neither a mark nor a space logic level.

The purpose of the phase detector is to produce a DC voltage (the PLL's error voltage) whenever there is a difference between the phases of fc and input sine wave, fin. Figure 5.25 is shown to aid in the understanding of a phase detector for this application. A carrier square wave is applied to the fc input. The input sine wave is applied to the fin leads. The positive half of the fc input biases the diodes on and passes fin to the peak detector RC (resistor-capacitor) circuit at the output. The amount the capacitors charge to is dependent on the phase relationship between fc and fin. The long RC time constant of the peak detector maintains the charge on

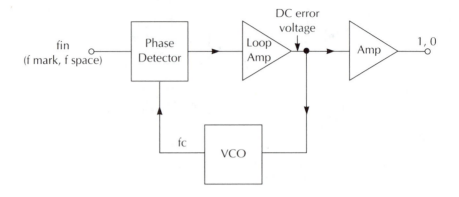

Figure 5.24 Phase Lock Loop

the capacitors during the negative alternation of fc when the diodes are biased off. For a PLL the voltage developed across the output is called an error voltage, which is fed back to the VCO input. The VCO frequency changes in response to this error voltage, causing a change in the phase relationship between fc and fin. This in turn changes the error voltage from the phase detector, causing further change in the VCO frequency. This continues until a "lock" condition occurs. The lock is the point at which the error voltage produces a frequency from the VCO that causes a constant 90° phase difference between the VCO signal and the incoming sine wave from the filters. Since the carrier frequency lies between the mark and space tone, the 90° phase shift will be leading the carrier frequency for one tone and lagging the carrier for the other. The DC error voltage will be positive for the leading 90° phase shift and negative for the lagging 90° shift. This error voltage is fed to another buffer amplifier, which establishes RS232C voltage levels that are sent to the UART.

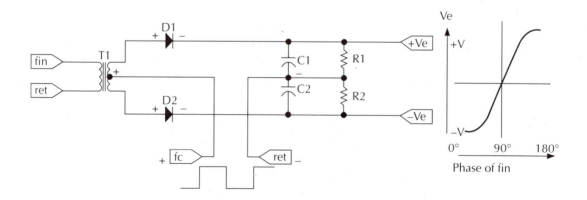

Figure 5.25 Phase Detector

● 5.18 LOW-SPEED MODEM OPERATING MODES

Different systems require different modes of data communications operation. Systems operating in full duplex have different needs than those operating in half duplex or simplex. A primary controlling station that initiates the call in a full-duplex data link uses an **ORIGINATE MODEM,** while the secondary remote station requires an **ANSWER MODEM.** The main difference in these modem types is the mark and space frequencies used to represent data logic 1s and 0s.

The Bell Telephone 103 modem is a full-duplex 300-bps modem that can be manufactured as an originate or an answer modem. As an originate modem the transmit mark and space frequencies are 1,270 and 1,070 Hz, respectively. The receive frequencies are 2,225 Hz for a mark and 2,025 Hz for a space. To operate with a 103 originate modem, a compatible answer modem must transmit mark and space tones at the originate's receive frequencies and be able to demodulate the originate's transmit mark and space tones. The use of different frequency pairs for each transmission direction (to the originate and to the answer station) allows the data transmissions to occur simultaneously. Secondary tones produced from any mixing of these signals on the transmission medium are rejected by the receive filters in each modem.

Some modems operating in half duplex echo back the transmitted data to the sending station's video screen or printer. Called **ECHO-PLEX,** this allows the sender to check the information being sent at the time it is being transmitted. This echo is permissible because, in half duplex, actual data transmission direction is one way at a time. A primary can send to a secondary, but the secondary must wait until the primary is finished before it can send data to the primary. Since the receive side is not being used by a station while it is transmitting, the data sent out are routed through the station's receiver and echoed back.

● 5.19 A MODEM ON A SINGLE CHIP—INTEL'S 2970

A number of integrated circuit (IC) manufacturers produce modems that have most of the required modem functions on a single chip. Among this group is the Intel 2970 Single Chip FSK Modem. The 2970 is designed to meet or exceed several popularly used modem specifications. They are for the Bell 103 and 202 and International Consultative Committee for Telegraphy and Telephony (CCITT) V.21 and V.23 modems. These specifications detail data rate **(BAUD),** duplex mode, and operating frequencies for their respective modem. Table 5.5 details the frequency and duplex mode assignment for each mode. The ''B'' prefix on signal names in the

**Table 5.5 Mark and Space Frequencies Specified by
Bell 103 and 202 and CCITT V.21 and V.23 Specifications**

Modem Type	Baud Rate (bps)	Duplex Type	Transmit		Receiver		Answer Tone (Hz)
			SPACE (Hz)	MARK (Hz)	SPACE (Hz)	MARK (Hz)	
103 Originate	300	Full	1,070	1,270	2,025	2,225	N/A
103 Answer	300	Full	2,025	2,225	1,070	1,270	2,225
202	1200	Half	2,200	1,200	2,200	1,200	2,025
202 Back	5	N/A	387*	387*	387**	387**	N/A
V.21 originate	300	Full	1,180	980	1,850	1,650	N/A
V.21 answer	300	Full	1,850	1,650	1,180	980	2,100
V.23 mode 1	600	Half	1,700	1,300	1,700	1,300	2,100
V.23 mode 2	1200	Half	2,100	1,300	2,100	1,300	2,100
V.23 back	75	N/A	450	390	450	390	N/A

*$\overline{\text{BRTS}}$ = 0; BTD is a MARK.
**Makes $\overline{\text{BCD}}$= 0.
For Bell 202 Back, an inactive $\overline{\text{BRTS}}$ forces BTD low. The absence of one 387-Hz tone on RC forces $\overline{\text{BCD}}$ high (inactive).

table denotes functions for a back secondary channel used to convey control or acknowledge information.

A functional block diagram with pin numbers for the 2970 is shown in Figure 5.26. The modem can be considered to be a three-section device consisting of the transmit, receive, and control sections. The transmit section takes the digital data from the UART on TD (transmit data) and on BTD if a back channel is used. These data are digitally synthesized into a sine wave using a digital to analog converter and digitally filtered to reject any unwanted side tones. The digitized sine wave is fully converted into the correct analog sine wave using a digital to analog converter (DAC) and filtered one last time before the tones are put out on the TC (transmit carrier) line.

On the receive side, analog data are brought into the modem on the RC (receive carrier) input and filtered to remove all but the mark, space, or back channel tones. These are passed to an analog to digital converter (ADC) and digitally filtered. The resulting digitized sine wave drives receive line signal detect and demodulator circuits. The detect circuit produces an active $\overline{\text{CD}}$ or $\overline{\text{BCD}}$ (carrier or back channel carrier detect) as long as a correct signal remains on the RC line. The demodulator produces the correct 1s and 0s in response to the mark and space data tones or back

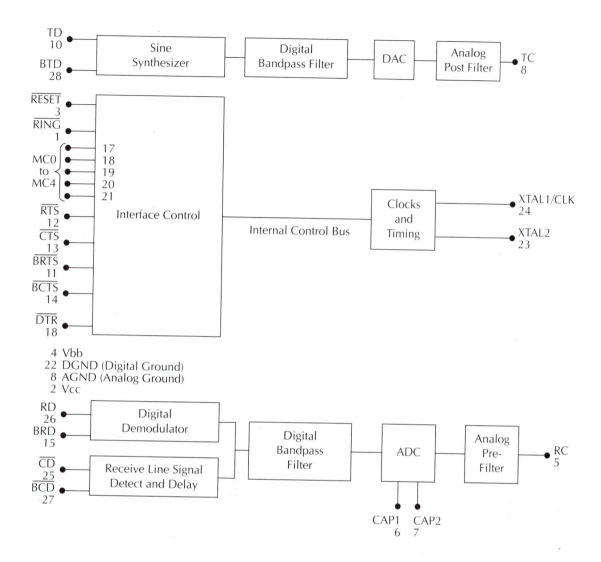

Figure 5.26 2970 Modem Functional Blocks

channel tones applied to it. These digital data are sent out to the DTE on the RD (receive data) or BRD (back channel receive data) as is appropriate.

The frequency of the tone used for the back channel is in the lower portion of the voice channel. Additionally, back channel data are sent and received at a low rate (5–75 bps) since they are primarily used to indicate an on (mark) or off (space) condition. Back channel tones (387 Hz for Bell 202 modem) present on TC indicate the on condition, while the absence of the tone indicates an off state for the 202 back channel. The CCITT modem back channel uses two different tones for mark (on)

Table 5.6 2970 Mode Control Truth Table

MC0–MC4 (in hexadecimal)	Mode Selected	Modem Type
0	300 bps full duplex originate	Bell 103
1	300 bps full duplex answer	Bell 103
2	1200 bps half duplex	Bell 202
3	1200 bps half duplex with equalizer	Bell 202
4	300 bps full duplex originate	CCITT V.21
5	300 bps full duplex answer	CCITT V.21
6	1200 bps half duplex—mode 2	CCITT V.23
7	1200 bps half duplex with equalizer	CCITT V.23
8	600 bps half duplex—mode 1	CCITT V.23
9–15	Not used—reserved by Intel	
16–24	As above with main loop-back capabilities	
25	Back loop-back	CCITT V.23
26–31	Not used—reserved by Intel	

and space (off) indications. Both back channels can still be used to interpret digital characters, but to do so would require some patience (at 5 bps for the 202 and 75 bps for the V.23).

The third section is the control section. It is responsible for recognizing $\overline{RTS}$ and producing $\overline{CTS}$, setting the $\overline{CD}$ delay time between detecting the mark tone and asserting the $\overline{CD}$ line, and selecting the modem specification that the 2970 will operate under. Table 5.6 lists the input states of MC0–MC4 (mode control) and the corresponding operating mode of the 2970.

Separate RTS and CTS control lines are present for the main channel ($\overline{RTS}$ and $\overline{CTS}$) and for the back channel ($\overline{BRTS}$ and $\overline{BCTS}$). Along with the carrier detect lines ($\overline{CD}$ and $\overline{BCD}$), these allow for individual data link control for main and secondary back channels.

A $\overline{DTR}$ signal is provided as a signal from the terminal to the modem to indicate that the DTE is connected and ready to start the communication link process. A RING input is used for an auto answer function, which allows the modem to automatically answer a call from a distant station.

An active RESET causes the following 2970 signals to be placed into their inactive states: $\overline{DTR}$, $\overline{RTS}$, $\overline{CTS}$, $\overline{BRTS}$, $\overline{BCTS}$, $\overline{RING}$, $\overline{CD}$, and $\overline{BCD}$. Data on TD and BTD are ignored, and a logic 1 is forced on the RD and BRD lines. In effect, the modem is placed into an inactive state with idle line 1s returned to the UART.

Figure 5.27 shows a data link with the IC devices discussed in this chapter. The main computer system supplies the parallel data to the ACIA, which supplies the serial digital data stream to the 2970 modem. The ACIA also supplies $\overline{RTS}$ while detecting $\overline{CTS}$ and $\overline{CD}$ inputs. Both the ACIA and the 2970 are manufactured using NMOS (N channel MOS) technology. The 2970, further, uses high-speed MOS (HMOS) technology. The main advantage is that both devices use 5V for Vcc and 0V for ground. Logic levels are directly compatible. This omits the need for an RS232C interface since the ACIA can be directly connected to the 2970. A data access arrangement (DAA) unit is placed between the modem and the telephone lines to supply interfacing between the modem data and telephone lines. The DAA is responsible for assuring line isolation between the telephone lines and user equipment. The purpose of this isolation is to protect the telephone lines and equipment from any electrical problems caused by faulty or poorly designed user data communications devices. The DAA responds to either an off-hook condition or an incoming ringing condition by connecting the telephone lines to user data equipment through a line transformer. Today, a number of modems incorporate the DAA functions as part of their internal circuitry. The conditions that DAAs (external or internal) must satisfy before user equipment can be interfaced with the telephone system are detailed in Federal Communications Commission (FCC) Rules and Regulations, part 68, which describes both electrical and physical specifications that must be met by makers of DAA circuits. An additional requirement is that all DAAs are to be registered with the FCC to assure they meet the requirements of part 68.

Example 5.6

What is the process required to use the system in this section to transmit ASCII data with even parity and 2 stop bits at 1200 bps using a half-duplex communication? Back channel functions are not used, and data transfers between DTEs and DCEs are performed by interrupt programs.

Solution: An application program at both local and remote stations in the communication link is required to program both the ACIA and 2970 modem to manage the type of character data specified. For the ACIA, its control register must be configured as follows:

$b7 = 1$ to enable receive interrupts

$b6 = 0$ to assert $\overline{RTS}$

$b5 = 1$ to enable transmit interrupts

$b4 = 0$ to select 7 bits per character

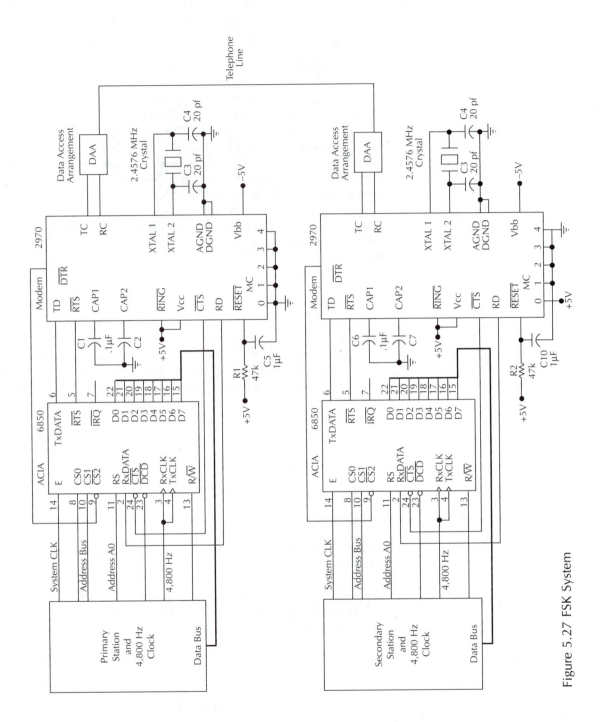

Figure 5.27 FSK System

b3 = 0 to select 2 stop bits

b2 = 0 to select even parity

b1, b0 = 01 for a divide ratio of 16

This translates into a hex data word = A1, which is sent by the CPU, in response to application program instructions, to the ACIA control register. TxClk and RxClk inputs are hardwired to a 19.2-kHz clock (1200 bps times 16).

The 2970 is hardwired to select its configuration. The mode control lines' (MC0–MC4) states are set according to Table 5.6. They are all set low except MC1 to select Bell 202 half-duplex operation at 1200 bps. After power is applied to the system and the applications program that configures the ACIA has been run, an active $\overline{RTS}$ from the ACIA to the 2970 starts the local handshake process.

The 2970 responds to an active $\overline{RTS}$ by sending a mark tone on the TC line to the data access arrangement (DAA). After a short delay the 2970 returns an active $\overline{CTS}$ to the ACIA. At the originating station, a call is placed to the remote station. The establishment of the off-hook condition enables the DAA, connecting the 2970 to the telephone lines. At the remote station, answering the call enables its DAA, and the mark tones from the 2970 modems are sent by both stations on the telephone lines.

Reception of the mark tones on the RC lines is detected by the 2970's carrier sense logic, which causes the $\overline{CD}$ lines to be asserted. These lines are connected to the ACIA $\overline{DCD}$ lines.

The terminal polls (reads) the ACIA status register, waiting for an active $\overline{DCD}$ indication (b2). When the CPU detects this bit going low, it sends the first character to the transmit buffer register in the ACIA. The ACIA, in turn, sets the transmit data register empty (TDRE) bit (b1) in the status register low, computes and inserts the even-parity bit, and adds start and stop framing bits to the character data. The completed character is transferred to the transmit shift register, and the ACIA commences to shift the data out on its TX data line. TDRE returns high as soon as the data are transferred into the shift register, generating a transmit interrupt request on the ACIA interrupt request ($\overline{IRQ}$) line. The central processing unit (CPU) responds to this signal, beginning a process that selects a transmit data program to send the next character to the ACIA while the ACIA is shifting out the current data character. The ACIA accepts the next data character, sets its TDRE bit low, and holds that character until it is finished sending the current data. Once the first character is fully shifted out, the ACIA automatically transfers the next character from its buffer register to its transmit shift register and repeats the preceding process.

The serial data from the ACIA's TX data line are fed to the 2970's TD input line. The modem accepts the incoming data stream, converts the digital data into mark (1,200 Hz) and space (2,200 Hz) tones and sends those tones, via the data access arrangement (DAA), onto the telephone lines. Between transmitted characters the CPU can perform other processing, since it can execute a number of instructions in the time it takes the ACIA to shift out the completed character. The

CPU will know when to fetch the next data character and send it to the ACIA by the reception of the interrupt request signal from the ACIA.

At the remote station the data tones are passed from the telephone lines to a 2970 through another DAA. The modem demodulates these tones, supplying a digital data stream to the ACIA via the RD and RX data lines. The ACIA detects the presence of a start bit and verifies it is a good start bit as described in section 5.6. The ACIA then shifts in the character data, parity, and stop bits. Once the data are completely shifted in, they are transferred to the receive buffer register and the receive data register full (RDRF) bit (b0) in the ACIA register is set high. A receive interrupt request is generated by the ACIA, asserting its $\overline{\text{IRQ}}$ line. At this time the internal logic of the ACIA checks for parity and framing errors. If any of these is present, its corresponding bit in the status register is set.

Upon receiving the interrupt request, the CPU at the remote station's terminal diverts to the receive interrupt program. This program begins by reading the ACIA status register to determine the cause for the interrupt request. Once it is determined that the interrupt was generated by the reception of the received data and that parity and framing indications are all right (both low), the CPU is directed to read the receive data buffer. This register must be read before the receive circuitry in the ACIA shifts in the next data character. Failure to do this in time causes the receive overrun bit in the status register to be set and would generate a second interrupt request. Reading the status register causes the error flags to be returned to their low or inactive state. Reading the receive buffer register forces the RDRF bit in the status register to return to a low state.

If an error is detected (parity, framing, or overrun bit set), the receiving station application program causes a retransmit request message to be sent back to the sending station. If the data word is received without errors, it is stored into the secondary station's memory and the secondary station returns to other processing until the next received data character is fully entered into the ACIA (indicated by an active RDRF bit and $\overline{\text{IRQ}}$ line). The process of sending and receiving data continues until an end of message character is sent and detected by the remote station. At this point the data link would be disconnected and both stations return to normal processing of local programs.

The use of interrupt control allows local programs to be processed in between the reception of data similar in concept to the process at the transmit station. The remote station would process regular programs until it received the receive interrupt. Upon reception of the interrupt, the regular program is suspended while the secondary station processes the receive interrupt program. The regular program is resumed after the receive program is completed.

SUMMARY

The development of coded data from the simple dots and dashes of Morse code to using character codes transferred through the telephone system required those character codes to go through many transformations. First they leave the terminal in parallel format and are fed to a UART. In the UART extra bits are added to frame the characters and perform error detection (parity). The data are then shifted out of the UART in serial form to a modem. The modem further translates the binary data into pairs of sine wave signals that can be sent on the telephone lines to a remote sight. At the receive end the signals go through a reverse process to be converted back into parallel character codes for display or use. Low-speed communications that perform this type of data link utilize frequency shift keying modulation techniques. Data rates for these systems usually lie between 100 and 1200 bps. Numerous medium- to large-scale integrated circuit devices have been developed to fill the electrical functions of UARTS, modems, and RS232C interface drivers and receivers. These systems can be monitored and tested using specialized equipment such as datascopes and breakout boxes.

GLOSSARY

Answer modem—Modem which responds to signals sent by an originating station.

Answering station—Secondary station in a data link.

Back channel—Channel that is used to convey control data between two stations in a data link.

BAUD—Binary data rate.

Breakout box—Test box inserted into an RS232C or other interface line.

Clock skew—Difference in frequency between transmitter and receiver clock frequencies.

DAA (data access arrangement)—Interface between modem and telephone lines.

Datascope—Test equipment used to monitor a serial data stream.

Data set—Modem.

Double buffered—Device which uses two registers to hold incoming data.

Echo plex—Modem with a feature that allows transmitted data to be "echoed" back to the sender.

FSK (frequency shift keying)—Modulation technique in which the frequency of a tone is altered by the state of binary bit data.

Handshake—Term used to describe control signaling between two units prior to transferring information.

IRQ (Interrupt request)—signal informing the CPU to divert from its current program processing to execute a new program.

Loop-back—Test condition in which the transmitted data are looped back into their own receiver.

Modem—Modulator/demodulator used for converting between digital data and analog data forms.

Originate modem—Used with primary station to establish communications link.

Originating station—The primary station responsible for originating the call prior to establishing the data link.

PLL (phase lock loop)—Circuit that detects the difference between an incoming tone and a reference tone. It produces an error DC voltage corresponding to the difference.

VCO (voltage controlled oscillator)—Circuit that produces an AC signal whose frequency is determined by the DC voltage at its input.

QUESTIONS

5.1. What are the main functions of the chief components (UART and modem) of a low-speed system?

5.2. Define mark, space, and idle line 1s.

5.3. What is the general purpose of each of the following handshake signals? Which functional unit generates each signal?
   a. RTS
   b. CTS
   c. CD

5.4. What do DSR and DTR signify? Which units generate each signal?

5.5. Which bits in a serially transmitted character word from a UART are optional? Which are always the same amount regardless of the character code used?

5.6. An ASCII letter A with even parity and 1 stop bit is sent and received at 1200 bps. The receive clock has drifted and become higher in frequency than the maximum allowable frequency. Which error flag in the status register, if any, is set? Why? If none are set, why not?

5.7. What are the errors detected by the ACIA? What causes each to be set?

5.8. What is the function of the ratio factor between transmit and receive clocks and the data rate?

5.9. What are the data size, stop bit, and parity options of the ACIA? Of the 8251?

5.10. What is the major difference between the ACIA and the 8251?

5.11. What are the options available in the synchronous data mode of the 8251?

5.12. What are the operating voltage ranges for a mark and space level on a RS232C interface from the driver (transmitter)? What is the minimum voltage that is expected at the receive or terminal end for RS232C data lines?

5.13. What is the maximum voltage that may be applied to the transmit or receive data line without causing damage to the equipment?

5.14. Which RS232C control signal indicates the presence of a tone on the telephone lines?

5.15. What are the maximum recommended cable length and total capacitance for an RS232C cable?

5.16. List the minimum control signals used to interface a modem to a UART using an RS232C interface.

5.17. List at least three changes incorporated in the RS449 standard when compared with the RS232C.

5.18. What are the main transmit functional blocks of a modem? Explain the function of each unit.

5.19. What are the main receive functional blocks of a modem? Explain the function of each unit.

5.20. Which tone is initially used to cause CD to be asserted by a modem? What logic level does that tone represent?

5.21. What are the functional and electrical differences between an originate and answer modem?

5.22. What type of communications link requires originate and answer modems?

5.23. Secondary and/or back channels in a modem normally carry what type of information?

5.24. Why is there no requirement for a RS232C interface between the ACIA and the 2970 modem?

5.25. What is the purpose of a data access arrangement (DAA)?

5.26. What agency is responsible for policing use of DAAs? What is the rule governing the requirements for a DAA?

PROBLEMS

5.1. What is the ACIA control word required to configure an ACIA to handle EBCDIC character transmissions using a divide ratio of 64? Both transmit and receive interrupts are to be enabled.

5.2. A parity error is detected by an ACIA on a received word. What is the content of the status register read as the first part of a receive program? Assume no other error has occurred.

5.3. Using a divide ratio of 64, what are the transmit and receive clock frequencies for a 1200-bps data rate?

5.4. What are the lowest and highest possible clock frequencies usable without loss of any bit information for the following character: 7-bit character code with even parity and 2 stop bits transmitted at 1200 bps? Use divide ratio of 65.

5.5. The ASCII character H is sent using odd parity and 1 stop bit. A data storage scope is used to read the transmit line. What is the bit pattern that appears on the CRT screen of the datascope?

5.6. What are the contents of the 8251 mode word to configure the USART to handle the following data links:
   a. Asynchronous ASCII with 1 stop bit and odd parity. A clock to data rate ratio of 16 is used.
   b. Synchronous EBCDIC with two SYN characters. When SYN is detected, the SYNDET will be asserted by the USART. Parity is not used with the transmission.

5.7. Give the hexadecimal equivalent of the 8251 command word for each of the three conditions below:
   a. Place the USART in a hunt mode to start data reception.
   b. Internally clear the USART and error flags. While this is happening, a break is sent to the receiver.
   c. Prepare to send an asynchronous data transmission following a reset.

5.8. Which bits in the receive 8251 USART status register are set for each of the conditions in problem 5.7? Assume data link is established and the transmitting station has reacted to the command word sent to the 8251.

LAB PROJECT

5.1. Design an FSK system using selected component chips suggested by your instructor. These chips should include a UART, a modem, RS232C drivers and receivers, and support chips. Once the system is designed and breadboarded, interface it with a microprocessor test set (a 8080, 8085, Z80, or 6800 8-bit microprocessor trainer, for instance). Write the programs for the trainers to send and receive data through the communications system.

RESEARCH ASSIGNMENT

5.1. Research and write a paper (or give a class presentation) on one of the following topics:
   a. Two-port UART used with 16/32 bit microprocessors (68000 family or 8086 family)
   b. Auto answer modems
   c. Commercial 1200- or 2400-bps modem
   d. Other related up-to-date interface standards, UARTs, and modems

Answers to Odd-Numbered Questions

5.1. UART's main function is serial/parallel data conversion. Modem's main function is digital/analog data conversion.
5.3. a. RTS—request to send sent by the UART signals the modem that the UART wants to send and receive data.
   b. CTS—clear to send sent in return by the modem informs the UART that the modem is ready to accept and send data.
   c. CD—carrier detect; sent by the modem, informs the UART that the modem has detected a signal from the remote station.
5.5. Number of character bits, number of stop bits, parity are optional. One start bit is required.
5.7. ACIA detects parity error for data bit errors; overrun error when CPU fails to read receive data in time; framing error—a low is detected as stop bit.
5.9. ACIA: 7 or 8 data bits; 1 or 2 stop bits; odd, even, or no parity
   8251: 5, 6, 7, or 8 data bits; 1, 1.5, or 2 stop bits; odd, even, or no parity.
5.11. Single or double SYN character, internal/external SYN detect, odd/even/no parity, number of data bits
5.13. Maximum on a pin: $+/-25V$
5.15. 50 foot; 2,500 pf
5.17. Two connectors, longer length, added functions, higher data rates
5.19. Mark and space filters—reject unwanted signals
   PLL—convert tones to digital data
   CD full wave rectifier—remote handshake detection
5.21. The originate modem is used by the station that initiates a call to establish a data link. The answer modem is used by the receiving station. Electrically,

the two modems differ in the frequencies used for MARK and SPACE tones. The originate modem's transmit frequency pair is the answer's receive frequency pair. Likewise, the frequencies used by the answer modem's transmitter are the same as the originate's receiver.

5.23. Control and acknowledge data

5.25. The DAA interfaces non–telephone company equipment to the telephone lines.

Solutions to Odd-Numbered Problems

5.1. 1011 0110, or B6 in hexadecimal

5.3. 76.8 kHz

5.5. 0 1 0 0 1 0 0 0 1 1 1

5.7. a. 1 0 1 0 0 1 1 1 = A7H
b. 0 1 1 1 1 1 1 1 = 7FH
c. 0 0 1 0 0 1 1 1 = 27H

Lab Project Guide

5.1. Very dependent on components selected for use for the lab.

Research Assignment Direction

5.1. As the field continues to grow, more and more topics are available for student research.

Six • Medium-Speed Modems and Systems

After studying this chapter, the student will be familiar with modems and systems transferring data at rates equal to and above 2400 bps. Topics include the following:

1. Balanced modulators
2. Phase shift keying modems
3. Quadrature amplitude modulation modems
4. Clock recovery circuits

ACRONYMS

ADC—**A**nalog to **D**igital **C**onverter

ASK—**A**mplitude **S**hift **K**eying

DAC—**D**igital to **A**nalog **C**onverter

DPSK—**D**ifferential **P**hase **S**hift **K**eying

FSK—**F**requency **S**hift **K**eying

MODEM—**MOD**ulator/**DEM**emodulator

PLL—**P**hase **L**ock **L**oop

PSK—**P**hase **S**hift **K**eying

QAM—**Q**uadrature **A**mplitude **M**odulation

QPSK—**Q**uadrature **P**hase **S**hift **K**eying

VCO—**V**oltage **C**ontrolled **O**scillator

A **SYMBOL** element is any electrically measurable characteristic of a waveform that can be used to hold information. For instance, in the previous chapter, **FREQUENCY SHIFT KEYING (FSK)** modulation was used for the transfer of data between two stations. Each binary bit was represented by a number of cycles of a sine wave signal on the telephone lines. One frequency represented a 1 logic bit and another the 0 logic level. In this case the frequency of the signals is the symbol used to represent the data.

The data rates used with FSK modems are limited by the bandwidth of the telephone lines (300 Hz to 3 kHz). Each signal symbol transmitted by an FSK modem represent a single data bit, so that the **SYMBOL RATE** (symbols per second, or sps) for data is identical to the data rate and cannot exceed 3000 sps. Ideally the symbol rate should be at a rate closer to the center of the telephone line bandwidth (approximately 1,500 Hz). A symbol rate of 2400 sps approaches the high side of that bandwidth. If the same data rate could be transferred at one-half of the data rate (1200 sps), its transmission would fall nearer the middle of the bandwidth. The problem is trying to squeeze the same information at 2400 sps into a switching rate of 1200 sps.

Amplitude can also be used to represent digital data. A digital data stream can be fed to a **DIGITAL TO ANALOG CONVERTER (DAC)** to produce an output voltage that varies with the input digital information. A group of bits (for instance, 4 bits per group) from the data stream are "collected" and presented to the DAC. A different analog voltage level out results from each of the sixteen combinations of 1s and 0s of the four bits. In this manner, each voltage level is a symbol representing four data bits. This form of digital coding is called **AMPLITUDE SHIFT KEYING (ASK)** modulation.

Another electrical aspect of an analog signal that can be used as a symbol is the phase of a sine wave voltage. Using a group of 4 bits again, a separate phase can be generated for each of the 16-bit combinations. One advantage of **PHASE SHIFT KEYING (PSK)** is that the frequency and amplitude of the signal used remain the same for each symbol. The frequency can be selected to fit comfortably in the middle of the telephone line bandwidth. Further, translating 4 bits at a data rate of 4800 bps into a symbol rate of 1200 sps keeps the switching rate close to the center of the bandwidth as well.

Amplitude and phase aspects of a sine wave can be simultaneously altered to form symbols. This type of modulation, called **QUADRATURE AMPLITUDE MODULATION (QAM),** is used for data rates of 9600 bps and higher. Use of PSK at these rates causes the difference between the phases to become too small to meet phase hit limits specified for the telephone lines. Recall from chapter 2 that a phase hit is a sudden change in phase that lasts for a short period of time. The telephone company specifies phase hits of $+/-10°$ as acceptable on the telephone lines. To avoid this problem, the combination of amplitude and phase modulation increases the phase differences while introducing amplitude variations. The circuits used to develop these forms of modulation consist of balanced modulators, bit splitters, digital to analog converters (DACs), linear summers, and various filters.

● 6.2 THE BALANCED MODULATOR

The heart of most PSK modulators is the balanced modulator. This circuit allows the phase of a carrier sine wave (fc) to be altered by a modulating digital signal. Figure 6.1, an example of a transformer balanced modulator, illustrates the concept behind balanced modulators. By and large, most balanced modulators today are made from semiconductor devices to avoid the expense, bandwidth limitations, and associated magnetic distortions and interference inherent in transformers.

The reference frequency (fc) is applied to T1 and is coupled through the secondary winding to the diodes D1 and D2 on the high side and D3 and D4 on the return side. The digital data stream is applied to the center taps of T1's secondary and T2's primary. The current level supplied by the digital circuits is enough to cause the diodes to turn on when the correct polarity is applied. As a point of reference, a logic 1 is selected to be positive at input A and negative at input B. This forward biases diodes 1 and 3 and turns off diodes 2 and 4. The signal coupled from T1's secondary is not large enough by itself to turn the diodes on, but once the diodes are on (from the digital input), this signal easily passes through to the primary of T2. The logic 1, in this case, causes fc to be passed to T2 and coupled so that the phase of the output signal is the same as the input signal.

Reversing the polarity at inputs A and B to represent a 0 turns on D2 and D4 while backbiasing D1 and D3. This time fc is directed to the opposite end of T2. The output signal coupled to T2's secondary is 180° out of phase with fc at the

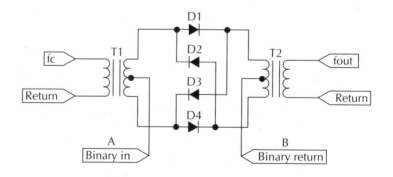

Figure 6.1 Diode Balanced Modulator

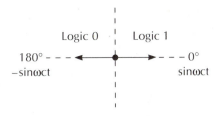

Figure 6.2 Balanced Modulator Phasor Diagram

input. A vector diagram (Figure 6.2) can be drawn to illustrate the phases representing a logic 1 and a logic 0. Zero degrees of phase lies on the positive side of the x-axis and is used as a reference for any phases generated by the modulator. It is important to note that the 0 reference is arbitrary and that the interpretation of data is done by relative rather than absolute means. The sine wave form of an AC signal is usually designated as the 0° reference lying on the x-axis. When using this designation, the y-axis, which is 90° from the x-axis, becomes the cosine axis; 180°, or negative x-axis, is denoted −sine, and 270°, or −90°, is the −cosine (see Figure 6.2).

Logic 1 in our example generated a signal that was the same phase as the input. This is represented in Figure 6.2 as a vector at 0°. This does not imply that the output signal is at 0° but that the difference between the reference (fc at the input) and the output is 0°. Similarly, the phase of the signal for a logic 0 (180° out of phase with reference, fc) is shown lying on the negative x-axis. The magnitude of the vectors used on this and subsequent phase diagrams (except QAM) is of unit length. This is easily acceptable since these devices are modulating the phase while keeping amplitude and frequency constant.

Since diodes are being used as the switching elements in the balanced modulator, a certain amount of mixing action occurs. This causes the generation of sum and difference frequency signals at the circuit's output. These signals are small in amplitude compared with the carrier signal that is being passed by the diodes, which are biased on. A bandpass filter at the output of the modulator is often used to prevent these unwanted signals from being passed on. The bandwidth of the circuit driven by the modulator is therefore large enough to pass the carrier frequency whose phase is shifted at the modulating rate. As an example, a data rate of 2400 bps, using a non-return to zero signal format (described in chapter 1), has a fundamental sine wave of 1,200 Hz. If the carrier frequency to be passed by the modulator is 1,650 Hz, then the bandwidth of the system driven by the modulator must be sufficient to pass both the 1,200-Hz switching rate and the 1,650-Hz carrier signal. For this example, the minimum bandwidth is 1,650 − 1,200, or 450 Hz. The telephone company system, with its 2,700-Hz (300 to 3,000 Hz) bandwidth, is more than adequate to handle the modulator's signals in this example.

One aspect of frequency shift keying (FSK) modems is that in order to use them, especially full-duplex modems, the lines they are connected to are required to possess sufficient bandwidth to handle the mark and space signal pairs from the originate and answer modems. To reduce the bandwidth requirements for the connecting medium, it becomes desirable to modulate a characteristic of a signal other than its frequency. Holding the frequency of the signals delivered by a modem constant reduces the amount of bandwidth needed for that signal. One option is to modulate a signal's phase instead of the frequency.

A **DIFFERENTIAL PHASE SHIFT KEYING (DPSK)** modem accepts a serial data stream and produces a single carrier signal whose phase changes according to the digital values of the incoming data stream. As the data stream is applied to the modem (Figure 6.3), it is fed to a balanced modulator (M1) and a 1-bit time delay circuit. The output of the delay circuit is fed to another balanced modulator (M2). A carrier frequency (fc) is applied to the first balanced modulator (M1) with a starting reference of 0 (sinωct). This carrier is also shifted 90° (cosωct), which is then applied to the second balanced modulator (M2). Each binary data bit causes its respective modulator to output a signal that has the same phase as the fc reference input to the modulator or a signal that is 180° out of phase with that input. In the diagramed example of Figure 6.3, a 0 bit into M1 causes that modulator to produce a sinωct (0°) signal. A 0 bit into M2 produces a cosωct (90°) out.

The two signals from the balanced modulators are summed linearly (that is, they are not mixed), and the resulting data symbol is sent to the connecting medium. Effectively, the linear sum of the two signals produces a vector sum of those signals. In the example of 0 bits applied to M1 and M2, the sinωct and cosωct signals have a vector sum of 45° from the 0° (sinωct) reference. Therefore, the symbol representing a 00 combination to the modulators is 1,800-Hz carrier signal whose phase is 45° from the sinωct fc reference.

Figure 6.3 shows a truth table for the bit combinations applied to the modulators and resulting phase of the symbols. The differential bits applied to the modulators are called **DIBITS,** since two (di) bits—the current and preceding bits—are grouped for each symbol. The phases specified by standards organizations in the United States (U.S.) and Europe (CCITT) are also illustrated.

The key behind differential phase shift keying is that the inputs to M1 and M2 are the current data bit (M1) and the preceding data bit (M2). The symbols produced at the output result from comparing the current data bit with the previous data bit. The reference clock (fc) is recovered from the data stream using a **PHASE LOCK LOOP (PLL)** to reproduce the clock. This is frequently called **CLOCK RECOVERY.** The receiver will then establish its own phase for fc reference based on a digital training sequence it receives from the sending station. This sequence is an established bit pattern that the receiver can decipher to establish a correct reference for its recovered fc clock.

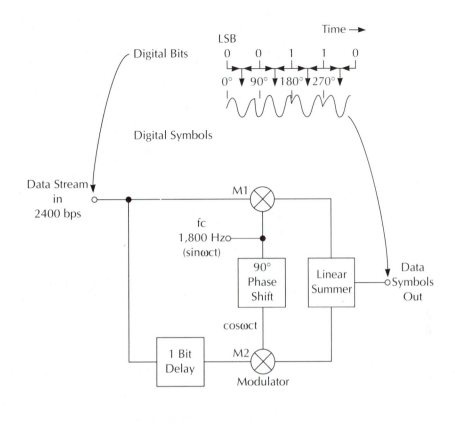

Dibit	CCITT	U.S.
0 0	0°	45°
0 1	90°	135°
1 1	180°	225°
1 0	270°	.315°

Figure 6.3 DPSK Modulator

● 6.4 BIT SPLITTERS, PSK MODULATOR INPUT CIRCUITS

To translate a number of bits from a serial data stream to a single symbol requires presenting all the bits in a group simultaneously to the circuits that are doing the translating. This job is done in PSK modems using a circuit called a **BIT SPLITTER.** Bit splitters are chiefly shift registers with controlled outputs. The data are shifted in at the same rate as the data rate. Output registers are loaded at the

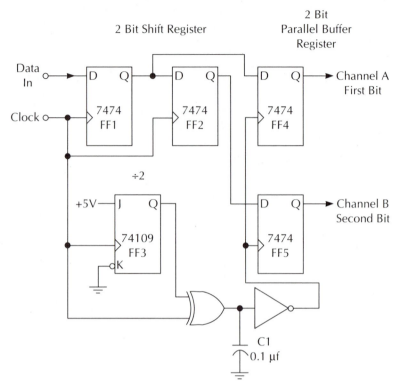

2 Bit Shift Register

2 Bit Parallel Buffer Register

Note: All set (preset) and reset inputs connected to +5V (inactive state)

(a)

Figure 6.4 Bit Splitter

symbol rate. This causes the bits to be presented to the translating circuits at the symbol rate. Phases of the output symbols will change at the same rate as data changes are presented to the translating circuits.

Figure 6.4a is the bit splitter used for a **QUADRATURE PHASE SHIFT KEYING (QPSK)** modem. Quad signifies that there are four different symbols created by this modulator, which uses 2 bits per symbol. Note that the clock signal (recovered from and equal to the data stream rate) is used to shift the input data in and is divided by two for clocking the two data flip-flops. These flip-flops are updated at a rate equal to one-half of the data input rate. The exclusive OR following the divide-by-two flip-flop sets up the correct clocking edge for timing purposes. The operation of the circuit begins with the presentation of the serial data to the data input of flip-flop 1. Flip-flops 1 and 2 act as a 2-bit shift register, which is loaded at the data rate (by the clock recovered from the data stream—clock recovery is discussed later in the chapter). Flip-flop 3 is configured as a toggle flip-flop. The output from this flip-flop is a square wave at one-half the clock (data)

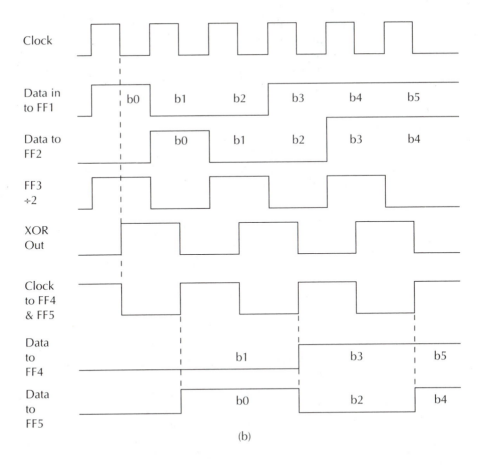

(b)

Figure 6.4—Continued

rate. The exclusive OR combines the original clock with the output of flip-flop 3. This presents a single pulse in coincidence with the negative half of the clock cycle (see the timing signals in Figure 6.4b). The 0.1-μf capacitor is added to cause a slight delay in the clocking of flip-flops 4 and 5 to avoid a "race" condition between the shifting in and parallel transfer of the data bits.

The inverted pulse is used to clock flip-flops 4 and 5, which are updated with new data for each pair of bits presented on the serial data in line. In essence, this circuit is a 2-bit serial to parallel converter. Each 2 bits presented to the circuit are present at its output. Because of the divide-by-two flip-flop, the output bits change at a rate that is one-half the data rate.

Example 6.1 For a data rate of 2400 bps, what is the symbol rate at the output of a QPSK bit splitter?

Solution: Because of the divide-by-two circuit, 2 bits from the data stream are present at the output at a symbol rate one-half that of the data rate, or 1200 sps. It should be noted that despite the fact that data at flip-flops 4 and 5 are still binary bits, they are no longer referred to as bits but as dibits. This is because they are no longer presented as single individual bits but are grouped as pairs of bits.

Other PSK modulators used for still higher data rates require bit splitters capable of converting serial data into groups of 3, 4, or more bits per symbol. These circuits are presented as the need arises.

● 6.5 QPSK MODEM

Figure 6.5 shows a QPSK modulator. The input is the 2-bit splitter of Figure 6.4. The digital symbol outputs of the bit splitter are fed to digital modulating inputs of two balanced modulators (Figure 6.1). The first bit into the bit splitter (at the output of flip-flop 4) is presented to the upper modulator, and the second bit (flip-flop 5), to the lower one. The reference sine wave (fc) is applied to the input of the top modulator and to a 90° phase shifter. The phase shifter uses the lag characteristics of a resistor-capacitor circuit to generate a signal that is designed to lag the input by 90°. In effect, if fc is referenced as a sine signal, then the output of the phase shift network is considered a negative cosine signal.

The outputs from the balanced modulators, which will be in phase with their respective fc signals for a logic 1 binary input and 180° out of phase for a logic 0, are fed to a linear summer. The resistors in this op amp circuit are selected for unity gain so that the output is the linear sum of the inputs. Linear summing avoids developing sum and difference frequencies normally generated by a nonlinear circuit used for standard radio modulators. A bandpass filter rejects unwanted signals produced by the summer. The result, at the output, is a vector sum of the outputs of the two balanced modulators. Again, recall that the output amplitude and frequency of the balanced modulators are identical. The bandwidth requirements for the QPSK modulator are similar to those discussed in the section on balanced modulators. The linear sum of the two modulator signals produces a signal whose frequency is that of the carrier frequency. The rate at which the phase of that signal changes is one-half that of the data rate applied to the input of the modulator.

As an example, a carrier frequency of 1,650 Hz is used with a 2400-bps QPSK modulator. The dibits cause the phase of the carrier to switch at a rate of 1,200 Hz (2 bits required for each phase signal generated). The bandwidth of the system driven by the QPSK modulator is 1,650 − 1,200 or 450 Hz. Attempting to send the

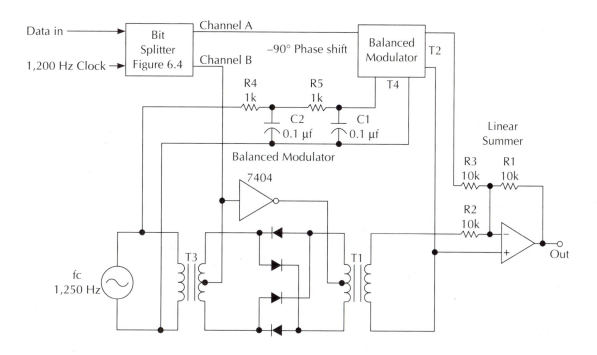

Figure 6.5 QPSK Modulator

same data using the frequency shift keying methods described in chapter 5 requires at least two different signals (one for space and one for a mark in a half-duplex scheme), which are to be switched at a 2,400-Hz rate. Depending on the actual mark and space frequencies, this will require a lot more than 450 Hz of bandwidth.

For this chapter the following conventions are used. Logic 1 inputs to the balanced modulator are positive since they produce outputs that are in phase with the fc input. Logic 0 inputs are negative since they produce a signal 180° out of phase with the input. Keep in mind that this is an arbitrary selection. The logic influences could be reversed if we so desired. To keep life easy, this text will retain positive logic convention. Equally as arbitrary is the selection of sine wave as a reference for fc. It could as easily be −sine wave or +/− cosine wave. The text will retain the sine wave as the signal that begins with 0° phase at time 0 (t0), as a reference for purposes of discussion. The phase shifter could also as easily have been designed to yield a positive or negative 90° shift. The shifters used in this chapter will always employ a −90° lag. This makes the fc input to the lower balanced modulator −cosine wave in reference to the sine wave input to the upper modulator.

Figure 6.6 is the vector diagram for all the possible results at the output of the QPSK modulator. Two binary bits can form four different combinations of 1 and 0 states. Each combination generates a separate vector symbol at the output. As an

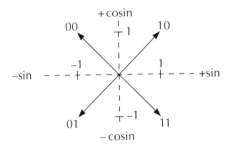

Figure 6.6 QPSK Phasor Diagram

example, if both bits are low, the outputs of the balanced modulators will be 180° out of phase with their respective fc inputs. This results in a negative sine signal summed with a positive cosine signal. The amplitudes of these signals are considered to be of unit length, so the vector sum becomes a signal that lies halfway between both signals (at 135°) at a length of 1.414 times the unit length. The vector representing this symbol, shown on Figure 6.6, is created by starting along the sine axis for a distance of negative unit length (to the left of the cosine axis intercept). From that point on, the sine axis goes up a positive unit length in parallel with the cosine axis. This is the endpoint of the vector that represents the symbol created by both bits being low. Draw a line from the origin to this point. The triangle formed by this vector and the unit lengths along the sine and parallel to the cosine axis can be used to determine the magnitude and angle of the vector using standard trigonometric mathematics.

The process is repeated for each combination of 1 and 0 states of the 2 bits from the bit splitter. Note that there are four distinct phase symbols, one for each combination, and that they are separated from each other by 90°.

6.6 SOME PSK GENERALITIES

Observations from the QPSK circuit and subsequent PSK circuits produce these mathematical relationships:

1. The symbol rate (S) equals the data rate (D) divided by the number of bits per symbol (N):

$$S = \frac{D}{N} \qquad (6.1)$$

2. The number of symbols (M) possible at the output is equal to the radix 2 raised to the Nth power:

$$M = 2^N \qquad (6.2)$$

DIGITAL & DATA COMMUNICATIONS

3. The size of the bit splitter register and the frequency division factor is N.

4. The amount of phase difference between symbols (P) is 360° divided by M:

$$P = \frac{360°}{M} \qquad\qquad (6.3)$$

Example 6.2   For a QPSK modulator, what is the number of symbols possible at its output? What is the phase difference between each symbol? For a data rate of 3600 bps, what is the symbol rate at the output?

Solution: Applying equations 6.1 and 6.2, the number of symbols possible for a QPSK is $2^2 = 4$. The phase difference between each symbol is $360°/4 = 90°$. Finally, the symbol rate is the bit rate divided by the number of bits per symbol, or $3600/2 = 1800$ sps. This last piece of data says that the phases of the symbols will change at a rate of 1800 sps.

• 6.7 QPSK DEMODULATOR

In a QPSK demodulator, the reference frequency (fc) is recovered from the incoming bit stream. A simple clock recovery circuit is shown in Figure 6.7. Before actual data are sent to a PSK receiver, they are preceded by a string of an alternating pattern of 1s and 0s called a preamble. The purpose of this preamble is to allow the clock recovery circuit to "lock in" on the clock frequency. The PSK demodulator crudely demodulates the incoming preamble using the free-running clock of the phase lock loop's (PLL) **VOLTAGE CONTROLLED OSCILLATOR (VCO).** The digital-type data are then fed into the clock recovery circuit. This consists of a ½-bit delay and an exclusive OR as shown in Figure 6.7. The timing diagram illustrates how the preamble of alternating 1s and 0s creates a "clock pulse," which is sent to the PLL. The preamble is long enough to allow the PLL to capture and hold lock between the data stream pulses and the VCO of the PLL. At this lock condition the frequency of the VCO output is the same as the pulse rate generated by the exclusive OR. Since the PLL is locked at this point, the loss of pulse information for a limited time period does not cause the PLL to unlock, thus maintaining correct clock frequency. That is, once actual data replace the preamble, clocking is not lost for short consecutive strings of 1s or 0s.

Clock recovery is enhanced by any scheme that increases the number of clock pulses presented from the exclusive OR. Strings of 1s and 0s cause the inputs to the exclusive OR to remain the same and thereby produce no pulses. If the string is too long, the PLL may lose capture and cause the VCO's frequency to change. One

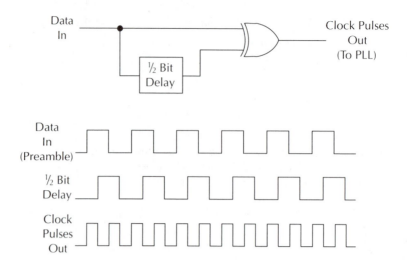

Figure 6.7 Clock Recovery

circuit developed to assist in clock recovery by increasing the number of clock pulses is called a **SCRAMBLER,** which is illustrated in Figure 6.8 along with the receiver's **DESCRAMBLER.** As data are being sent to the modem, they first shift through the scrambler's shift register. AND gates on the outputs of that register monitor the Q and $\overline{Q}$ outputs of the register's flip-flops. As long as there are fewer than four consecutive 1s or 0s, at least one input to each AND gate is low. The outputs of the AND gates feed an OR gate. As long as one input to each AND gate is low, their outputs will be low, keeping the OR gate output low. The OR gate, in turn, feeds into one input of an exclusive OR, acting as a controlled inverter. A low input to the exclusive OR control input maintains its output at the same state as the other exclusive OR (data) input.

If a string of four consecutive 1s or 0s is shifted into the register, the scrambler will cause the next bit to be inverted. As an example, suppose the ASCII character A (1000001) is sent as data to the scrambler. The first 1 and four following 0s shift through the register and proceed through the exclusive OR unchanged. However, once the four consecutive 0s are shifted into the scrambler register, the AND gate connected to the $\overline{Q}$ outputs senses four high inputs. Its output goes high, causing the OR gate to follow suit. A high state on the control input of the exclusive OR causes the fifth consecutive 0 to be inverted. The inverted bit is shifted out the data line to the receiver and into the scrambler shift register as well. The shift register receives its inputs from the output of the exclusive OR to prevent the circuit from continuing to invert data if the number of consecutive 0s (or 1s) continues. It would take another consecutive four 0s following the inverted bit to cause another 0 to be inverted.

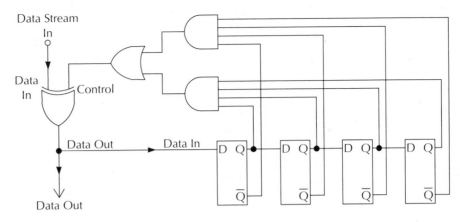

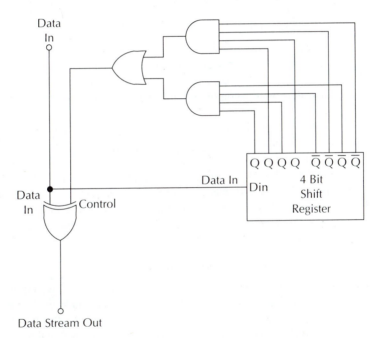

(a) Scrambler

(b) Descrambler

Figure 6.8 Data Scrambler and Descrambler

Example 6.3   How does the following data stream appear as it is sent using a scrambler (LSB on the right)?

110001111111000000001100111110100001

Solution: Starting with the LSB first, the data are unchanged until four consecutive 1s or 0s are experienced:

00001

The next bit is inverted (and sent into the scrambler shift register as a 0). This now places four consecutive 0s into the scrambler again (the 1 bit following the original four 0s has been converted to 0 also). This results in the next bit (a 0) being converted to a 1. The data continue until the next four consecutive 1s or 0s occur. Note that the inversion of the 1 bit following the first four consecutive 0s causes the scrambler to receive a fourth 0 and will invert the bit as well.

1000001

This appears to defeat the intent of the scrambler by actually increasing the number of consecutive 0s in this data stream. Such is the case and the chief limitation of the circuit. Given occasional strings of four consecutive 1s or 0s followed by an alternating combination of 1s and 0s will generate a longer stream of consecutive 1s or 0s.

The process in the example will continue until the data out appear as:

110001<u>1</u>0111100<u>1</u>00001100<u>1</u>01111<u>1</u>000001

Compare this with the original data stream and verify that the bit following each string of four 1s or 0s has been inverted. Inverted bits are underlined to assist in the comparison. Do not forget that if an inverted bit enters the scrambler in the same state as the three preceding consecutive bits, the next bit is also inverted.

At the receiver end is a descrambler (Figure 6.8), which unscrambles the data after the clock has been recovered. The digital data stream enters the descrambler and the exclusive OR. AND gates again monitor the flip-flop outputs. The circuit operation is essentially the same. The main difference is that the descrambler shift register is fed from the data stream directly, while in the scrambler its shift register was driven from the output of the exclusive OR.

The clock is fed into one balanced modulator and a 90° phase shifter. The shifted signal is presented as the f_c input to a second balanced modulator. This time, the amplitudes of these signals are large enough to turn the diodes on. The incoming data stream is applied as the other input to the balanced modulator. The output of the modulator is altered to perform the function of an FM detector or phase detector (Figure 6.9).

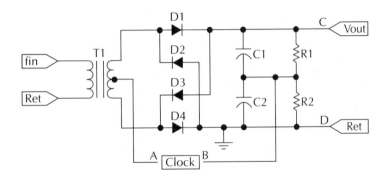

Figure 6.9 Phase Detector

Notice that the phase detector is similar to the modulator (Figure 6.1). The capacitors and resistors that replace the second transformer found in the modulator form a peak detector circuit. The capacitors charge to the peak value of the applied sine wave signals and, in combination with the resistors, filter out the AC sine wave components. Signal fc is applied to inputs A and B with sufficient amplitude to bias alternating pairs of diodes on for the positive and negative alternations of fc. The positive alternation of fc turns on D1 and D4, while the negative half turns on D2 and D3. The effect is to pass the incoming signals (fin) to the RC circuit. The amount the capacitors will charge to depends on the phase relationship between fin and fc. Figure 6.10 shows four cycles of fc and their corresponding fin cycles.

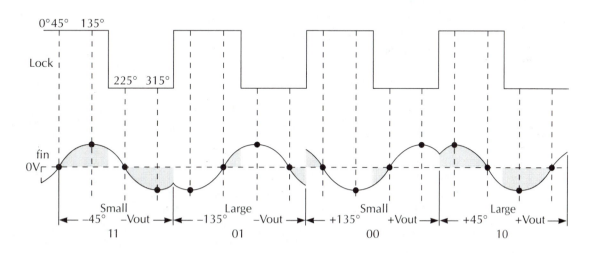

Figure 6.10 Phase Relationship between fin and Clock for QPSK Demodulator

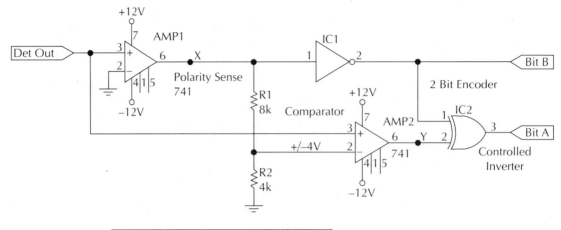

Det Out	X	Bit B	Y	Bit A
−3V	−12V	1	+12V	1
−6V	−12V	1	−12V	0
+3V	+12V	0	−12V	1
+6V	+12V	0	+12V	0

Figure 6.11 2-Bit Analog to Digital Converter

These fin cycles represent the four phases produced by the QPSK modulator. The shaded area indicates the amount of the fin signal that causes a ''positive'' current through capacitors C1 and C2.

When the fin cycle begins with a shaded area, the capacitors charge to a positive voltage with respect to ground. Resistor values are sufficiently large to prevent discharge during the nonshaded time periods. If the fin cycle begins with a negative alternation, then the capacitors charge negatively. The amount of capacitor charge time and hence the value of capacitor voltage is directly related to the time that fin has the same polarity as it began with. As a result, the voltage from C to D at the output is different for each input phase. There are two positive and two negative voltage levels possible at the detector's output. An **ANALOG TO DIGITAL CONVERTER (ADC),** as shown in Figure 6.11, is required to translate the four levels into binary bits.

The output from the detector (DET OUT) is fed to polarity sense amplifier AMP1. The polarity of DET OUT causes this amplifier to saturate to +12 or −12V. A voltage divider composed of R1 and R2 supplies +/−4V to comparator amplifier, AMP2. IC1 inverts the output of the polarity sense amplifier to produce the correct logic level for bit B. DET OUT is also fed to comparator AMP2. This circuit produces positive or negative saturation based on the value of DET OUT compared with +/−4V reference from the voltage divider. The output of the comparator is exclusive ORed with bit B to create bit A. A table of the values and results through the ADC is provided with Figure 6.11. IC1 and IC2, besides performing the

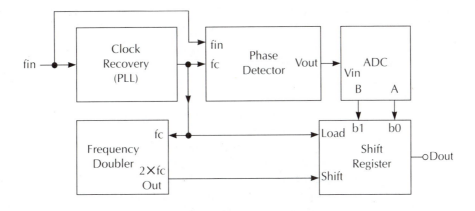

Figure 6.12 QPSK Demodulator Functional Blocks

indicated logic functions, also must perform voltage level translation to produce the correct logic family (TTL, MOS, etc.) for bits A and B.

The components of the QPSK demodulator are summarized in the block diagram of Figure 6.12. The incoming data stream (fin) supplies the clock signal (fc) through a clock recovery circuit as well as data input to the phase detector. The recovered clock supplies the other input to the phase detector and to a frequency doubler (Figure 6.13). The doubler works by combining pulses from two differentiators, which each produce a positive pulse for alternating clock edges. The output from the phase detector is the DET OUT input to the ADC. Digital bits A and B from the ADC feed a bit-combining shift register. The bits are loaded into the shift register at the symbol rate and shifted out at twice that rate to match the data

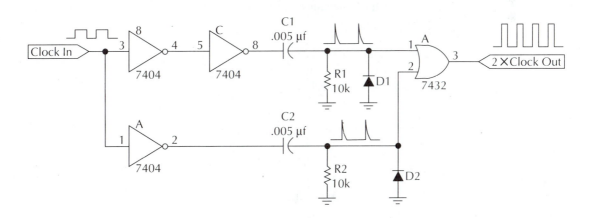

Figure 6.13 Clock Frequency Doubler

rate used for input data to the QPSK modulator that sent the information. Two bits are then shifted out for every symbol input to fin.

● 6.7.1. An Alternate QPSK Demodulator

Another type of QPSK demodulator (Figure 6.14) literally reverses the process used in the QPSK modulator. The recovered clock supplies the reference clock to a balanced modulator and 90° phase shifter. The output of the phase shifter is fed to a second balanced modulator. Signal fin and the clock inputs to the balanced modulators cause these signals to be "multiplied," producing second harmonic signals of the original clock input. These second harmonics are filtered at the output of the balanced modulators, leaving DC voltage levels equivalent to amplitude results of the "multiplication." These DC levels are fed to ADCs and bit-combining shift circuits.

The mathematics supporting the QPSK demodulator involve these trigonometric product relationships:

$$x(\sin\omega ct)(\sin\omega ct) = 1/2(1 - x\cos2\omega ct) \tag{6.4}$$

$$x(\sin\omega ct)(\cos\omega ct) = 1/2(x\sin2\omega ct) \tag{6.5}$$

$$x(\cos\omega ct)(\cos\omega ct) = 1/2(1 + x\cos2\omega ct) \tag{6.6}$$

where x is the symbol's amplitude and ωct is the symbol's frequency.

Example 6.4

Analyze the QPSK demodulator of Figure 6.14 for detection of the phase generated by binary bit combination 00 into the QPSK modulator (Figure 6.5). The symbol for this combination produced a phase equivalent to the sum $-\sin\omega ct + \cos\omega ct$. Channel A of the demodulator uses a clock reference of $\sin\omega ct$ as it did in the modulator.

Solution: The lower balanced modulator has a clock reference shifted $-90°$ or $-\cos\omega ct$. fin is multiplied by both references, producing these results:

For the upper balanced modulator (channel A):

$$A = x(\sin\omega ct)(-\sin\omega ct + \cos\omega ct)$$

$$A = x[(-\sin\omega ct)(\sin\omega ct) + (\sin\omega ct)(\cos\omega ct)]$$

$$A = \frac{-1 + x(\cos2\omega ct + \sin2\omega ct)}{2}$$

After filtering out the second harmonic, only the constant DC level remains:

$$A = -\frac{x}{2}$$

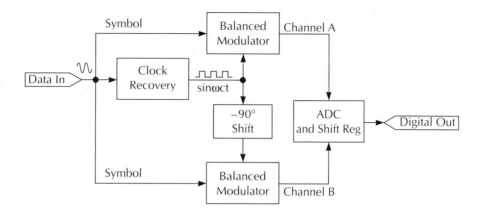

Figure 6.14 QPSK Demodulator

And for the lower B channel:

$$B = x(\cos\omega ct)(-\sin\omega ct + \cos\omega ct)$$

$$\frac{x(-\sin2\omega ct + 1 + \cos2\omega ct)}{2}$$

which, after filtering out $2\omega ct$ signals, leaves $x/2$.

Repeating the trigonometric calculations for the remaining three symbols produced by the QPSK modulator produces the results listed in Table 6.1. An ADC and binary encoder are used to convert these four different voltage combinations

Table 6.1 QPSK Demodulator States

DC out		Logic states	
A	B	A	B
$-\dfrac{x}{2}$	$-\dfrac{x}{2}$	0	0
$-\dfrac{x}{2}$	$\dfrac{x}{2}$	0	1
$\dfrac{x}{2}$	$-\dfrac{x}{2}$	1	0
$\dfrac{x}{2}$	$\dfrac{x}{2}$	1	1

into the binary bit combinations that produced the symbols from the QPSK transmitter. These bits are loaded into a 2-bit shift register and sent to the USART in the data terminal equipment.

• 6.8 HIGHER DATA RATE MODEMS

One major limiting factor to the data rate that can be used with any modem on any system is the bandwidth of that system. For binary, essentially noiseless communications, a form of the **SHANNON CHANNEL CAPACITY** (C) is used to determine the relationship between the number of different symbols (n) a modem can produce and the maximum data rate that modem can be used for when connected to a specific system. This relationship is stated mathematically as:

$$C = BW(\log_2 n) = BW \times m \qquad (6.7)$$

where C is the channel capacity in bits per second, BW is the bandwidth of the channel (or medium), n is the number of different symbols, and m is the number of bits per symbol.

For FSK modems where n = 2 (mark and space signals) and the BW = 3000 − 300 or 2,700 Hz, the theoretical maximum data rate is:

$$C = 2,700 \log_2 2 = 2700 \times 1 = 2700 \text{ bps}$$

Above 2700 bps it is necessary to transmit more bits per symbol to use the same bandwidth. QPSK uses two bits per symbol, increasing the maximum data rate to $2 \times 2700 = 5400$ bps. Higher data rates require still more bits per symbol from the transmitter.

Figure 6.15 is an 8PSK modulator. The number 8 relates to the total number of different symbols produced by this circuit. Each symbol is separated by 360°/8, or 45°. The bit splitter is expanded to accommodate 3 bits at a time from the data stream. Bit A, the first bit into the bit splitter, and bit B, the second, are sent to 2-bit DACs. The third bit (bit C) is sent to one DAC directly, inverted, and sent to the other DAC. Bit C or $\overline{C}$ (depending on the DAC) controls the DAC amplitude voltage required to eventually produce the desired symbol vectors. These values are 0.383V when the state of the C input is low and 0.1585V when it is high. The states of bits A and B determine the polarity of the voltage from the DAC. A 1 on either bit causes its corresponding DAC to output a positive voltage. A 0 produces a negative voltage value. The polarity of this voltage determines whether the outputs of the balanced modulators are in or out of phase with the reference (fc) inputs. The amplitude of the voltage determines the amplitude of the balanced modulator signals. While actual voltages used may not be the same used here, the ratio between the two levels remains the same. This ratio is required to develop the phase relationships between symbols.

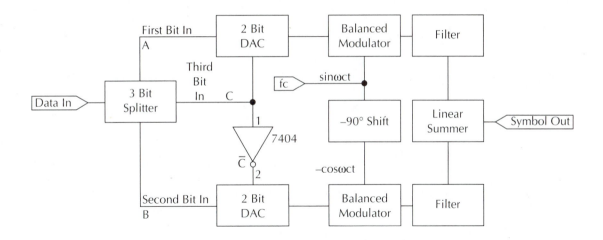

Figure 6.15 8PSK Modulator

Figure 6.16a is the vector diagram for an 8PSK modulator. A similar type of diagram (Figure 6.16b) showing just the endpoints of the vectors, called a **CONSTELLATION** diagram, is used for purposes of clarity, particularly as the number of symbols produced by the modulator increases. Using the ''triangle'' in quadrant I, consider the symbol vector as the hypotenuse and the individual vectors representing the transmitter balanced modulator outputs as the adjacent and opposite sides of the triangle. The angle between the symbol vector and the axis (which is the phase of the symbol) is found by calculating the inverse tangent using the values of the triangle's ''sides.''

$$\tan^{-1}(.1585/.383) = 22.5°$$

This makes the angle between each vector twice 22.5°, or 45°, which agrees with the earlier computation of 360°/8, where 8 is the number of symbols produced by an 8PSK modulator.

The magnitude of each symbol is calculated using the Pythagorean theorem:

$$A^2 + B^2 = C^2 \qquad (6.8)$$

where A and B are the adjacent and opposite sides and C is the hypotenuse representing the symbol vector.

$$C = \sqrt{[(.383)(.383) + (.1585)(.1585)]} = .4145V$$

Example 6.5   Show the steps needed to derive the vector symbol for a binary data stream of 000.

Solution: The 3 bits enter the bit splitter and are directed to the 2-bit DACs. The output of channel A's DAC (A = 0 and C = 0) is −.383V, and the lower DAC

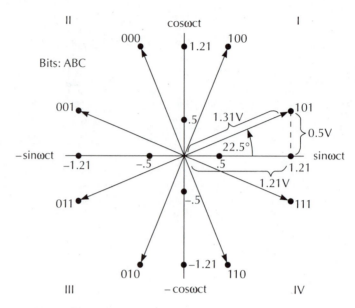

(a) Vector Diagram

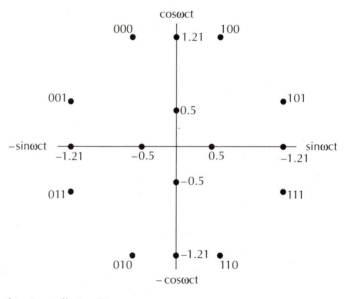

(b) Constellation Diagram

Figure 6.16 8PSK Output Symbol Diagrams

(B = 0 and $\overline{C}$ = 1) is −.1585V. Note, since bit C controls the amplitude of the DAC circuit, the amplitudes of the outputs of these two DACs are always different. The DAC outputs are fed to the balanced modulators, producing −.383 sinωct on channel A and .1585 cosωct on channel B. These are linearly summed to form the appropriate symbol in the second quadrant on the vector diagram.

Note that the actual voltages and polarities, as well as the lead or lag condition of the phase shifter, are selected by the circuit designer and can be any combination. The benefit of using 8PSK modems is to allow for an increased data rate without an increased bandwidth. A 4,800-Hz modem using 1,850 Hz for a carrier frequency causes the phase of the output signal to change at a 1,600-Hz rate. The bandwidth required for this example is 1,850 − 1,600, or 250 Hz. While the values have been selected arbitrarily in this case, as in earlier examples involving bandwidth considerations, the results illustrate the types of bandwidths required for circuits driven by these modulators. For the 8PSK modem, the Shannon law (2,700 $\log_2 8$) determines that the upper limit for the data rate using an 8PSK modem is 8100 bps. Data rates above 8100 bps are transferred using a different form of modulation.

• 6.9 A 9600 bps MODEM

To satisfy the Shannon law, a system operating at 9600 bps would require 4 bits per symbol. The circuit produces 16 different symbols at its output. Using standard PSK for modulation creates symbols that are 360°/16, or 22.5° apart. This phase difference comes close to the +/−10° of phase hit specified as permissible by the telephone company. An alternate method of modulation has been developed to reduce the phase error risk inherent in PSK modems that begin to approach the phase hit limit. This method incorporates amplitude and phase modulation within a single modulator.

Quadrature amplitude modulation (QAM) causes the phase and amplitude of a signal to be modulated by the binary bit stream. Figure 6.17 is the block diagram of a QAM modulator. It uses the circuits we are now familiar with—bit splitters, DACs, balanced modulators, and linear summers.

Four bits are shifted into the bit splitter. These bits are directed to the DACs, with bits A and C sent to the upper-channel DAC and bits B and D to the lower-channel DAC. As with the 8PSK modulator, these bits create one of four output levels, two of which are a positive voltage and two a negative voltage. The amplitudes are selected to produce phase angle differences between symbol phases of 30° as shown on the vector and constellation diagrams of Figure 6.18. Note that there are two vectors in each quadrant that have no phase difference but do exhibit amplitude differences. This occurs because, unlike the 8PSK modulator, it is now

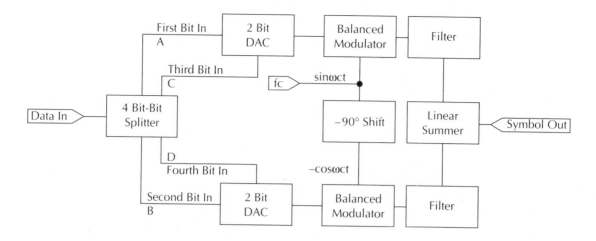

Figure 6.17 16 QAM Modulator

possible for the DACs to produce the same voltage magnitude. There are twelve different phases and three different amplitudes for the symbols shown on the vector diagram for the QAM modulator.

Depending on the inputs to the DAC, the outputs can be multiples of 1.34V, 5V, −1.34V, or −5V. Combining these with the different phase possibilities from the balanced modulators (sinωct, −sinωct, cosωct, and −cosωct) results in sixteen possible vectors (eight per channel). These are summed to form sixteen possible vectors at the modem's output as shown in Figure 6.18.

Example 6.6   Show how the vectors representing 0000 and 1001 are created.

Solution: Assuming the DACs work the same as they did for the 8PSK modulator, bits A and B determine the polarity (0 = − and 1 = +) and bits C and D, the amplitude (0 = 5V and 1 = 1.34V), then the bit combination 0000 puts out −5V from both DACs. This reverses the phase outputs of the balanced modulators. When summed, the output symbol is −5Vsinωct + 5Vcosωct, which is the middle long vector in quadrant 2 II the vector diagram, Figure 6.18.

For 1001, bit A = 1, bit B = 0, bit C = 0, and bit D = 1. Channel A's DAC is 5V, and channel B's is −1.34V. This makes channel A's balanced modulator's output to be 5Vsinωct and B's, 1.34Vcosωct. These are summed to produce the symbol vector indicated in quadrant I of Figure 6.18.

A 9600-baud modem causes the symbols to change at a rate that is one-fourth of the data rate, or 2,400 Hz. If the carrier used in the modem has a frequency of 1,870 Hz, then the bandwidth required for this example 16 QAM modulator is

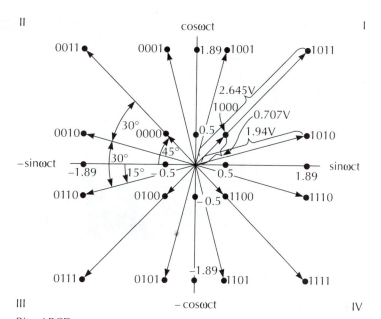

(a) Vector Diagram

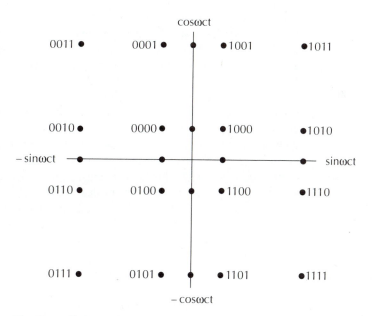

(b) Constellation Diagram

Figure 6.18 16 QAM Output Symbol Diagrams

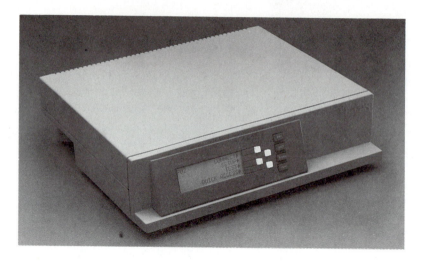

Figure 6.19 9600-bps QAM Modem—Hayes Smartmodem 9600. (9600 is a trademark of Hayes Microcomputer Products, Inc. Used by permission.)

2,400 − 1,870, or 530 Hz. Additional consideration must be given when using a QAM system since amplitude as well as the phase of the symbols is being changed for each quad bit combination.

A 9600-baud QAM modem manufactured by Hayes Smartmodem is pictured in Figure 6.19. A number of RS232C function status lights are brought to the front panel to assist the user or technician. This modem is an example of a stand-alone modem, which is installed separately from the terminal or personal computer. Many modems are manufactured onto printed circuit cards, which are inserted into one of a personal computer's expansion slots. It is only necessary to plug the telephone line into such modems to complete the installation.

As the data rate is increased, the need to transmit more bits during a single symbol period increases. Low-speed modems operate fine using FSK modulation and 1 bit per symbol. Moderate speeds are handled using PSK techniques allowing multiple bits to be included in each symbol. To maintain a lower number of bits per symbol but still increase the data rate requires that more than one element of a symbol be modulated. This is accomplished using QAM modulation. To further increase data rates, more sophisticated technology involving DACs and bit compression is required. In essence, a form of coding—that is, changing the original data to a form using fewer bits—is utilized to allow higher data rates to be sent on the telephone lines.

SUMMARY

The process of transferring data rates beyond the bandwidth limitations of the telephone lines requires translating the digital data rates into symbol rates that do conform to the bandwidth. To do this, groups of bits were used to create single symbols. These symbols are used to modulate the phase and, in the case of QAM,

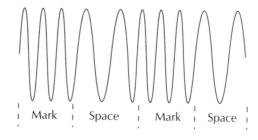

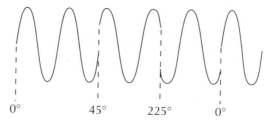

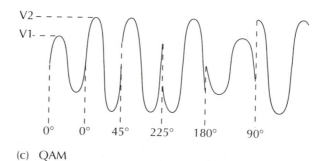

Figure 6.20 Comparison of FSK, PSK, and QAM Signals

the amplitude of the signal placed onto the telephone line. These transfers generally use synchronous data, which require a synchronizing clock signal to be regenerated, at the receiver, from the data stream itself. Various circuits are used to recover this clock signal, including scramblers and descramblers. The scrambler is used to reduce the possibility that a long stream of 1s or 0s will be transferred. These long streams would appear as a constant DC level to the clock recovery circuit, which would cause a loss of clock signal at the receiver.

Differences in the modulated signals for each of the primary low- and medium-speed modulators are illustrated in Figure 6.20. The FSK signal (Figure

6.20a), described in chapter 5, shows changes in the signal's frequency as the data bits vary between mark and space states. PSK signals (Figure 6.20b) show variations in phase caused by the modulating symbol information. Finally, QAM signals show changes in both phase and amplitude in Figure 6.20c. The number of actual amplitude changes have been minimized for this figure to reduce confusion.

GLOSSARY

Bit splitter—Accepts serial data stream and supplies individual bits to balanced modulators in a PSK or QAM modem.

Clock recovery—Restoring synchronizing clock from the data stream.

Constellation—Diagram of the vector outputs of a PSK or QAM modem.

Dibits—Two binary bits per symbol.

Quadrature amplitude modulation (QAM)—Phase and amplitude modulation technique.

Scrambler/Descrambler—Circuits which aid clock recovery by preventing long streams of 1's or 0's in the transmitted and received data.

Shannon channel capacity—Relates data rate to bandwidth and bits per symbol.

Symbol—A single electrically measurable signal that represents groups of one or more data bits.

Symbol rate—Rate at which symbols are transmitted in symbols per second, or sps.

QUESTIONS

6.1. What three measurable electrical characteristics of a sine wave can be used as a symbol? Which modulation types discussed in the chapter are used for each symbol type?

6.2. Which circuit is used to phase modulate a sine wave with digital data?

6.3. How many bits are grouped into a DPSK symbol?

6.4. How does a DPSK receiver decipher the incoming symbols into binary bits?

6.5. What is the main advantage of PSK modulation compared with FSK?

6.6. What is a drawback of PSK versus FSK modulation?

6.7. How many different symbols are produced at the output of a DPSK modulator? How do they differ?

6.8. Since a grouping of 2 bits is called a dibit, what would a group of 3 bits be called?

6.9. What type of digital circuit is used as a bit splitter?

6.10. How many bits are grouped to form a QPSK symbol?

6.11. Which functional circuit block is used to recover the clock from the input data stream in a PSK demodulator?

6.12. What is the purpose of a scrambler and descrambler circuit?

6.13. How many different symbols are possible at the output of a 16QAM modulator?

6.14. What are the advantages and disadvantages of a 16QAM modulator compared with a 16PSK modulator?

6.15. What standard data rate is associated with a QAM modulator?

PROBLEMS

6.1. What is the symbol rate for a 2100-bps data stream using a QPSK modulator? An 8PSK modulator?

6.2. What is the phase angle difference between symbols for a QPSK modulator? An 8PSK modulator? A 16PSK modulator?

6.3. Draw the vector and constellation diagrams representing the possible symbols at the output of the QPSK modulator of Figure 6.4 if the 90° phase shift is positive (leading) rather than negative (lagging).

6.4. Draw the vector and constellation diagrams for the 8PSK modulator shown in Figure 6.12 with a noninverting buffer in place of the inverter connected to bit C. What type of modulator does this actually create?

6.5. What is the noiseless telephone line channel capacity in bps of a modulator that generates a symbol for every group of 5 bits at the data input? For a group of 12 bits?

6.6. A QAM-type modulator converts groups of 6 bits into symbols. What are the values of the following parameters for this modulator if the data rate is 12000 bps?
a. Symbol rate   b. Number of different symbols
c. Number of flip-flops in the bit splitter

6.7. The following data are sent synchronously using the ASCII code. What is the binary sequence at the output of a scrambler connected before the modulator?

BAD ACE

6.8. Using equations 6.4, 6.5, and 6.6, demonstrate how the incoming symbol $\sin\omega ct - \cos\omega ct$ produces a DC level.

6.9. Draw the vector and constellation diagram for the 16QAM modulator shown in Figure 6.17 using $-\cos\omega ct$ as the reference fc in place of $\sin\omega ct$.

LAB PROJECTS

6.1. Design, construct, and test each of the circuit types below:
a. QPSK bit splitter
b. Solid-state balanced modulator
c. Linear op amp summer
d. Telephone line bandpass filter

6.2. Combine the circuits in project 6.1 into an operating QPSK modulator.

6.3. Design, construct, and test a QPSK demodulator. Upon completing this and project 6.2, interconnect both modulator and demodulator. Use a square wave generator to supply a square wave whose frequency is one-half the bit rate for your circuit. With an oscilloscope, observe the output of the demodulator to see if the square wave is reproduced.

6.4. Use a microprocessor-based test set (8085, 6800, or Z80, for example) to generate the serial data bit stream. If your test set is equipped with a UART such as an ACIA or 8251, write the program to utilize that device to send out the serial data. If your set is equipped only with a parallel interface such as a Motorola PIA, INTEL 8255, or Zilog PIO, or only with access to the data bus, write the program to output the serial data through one of the data lines. Connect your circuit of project 6.3 to the microprocessor test set. Use the oscilloscope to monitor the output of the QPSK demodulator for reproduction of the data generated by the microprocessor test set.

6.5. One last step: Connect the demodulator to a second microprocessor test set. Write the program that will read the incoming data from the QPSK modulator and store that data in the test set's memory.

RESEARCH ASSIGNMENT

6.1. Many modem manufacturers are producing 2400-, 4800-, and 9600-bps modems. Research one of these modems, using magazine articles and company data manuals. Your report should include the circuits used in the modem, which specifications the modem adheres to, the standardizing organization whose standards are being used, all of the specifications for the modem, and any other pertinent information.

Answers to Odd-Numbered Questions

6.1. Amplitude—ASK and QAM; frequency—FSK; phase—PSK and QAM
6.3. Two consecutive bits
6.5. Single line frequency reduces bandwidth requirements; less affected by line noise
6.7. Two different symbols that vary in phase
6.9. Shift register
6.11. Phase lock loop
6.13. Sixteen
6.15. 9600 bps

Solutions to Odd-Numbered Problems

6.1. 1050 sps; 700 sps
6.3. See Figure 6.21 for solution.
6.5. 13500 bps; 32400 bps
6.7 B = 1000010   A = 1000001   D = 100 0100   space = 0100000
     A = 1000001   C = 1000011   E = 100 0101

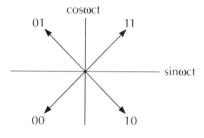

a) Vector Diagram

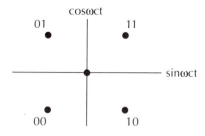

b) Constellation Diagram

Figure 6.21 Solution to Problem 6.3

Sent as unscrambled data, with lsb on the right:

E C A space D A B

1000101100001110000010100000100010010000011000010

After being scrambled (inverted bits are underlined):

1000101011111111000010110000100010100000100000010

Notice the limitation of the scrambler, here—four 0s in the second A caused a bit to be changed to 1, which, in turn, created four 1s followed by a set of 0s. As each 0 is inverted to a 1, another set of four 1s is produced, requiring the next bit to be inverted from a 0 to a 1. This continued until a 1 was inverted to a 0. This sequence actually created a more severe problem than was in the original data stream.

6.9. See Figure 6.22 for solution.

Lab Project Guides

Urge instructors to either set lab kits to have some ICs that include these circuits or to have students research and use such ICs. Most of today's modems are built

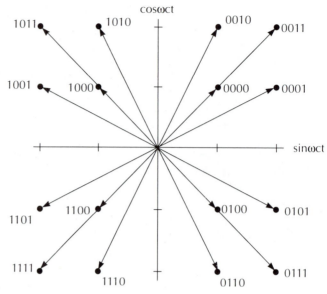

A,B (0 = –, 1 = +)
C,D (0 = .5, 1 = 1.3V)
ABCD

(a) Vector Diagram

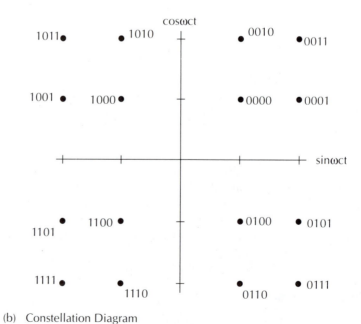

(b) Constellation Diagram

Figure 6.22 Solution to Problem 6.9

around these LSI and MSI chips. These projects are selected as a progressing lab assignment through a portion of the lab course. Some guidance may be required depending on the level of the students taking the course.

Research Assignment Direction

Many modems are manufactured, with new ones announced every month. The headache is to find any in-depth technical material on any of them. Specifications and the like are supplied by the manufacturers, and magazine articles show applications and tout each modem's abilities. The challenge is to gather the technical background on a selected modem. The depth of the assignment should reflect the availability of the material.

Seven • Multiple-Channel Digital Communications

OBJECTIVES

The objectives of this chapter are to familiarize the reader with high-speed digital communications systems. The chapter begins by illustrating multichannel multiplexing using frequency division multiplexing. It continues with a discussion of telephone company digital T1 carriers and a presentation of the methods of digital multiplexing and data encoding to facilitate the transmission of multiple channels on a single link. Concepts of pulse coded modulation incorporating sampling and quantization are examined. Lastly, methods to reduce noise and the number of bits required for coding are shown through a method called companding. Many of the later concepts are brought together by the use of a CODEC (coder/decoder) circuit, which also performs the functions of a modem discussed in chapter 5. Specific topics include the following:

1. Frequency division multiplexing
2. T1 and other digital carriers
3. Time division multiplexing
4. Pulse coded modulation
5. Sampling theorem
6. Quantization
7. Delta modulation
8. Companding
9. CODER/DECODER or CODEC data set

ACRONYMS

ADC—**A**nalog to **D**igital **C**onverter

ADM—**A**daptive **D**elta **M**odulation

AMI—**A**lternate **M**ark **I**nversion

ATDM—**A**synchronous **T**ime **D**ivision **M**ultiplexing

B8ZS—**B**inary **8** **Z**ero **S**uppression

BPRZ-AMI—**B**i**P**olar **R**eturn to **Z**ero-**A**lternate **M**ark **I**nversion

BPV—**B**i**P**olar **V**iolation

CCITT—International Consultative Committee on Telegraphy and Telephony

CODEC—**COD**er/**DEC**oder

COMPAND—**COM**press/Ex**PAND**

DAC—**D**igital to **A**nalog **C**onverter

DUV—**D**ata **U**nder **V**oice

ESF—**E**xtended **S**uper**F**rame

FDM—**F**requency **D**ivision **M**ultiplexing

FSK—**F**requency **S**hift **K**ey

FT1—**F**ractional **T1**

MUX—**MU**ltiple**X**er

PAM—**P**ulse **A**mplitude **M**odulation

PCM—**P**ulse **C**ode **M**odulation

RZI—**R**eturn to **Z**ero Mark **I**nverted

SF—**S**uper**F**rame

STATDM—**STAT**istical **T**ime **D**ivision **M**ultiplexing

STAT MUX—**STAT**istical **MU**ltiple**X**er

STDM—**S**ynchronous **T**ime **D**ivision **M**ultiplexing

TDM—**T**ime **D**ivision **M**ultiplexing

● 7.1 INTRODUCTION

The modems and communications systems that have been explored in previous chapters are utilized for low- to moderate-speed data rates. This data rate is limited by the system's bandwidth (300 Hz to 3 kHz for the telephone system). The various methods thus far discussed utilize a small portion of this bandwidth to manage a single communications link (called a **CHANNEL**) to send data. In these cases the digital data are converted to an analog signal and sent via modems onto the telephone lines. Another method of transmitting the digital data in analog form is called **FREQUENCY DIVISION MULTIPLEXING (FDM).** This method allows many voice or **BASEBAND** channels (bandwidth 300 Hz to 3 kHz) to share space in a much larger total band. This band of channels is then sent via radio, microwave transmitters to satellites, or along fiber-optic cables to a receiving station, where the band is demultiplexed into individual channels which are destined for specific end users.

The term *baseband* has evolved into two distinct definitions depending on whether the area under discussion is general in nature or applies to local area networking (chapter 8). In general terms, baseband refers to a single-channel transmission. Thus a single voice channel can be considered a baseband signal whose band is from 300 Hz to 3 kHz. Similarly, if a single channel is used to transport a baseband signal, it is called a baseband system. In the context of local area networks (LANs), the term baseband means "A transmission method— typically for shorter distances—in which the entire bandwidth of the LAN cable is required to transmit a single digital signal."[1]

The limiting aspect of the definition when applied to local area networks is the digital nature of the signal. A point-to-point system as discussed in chapter 1 is an example of a baseband system. A single channel's signal is the only signal occupying the data link. However, the signal is analog (output from the modem) and not digital.

In contrast, **BROADBAND** systems are those that allow many channels to share the system's bandwidth. By definition broadband is

1. [in general] Communications channel having a bandwidth greater than a voice grade channel and potentially capable of much higher transmission rates; also called wideband.

2. [in LAN technology] A system in which multiple channels access a medium that has a large bandwidth using radio frequency modems. Each channel occupies a different frequency slot on the cable and is demodulated to its original frequency at the receiving end.[2]

Baseband is a single channel (either analog or digital) that occupies the entire bandwidth (though it may not utilize all of that bandwidth) of a system's medium. Broadband, which is analog, refers to multiple-channel transmissions in which each channel occupies a portion of the medium's bandwidth.

---

[1]*Pocket Glossary of Data Communications and Computer Terms,* copyright 1988 by Black Box Corporation, p. 8.

[2]*Pocket Glossary of Data Communication and Computer Terms,* copyright 1988 by Black Box Corporation, p. 13.

In its simplest form, frequency division multiplexing (FDM) is represented by a **FREQUENCY SHIFT KEY (FSK)** modem. Here two frequencies were used to represent a 1 and a 0. In a full-duplex system, four sets of frequencies occupied the line, with a maximum of two signals on the line at any one time. In effect, the bandwidth of the telephone line was shared by each of the signal frequencies. This example falls short, since it deals with a single channel of communication and does not involve any form of true multiplexing. It does, however, illustrate frequency allocation usage of the telephone system bandwidth. Full-duplex operation is performed by allocating the transmit frequencies of the originate and answer modems into different frequency bands within the bandwidth. FDM applies itself to the increased usage of a network's capacity by utilizing as much available frequency bandwidth as is allocated to the system. Applications of FDM include fiber-optic links, microwave, and satellite communications that have bandwidths in the megahertz and gigahertz ranges.

A single voice or baseband channel occupies (ideally) a 0 to 4 kHz bandwidth as allocated for a telephone subscriber circuit. Actual operating bandwidth for the voice channel is 300 Hz to 3 kHz. The remaining band area (0 to 300 Hz and 3 kHz to 4 kHz) is unused and acts as a built-in **GUARD BAND.** Simply put, a guard band is an area that acts as a buffer between adjacent channels to avoid one channel's data from crossing over into another channel. It is similar in concept to the "dead" area between broadcast channels on an FM radio. The first channel in an FDM system occupies the frequency band from 0 to 4 kHz. A second channel could then begin at 4 kHz and end at 8 kHz and still have a 4-kHz bandwidth and contain the same guard band area. Putting these two channels together produces a total bandwidth of 0 to 8 kHz. These two channels, of course, are not going to be sent using unconditioned telephone lines, since the upper end of the telephone line bandwidth is only 4 kHz.

All voice-grade channels have a bandwidth of 0 to 4 kHz. In order to be transmitted as part of a broadband signal, the voice channel is mixed with a carrier frequency. A mixer (Figure 7.1) is a nonlinear circuit that produces the original signal and the sum and difference frequencies of two input waveforms. When a balanced modulator is used as the mixer, the voice channel and carrier signals are suppressed at the mixer's output. All frequencies above the difference frequency are filtered out, so that the mixer/filter circuit only produces the difference frequency.

Example 7.1   Illustrate the results from mixing a voice-grade channel (0 to 4 kHz) with a 64-kHz carrier signal.

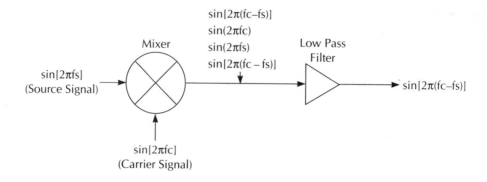

Figure 7.1 Mixer Blocks

Solution: Refer to Figure 7.2 to observe the frequency spectrum produced by the mixing action. The original voice-channel signal and the carrier signal are suppressed and do not appear at the mixer output. The difference of the carrier and the voice channel (60 to 64 kHz) is the lower sideband signal, while the sum (64 to 68 kHz) fills in the upper sideband area. The upper band is filtered using a low-pass filter following the mixer so that only the difference frequencies (lower sideband) are produced from the mixer. A second and third channel can be produced at higher frequencies to this example's channel by mixing them with higher carrier frequencies.

● 7.3 FREQUENCY DIVISION MULTIPLEX GROUPS

The telephone company has a set of standards that specify the frequency allocations for a broadband FDM system. The standard begins with twelve voice channels, each mixed with carrier signals that are 4 kHz apart. These carrier signals range from 64 to 108 kHz. The resulting difference frequencies are combined into a single *GROUP* channel by linearly summing the outputs of several mixers as shown in Figure 7.3. Each mixer mixes a voice channel (0 to 4 kHz) with a carrier signal. The carrier signals are 4 kHz apart, placing each channel adjacent to the next in the overall frequency spectrum of the group (Figure 7.4). The bandwidth of the group ranges from 60 kHz (the carrier of the group 12 minus 4 kHz voice channel) to 108 kHz (the carrier of group 1, 108 kHz, minus the lower end of the voice channel, 0 kHz), for a total bandwidth of bandwidth = 60 − 108 kHz, or 48 kHz. This 48-kHz bandwidth is verified by multiplying 12 channels times 4 kHz per channel, resulting in 48 kHz.

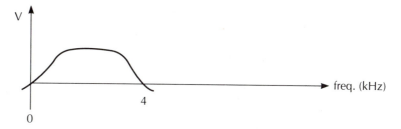

a) Original Voice Band Channel

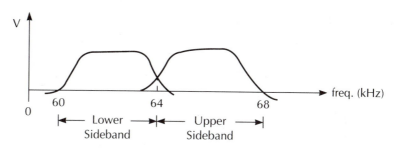

b) Voice Band Mixed with 64 kHz Carrier Signal

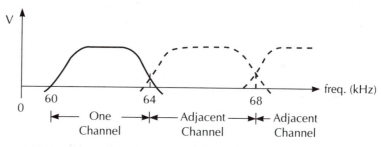

c) Filtered Channels (3 Adjacent Channels)

Figure 7.2 Results of Mixer/Filter Operation

Notice that the 60- to 108-kHz bandwidth is well above the telephone system's 300 Hz to 3 kHz bandwidth. This group channel is meant to be sent by mediums other than voice-grade telephone lines. It could be sent on a fiber-optic cable that has a larger bandwidth (discussed later), but it is more frequently used for radio, microwave, and satellite transmissions. Some of these have extremely large

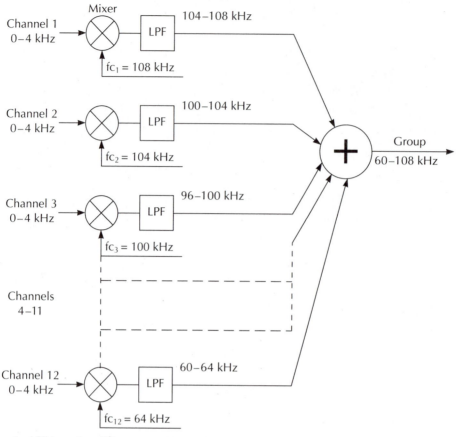

Figure 7.3 Formation of A Twelve-Channel Group

bandwidths compared with 48 kHz. As such, larger groupings are created in similar manner to take full advantage of these large bandwidths.

Example 7.2 What is the bandwidth range of channel 7 of a group transmission?

Solution: The carrier frequency (fc) for each channel is 4 kHz less than the previous channel. Channel 1's fc is 108 kHz, channel 2's is 104 kHz, and so on. This makes channel 7's fc equal to 84 kHz. Mixing 84 kHz with the 0- to 4-kHz channel range and then filtering produces a bandwidth of 80 to 84 kHz for channel 7.

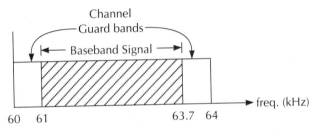

a) Channel 12 Frequency Spectrum After Mixing with
   64 kHz Carrier Signal

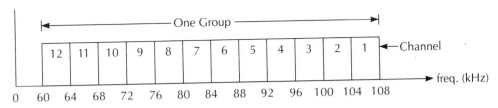

b) Frequency Spectrum of a Group (12 Channels)

Figure 7.4 FDM Group Formation

Each channel within the group may vary in amplitude since they are created from twelve separate sources. To regulate the amplitude of the group, a *CARRIER PILOT* signal at 104.08 kHz is added to the group. Like attenuation/gain distortion described in chapter 2, this carrier pilot is used to set the amplitude reference for the group. The amplitudes of the separate channels are adjusted to match the carrier pilot so that the group will have a uniform amplitude. The carrier pilot is also used to generate a carrier detect (CD) signal to indicate that a remote link has been established.

● 7.4 FREQUENCY DIVISION MULTIPLEX SUPERGROUPS AND BEYOND

By repeating the process described for creating groups, larger groupings are developed as actual transmission frequencies increase. Five groups (sixty channels) are assembled to form a *SUPERGROUP*, which fills a bandwidth between 312 and 552 kHz. The actual bandwidth (552 − 312 = 240 kHz) equals the five groups (5 × 48 kHz), which in turn is equivalent to the sixty channels (60 × 4 kHz) that make up the five groups. Group 1's carrier pilot is used as the supergroup's carrier

Table 7.1 Frequency Assignment for Telephone Company FDM

FDM Level	Number of Channels	Bandwidth Range	Bandwidth
Voice Channel	1	0–4 kHz	4 kHz
Group	12	60–108 kHz	48 kHz
Supergroup	60	312–552 kHz	240 kHz
Mastergroup	600	564–3,084 kHz	2,520 kHz
Jumbo Group	3,600	564–17,544 kHz	17 MHz*
Jumbo Mux	10,800	3.0–60 MHz*	57 MHz*

*Values approximate.

pilot for amplitude regulation of that supergroup. The next level formed from ten supergroups is the *MASTERGROUP*. An additional carrier pilot at 2,840 kHz is inserted at the mastergroup level for amplitude regulation at that level. Table 7.1 summarizes the bandwidth assignment of the telephone company's FDM standard allotment.

Various grouping levels are also categorized: A *JUMBO GROUP* is composed of six mastergroups; each mastergroup encompasses ten supergroups; one supergroup contains five groups, each with twelve channels. A *JUMBO MUX* combines three jumbo groups into a single channel. By performing a little multiplication, it is determined that $3 \times 6 \times 10 \times 5 \times 12 = 10,800$ channels are carried by a jumbo mux system.

An optional grouping specified is a *MASTERGROUP MUX*, which allows two to six mastergroups (1,200 to 3,600 channels) to be combined. Generally, use of a mastergroup mux transmission is terminal. That is, a mastergroup mux is used instead of a jumbo group or jumbo group mux.

Guard bands, as mentioned earlier, are used to prevent interference between adjacent channels. At the group and supergroup levels, the built-in guard band of the voice channels is sufficient to assure this. Starting with mastergroups, additional spacing is inserted into the frequency spectrum. Notice that the ten supergroups, each with a bandwidth of 240 kHz, should form a 2,400-kHz mastergroup. In reality, the mastergroup is specified to occupy a bandwidth of 2,520 kHz. The additional 120 kHz is assigned as ten additional 12-kHz guard bands between the supergroups that make up the mastergroup. Guard bands of 12 kHz also precede and follow the mastergroup (Figure 7.5). Similarly six mastergroups require a 15,120-kHz (15.12 MHz) bandwidth for a jumbo group. Jumbo groups are allotted bandwidths of 16,980 kHz, providing for additional guard bands of 310 kHz each between mastergroups in the jumbo group (Figure 7.6).

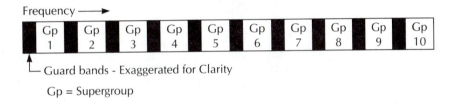

Gp 1 | Gp 2 | Gp 3 | Gp 4 | Gp 5 | Gp 6 | Gp 7 | Gp 8 | Gp 9 | Gp 10

Guard bands - Exaggerated for Clarity

Gp = Supergroup

Figure 7.5 Mastergroup with Guard Bands

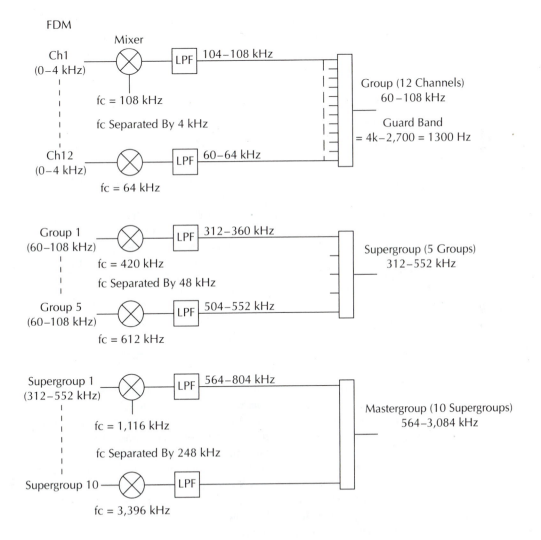

Figure 7.6 Forming of Multichannel Distribution Using FDM

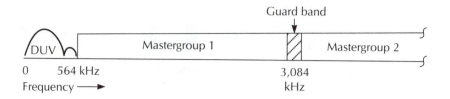

Figure 7.7 Jumbo Group with Data under Voice (DUV)

Essentially, the microwave and satellite transmitter carriers are modulated by the mastergroups or jumbo groups of channels and transmitted to receivers, where the signals are demodulated and remixed using the same carrier frequencies that formed the original groups. This process is repeated until each individual channel is recovered and sent on to its final destination.

● 7.5 DATA UNDER VOICE

The advent of FDM systems resulted in a secondary effect that has been capitalized on. The band below 564 kHz is not used by a mastergroup or above. That is, the band from 0 to 564 kHz is unused by these transmissions. It was soon discovered that nonvoice communications signals (control, test data, etc.) presented at high data rates could be sent along in this band, with a mastergroup transmission as shown in Figure 7.7. Signals of these type are referred to as **DATA UNDER VOICE (DUV)**.

● 7.6 DIGITAL T CARRIERS

As technology improved the speed at which electronic circuits could dependably operate, the need to increase the rate at which data are transferred also increased. To respond to the need for interconnecting computers and data base facilities, the telephone company turned to supplying lines for carrying data in digital form. In addition to supplying the lines, the telephone company also established the limitations for their use. These lines are designated T carriers, beginning with the basic T1 carrier specified for the Bell System Voice Network.

T1 CARRIERS are digital leased twisted-pair lines designed to handle twenty-four digitized voice-grade channels operating at a maximum data rate of 64 kbps each. Thus, the T1 line requires a bandwidth large enough to handle 1.544 Mbps (24 × 64 kbps). T1 lines are designed to interconnect stations that are from

5 to 50 miles apart, using regenerative repeaters every 6,000 feet. These repeaters reshape the digital data that have been distorted due to attenuation or pulse spreading caused by effects on the signals due to the reactive components of the cabling. The repeater closest to the central office station is placed 3,000 feet away instead of 6,000 feet.

Each channel is sent, using time division multiplexing (discussed in the next section), 8 bits at a time. Since twenty-four channels are sent at once, a total of 8×24 bits, or 192, data bits are sent per transmission period. An additional frame boundary bit (like a start/stop bit) increases the total to 193 bits per frame (transmission). Dividing the total data rate by the total number of bits per frame yields a frame rate of 8,000 frames per second.

The 193rd framing bits have several interpretations based on the frame format used. The basic frame format, called D1, is recognized by the alternating states of the 193rd frame bits of each succeeding frame. Two other common frame formats are the **SUPERFRAME (SF or D4)** and the **EXTENDED SUPERFRAME (ESF)** listed in Table 7.2. Every twenty-four channels make up a frame that ends with one of two types of framing bits, the TERMINAL FRAME (Ft) BIT or the SIGNAL FRAME (Fs) BIT. Furthermore, for a superframe, the states of the framing bits for

Table 7.2 Superframe (SF) and Extended Superframe (ESF) Formats and Use or 193rd Frame Bit

Frame Number	Superframe	Extended Superframe
1	Ft = 1	Data Link Control (DLC)
2	Fs = 0	Cyclic Redundancy Check (CRC)
3	Ft = 0	DLC
4	Fs = 0	Framing Bits (F) = 0
5	Ft = 1	DLC
6	Fs = 1	CRC
7	Ft = 0	DLC
8	Fs = 1	F = 0*
9	Ft = 1	DLC
10	Fs = 1	CRC
11	Ft = 0	DLC
12	Fs = 0	F = 1

Sequence is repeated for every twelve frames with the single exception for the state of the framing bit for the extended superframe frame 8(*). The state of this bit alternates between 1 and 0 for each group of twelve frames.

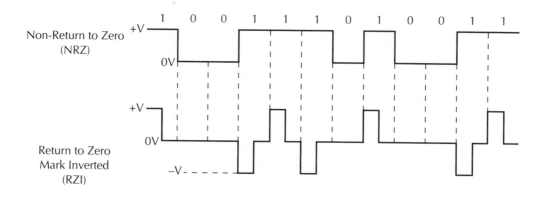

Non-Return to Zero (NRZ)

Return to Zero Mark Inverted (RZI)

Figure 7.8 Bipolar Return to Zero, Alternate Mark Inverted Digital Form

superframes repeat every twelve frames. The terminal and signal frame bits alternate each frame, with the odd-numbered frame bits being terminal bits and the even-numbered being signal bits. The pattern for the framing bits starting with frame number 1 is 100011011100, as shown in Table 7.2.

For the extended superframe, the framing bits take one of three meanings. They are regular framing bits (F) serving the same function of terminal and signal framing bits, data link control (DLC) bits used for control and diagnostic activities, and cyclic redundancy check (CRC) error-detection bits. Each of these bits appear as the 193rd bit of their particular frame. Framing (F) bits are related to every fourth frame starting with frame number 4 and have the following bit pattern: 001011 starting with frame 4 and ending with frame 24.

Data link control (DLC) bits are interpreted in the same manner as high-level data link control (HDLC) protocol or other bit protocol bits. They occupy every odd-numbered frame's 193rd bit position starting with frame number 1. Lastly, the CRC bits occupy the remaining frame bit positions, using a total of 6 bits for each twenty-four frames. CRC-6 is used, which is a shorter form of the CRC-16 error-detection method described in chapter 3.

The digital data form used on T1 lines is **BIPOLAR RETURN TO ZERO-ALTERNATE MARK INVERTED (BPRZ-AMI).** The format for this digital form is illustrated in Figure 7.8. It results from a combination of forms discussed in chapter 1. Zero logic levels are at 0V. Logic 1s are at $+V$ and $-V$, alternating between each level. Each logic 1 returns to a 0 level midway through the bit period. A string of thirty-two consecutive 0s indicates that a sender is no longer in service. To avoid misinterpreting a long string of 0s if they are required in the data string, a logic 1 bit is inserted after fifteen consecutive 0s. The receiving station is required to strip this added bit before interpreting the data. Another method frequently used to avoid the occurrence of a long string of 0s is an encoding scheme called **BINARY 8 ZERO SUPPRESSION (B8ZS),** which is illustrated in Figure

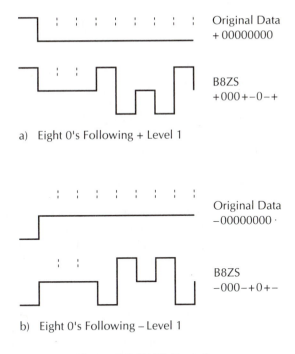

Original Data
+ 00000000

B8ZS
+000+−0−+

a) Eight 0's Following + Level 1

Original Data
−00000000 ·

B8ZS
−000−+0+−

b) Eight 0's Following − Level 1

Figure 7.9 B8ZS Encoding

7.9. When the transmitter detects eight consecutive 0s, it checks the level of the logic 1 bit that preceded them. If the level of that bit is positive, the 9 bits are encoded in accordance with Figure 7.9a. If the leading 1 bit is a negative level, then the encoding takes the form shown in Figure 7.9b. Notice that the difference between the two forms is that the logic 1 levels are inverted from one to the other. When using alternate mark inversion formats, two consecutive 1 bits with the same level (positive or negative) results in a condition called a **BIPOLAR VIOLATION (BPV).** Normally this condition indicates that an error has occurred. However, if the patterns shown in Figure 7.9 are detected by a receiver set to decode B8ZS format, then the double bipolar violation is recognized and decoded into a single 1 followed by eight 0s. Any other detected bipolar violations are assumed to be data errors.

The T1 carrier is similar to the telephone system digital signal-1 (DS-1) carrier. The Consultative Committee for International Telegraphy and Telephony (CCITT), a standards organization based in France, produced a similar digital carrier specifications to apply to European digital communications systems. The significant difference in the specifications is an increase from twenty-four to thirty-two channels (thirty voice-grade and two for frame synchronization and signaling), which increased the bandwidth requirements to handle 2.048 Mbps.

An addition to the T carrier system was made for users who wanted to have access to these lines but did not have sufficient data needs to fill twenty-four channels. This addition, called **FRACTIONAL T1 (FT1)** service, uses conventional T1 technology to send fewer channels of information. User's data are inserted into alternating channels of a T1 frame. Unused channels are filled with filler data. As an example, suppose a user had a need to send a 384-kbps video data sequence on a FT1 carrier. The user's data would fill six channels (64-kbps per channel × 6 yields 384-kbps). These data are placed in channels 1, 3, 5, 7, 9, and 11. Channels 2, 4, 6, 8, 10, and 12–24 are filled with filler data to complete the 24-channel frame. Since part of the frame is not being used for customer data, it is up to the customer to supply framing data so that a receiver can correctly interpret the data sent. The 193rd bit normally used for framing is still applied to the T1 frame, but it cannot supply specific framing information for the actual data contained in the fractional T1 frame.

T1 carrier lines are susceptible to all the impairments and problems inherent in telephone lines plus a few of their own. Among these additional problems is that of bipolar violation, which occurs when two consecutive logic 1s have the same voltage level (+V or −V). Since T1 carriers use alternate mark inversion, such an occurrence is generally in error. Shifts in clock frequency, jitter, and wander also cause data errors or framing misinterpretation to occur. A shift in the clock frequency by either the transmitter or the receiver in a T1 connection causes data and/or framing bits to be sampled at an incorrect time. This causes a condition called a *SLIP,* which is the loss or addition of a bit due to a slip in sampling time.

Excessive *JITTER* can also cause slips. Most jitter is introduced in clock recovery circuits as they regenerate the clock signal from the incoming data transitions. Most of the time, the jitter is very slight and causes no problem. However, as components in repeaters and customer premises equipment begin to degenerate with time, jitter becomes worse until data or framing errors result. *WANDER* is a form of low-frequency jitter (usually less than 10 Hz) and is caused by instabilities in the master timing source or from temperature variations experienced by the equipment. The effect of both jitter and wander is similar to frequency shifts. They cause data or framing bits to be incorrectly sampled, resulting in a data or framing error.

Additional T carrier lines are available for systems requiring expanded use. Table 7.3 charts these lines and the specifications associated with each carrier. T2 lines carry 96 channels of digital information at 6.312 MHz rate on lines that can reach 500 miles with the aid of repeaters. Instead of using the 1 bit insertion following fifteen consecutive 1s specified for T1, a false disconnect is avoided by using a method called binary six zero substitution (B6ZS). Any group of six consecutive 0s is replaced with one of two sequences. They are:

0 bit,   −V bit,   +V bit,   0 bit,   +V bit,   −V bit

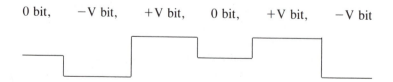

Table 7.3 T Carrier Specifications

Carrier Type	Alternate Designation	Number of Channels	Multiplex Format	Rate (Mbps)	Media Type
T1	DS-1	24	24 Analog	1.544	twisted pair
T1C	DS-1C	48	2 DS-1	3.152	twisted pair
T2	DS-2	96	4 DS-1	6.312	twisted pair
T3	DS-3	672	28 DS-1	44.736	3A-RDS 11 GHz radio
T4	DS-4	4,032	6 DS-3	274.176	T4M Coax WT4 Waveguide DR18 18 GHz radio

or

0 bit,   +V bit,   −V bit,   0 bit,   −V bit,   +V bit

Both sequences contain a bipolar mark inversion violation (occurrence of two consecutive +V or −V levels for logic 1 levels). When the receiver detects this violation, it replaces the sequence with 0s.

The remaining T carriers listed in Table 7.3 show the main difference between each level and the other levels. Each carrier specification contains details specific to that T level. Basically, each is used to handle an increased number of channels and each requires larger bandwidths to handle the increased number of channels and data rates they are operated at. Different techniques exist for multiplexing the channels onto a single T carrier. One of the most common methods is **TIME DIVISION MULTIPLEXING (TDM).**

● **7.7 TESTING T1 DIGITAL CARRIER LINES**

A number of instruments are available for monitoring and testing T1 lines. One of these, the TM1 T1 Carrier Monitor manufactured by Electrodata, Incorporated, is pictured in Figure 7.10. This hand-held unit is used to monitor active lines without disrupting service. It can also be used to test an off-line transmitter to check its operation. The monitor is capable of automatically detecting the type of framing

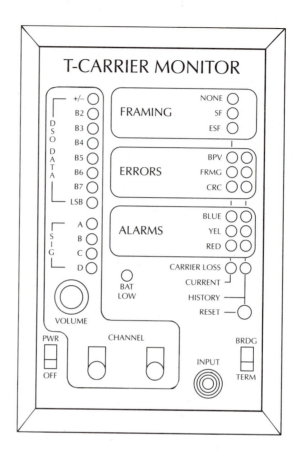

Figure 7.10 TM1 T1 Carrier Monitor Courtesy of Electrodata, Inc.

used (unframed, superframe, or extended superframe). Three LEDs in the upper right corner are used to indicate the framing type. To be useful, a T1 test set must be able to decipher digital data in either standard **ALTERNATE MARK INVERSION (AMI)** or B8ZS form. Additional indicators on the left side of the TM1 show the states of 8 current data bits and 4 signal bits. The user is capable of selecting which one of the twenty-four channels to monitor using two switches located at the bottom of the unit. Three types of errors can be detected by the monitor: bipolar violations (BPVs), framing errors, and CRC data errors. Current indicators will flash as they occur, while a second set, called the history indicators, latch on the first occurrence of an error. These latched indicators are manually reset by pressing the switch located directly below the history LED column.

The two remaining indicators tell the user if the carrier becomes lost (loss of signal) and if the battery charge on this portable unit is becoming too low. As an emulator, the TM1 Monitor is capable of generating a voice signal adjustable from +3 to −55 dBm DC coupled into 600 ohms termination. The frequency response of the output is held between + and− 0.2 dB of the original setting from 300 Hz to 3 kHz, the telephone line bandwidth and between + and− 0.3 dB overall from 200 Hz to 3.3 kHz. The device is small and easy to use, and helps the technician determine whether there are any problems on a T1 line.

● 7.8 TIME DIVISION MULTIPLEXING

Using the conventional means of transferring data on a baseband system previously discussed, sending three messages of varying lengths is accomplished in the following manner. Message A is transmitted first, followed in time by message B and then message C (Figure 7.11). This is a simple and straightforward process. The main drawback is that message C must wait until messages A and B are sent before it can be transmitted.

Expand the problem by applying it to a multipoint system (Figure 7.12) where messages are destined for different secondaries. Secondary station 3 in Figure 7.12

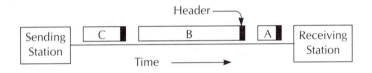

Figure 7.11 Conventional Transmission

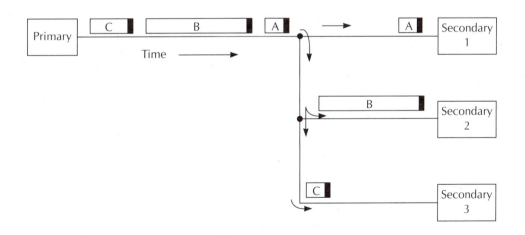

Figure 7.12 Conventional Multipoint Communication

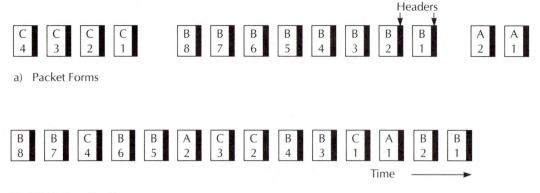

a) Packet Forms

b) TDM Time Slot Frame

Figure 7.13 Packet Time Allotment

would like to start receiving its message at about the same time that station 1 got message A and station 2 received message B.

To resolve the problem, all three messages are reformed into smaller parts called **PACKETS.** These packets are of equal length, resulting in more packets for longer messages (message B in Figure 7.13a). The packets constituting messages A, B, and C are interleaved and assigned time slots as shown in Figure 7.13b. A header (shaded area), containing address and packet number information, precedes each packet. The interleaved packets are transmitted and received by the secondaries. The appropriate packets (determined by destination address in the header) are extracted by each station as they are received and reassembled (by packet number, also included in the header) into their original message form. This is the essence of a time division multiplex (TDM) system.

Two basic forms of TDM are in use today—**SYNCHRONOUS TDM (STDM)** and **ASYNCHRONOUS TDM (ATDM),** also known as **STATISTICAL TDM (STATDM)** or **STAT MUX.** STDM systems assign time slots of equal length to all packets regardless whether or not anything is to be sent by each station with an assigned time slot. As an example, if message A was not included in the previous illustration, its allotted time would still be allocated. The TDM frame would appear as it does in Figure 7.13b, but time slots for message A would not contain information.

STDM systems are comparatively easy to implement once the software allocates the time slots. STAT MUX systems, on the other hand, are more complex, but they allow for a means of reassigning time slots that are not in use. Essentially, STAT MUX networks assign time slots only when they are to be used and delete them when they are idle. The total time used for a STAT MUX frame varies with the amount of traffic currently being handled.

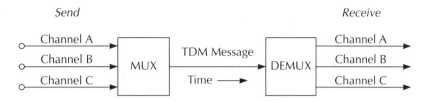

Figure 7.14 Multichannel TDM System

● 7.9 MULTICHANNEL TDM

Broadband systems manage many channels of communication on a single line at a given time by dividing the system's bandwidth so that each channel occupies a portion of that bandwidth. These channels are multiplexed onto a single carrier and demultiplexed at the receiving end (Figure 7.14). This requires the bandwidth of the system to be large enough to contain all the channels. Time division multiplexing, on the other hand, uses a baseband system requiring less bandwidth. Only a single channel's signal appears on the system at any given time. Multichannel use of a TDM system relies on sharing transmission time periods rather than a system's bandwidth.

Messages from channels A, B, and C can be sent one at a time (message A followed by messages B and C), but the same problem exists as in the multipoint system in Figure 7.12. A period of time elapses before message C is received by its destination station. This problem is compounded because each *channel* in this example is a multipoint system; that could delay a message from one station to another, using conventional methods, by a significant amount of time.

STAT MUX systems are most suitable for these high-density, high-traffic applications. The continuous messages are assigned time slots and interleaved as each channel on the send side becomes active and requires communications with another channel. If a channel does not have any traffic, its time slots are deleted and reassigned to an active channel. In this way the interconnecting media achieve a higher state of efficiency than with STDM systems.

● 7.10 SAMPLING THEOREM

Another technique for sending analog information on digital carriers involves translating voice or other analog signals into a train of pulses that are in turn coded into digital format. The digital data are then transmitted and the process reversed at

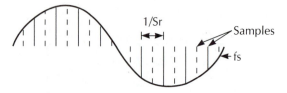

a) Sampled Signal

b) Reconstructed Waveform (So)

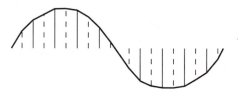

c) Reconstructed Waveform Using Twice
    as Many Samples

Figure 7.15 Sampled Signal

the receiving end. In effect, this process combines SAMPLING with ANALOG TO DIGITAL CONVERSION (ADC) and DIGITAL TO ANALOG CONVERSION (DAC). An advantage gained by this method is in the reduction of noise effects usually inherent in analog transmissions.

The process of digitizing an analog signal starts by dividing the original signal into uniformly spaced samples as shown in Figure 7.15. The amplitudes of the sample pulses rise and fall with the amplitude of the original signal. Specifically, this part of the process is a form of **PULSE AMPLITUDE MODULATION (PAM),** since the original signal is being separated into individual pulses (or samples), each of which has a different amplitude based on the amplitude of the original signal. At the receive end, these samples are used to reconstruct the original signal. The more frequently the samples are taken, the more accurate is the reconstructed waveform. Notice in Figure 7.15 that the bottom signal is sampled twice as much as the top signal. The recovered signals below the example show the difference caused by the differences in the sampling rates.

To determine the minimum number of samples to use to replicate the original waveform reliably, a sampling theorem developed by Nyquist is used. This theorem states that for a given signal, fs, the minimum sampling rate (known as the **NYQUIST SAMPLING RATE** [Sr]) to assure accurate recovery of the signal at the receiving end is twice the frequency of the highest sine wave element $\sin(2\pi fs)$ of the original signal, or as a formula:

$$Sr = 2[\sin(2\pi fs)] \qquad (7.1)$$

Original signals are sampled at rates at or above the minimum sampling rate to assure that the original signal is accurately replicated. If the sampling rate were less than twice the highest fundamental sine wave frequency, then a distortion called aliasing or fold-over occurs. To understand what this means, consider sampling the voice signals on the telephone lines, which contain signals from 300 to 3 kHz. Essentially, the sampling process causes mixing, which is similar to that used for regular amplitude modulation (AM) to result. This process creates the sum and difference frequencies as well as the original signals that were mixed. For a sampling circuit, these are Sr, $\sin(2\pi fs)$, $Sr - \sin(2\pi fs)$, and $Sr + \sin(2\pi fs)$. Filters are used to remove all but the difference and original fs signals. With an Sr = $2\sin(2\pi fs)$, the frequency range of the original signals resides next to the difference frequencies. If the sampling rate is higher than $2\sin(2\pi fs)$, there is a gap between one group and the other. The problem of fold-over occurs when Sr is less than $2\sin(2\pi fs)$.

Example 7.3   Show the differences between sampling a voice channel (300 Hz to 3kHz) using sampling rates at and below $2\sin(2\pi fs)$.

Solution: The minimum sampling rate is twice the highest frequency component of fs, or 2×3 kHz = 6 kHz. Mixing the voice band with 6 kHz and removing the higher-frequency elements produces the original voice channel (300 Hz to 3 kHz) and the difference band 6 kHz − (300 Hz to 3 kHz) = (3 to 5.7 kHz). These two bands are illustrated in Figure 7.16a. Using 4.5 kHz for Sr as an arbitrary value that is less than $2\sin(2\pi fs)$ results in the original voice band and a difference frequency band of 4.5 kHz − (300 Hz to 3 kHz) = (1.5 kHz to 4.2 kHz). Figure 7.16b illustrates these two bands. Note that there is a fold-over of the original band and the difference frequency band from (1.5 to 3 kHz).

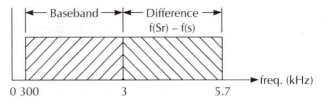

a) Sampling Rate (Sr) = 2 × Signal Rate (s)

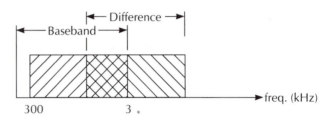

b) Sampling Rate < 2 × Signal Rate

Figure 7.16 Aliasing (fold-over) Distortion

• 7.11 SAMPLING USING TDM

Two or more signals can be sampled at different times. The samples are all sent and extracted at the receiver according to their time relationship. In essence, the samples of one signal will be assigned to specific time slots, while samples from other signals occupy different time slots. Figure 7.17 illustrates the concept of varying sampling times for different signals.

For simplicity, signals A and B have the same frequency and are sampled at the same Nyquist rate. Signal A is sampled beginning at time t0. Samples are spaced at 1/Sr time periods. The first sample for signal B is taken at a time of t0 + 1/2Sr, or halfway between the first two samples for signal A. Each succeeding sample for signal B occurs 1/Sr time period from the last sample. This places the samples for signal B in between the samples for signal A. All the samples are sent sequentially. The receiver extracts the sample pulses starting at t0 and occurring each 1/Sr following t0. From this the receiver recreates signal A. The receiver also extracts the samples starting at t0 + 1/2Sr and each succeeding sample 1/Sr away. From these samples signal B is reconstructed.

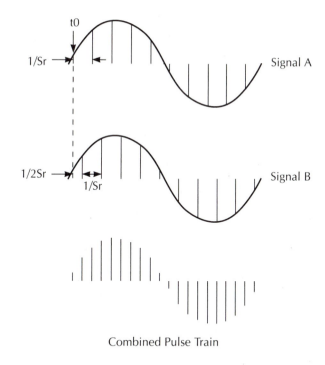

Figure 7.17 Sampling Two Signals at One Time (TDM)

● 7.12 NATURAL SAMPLING

Samples are done by generating a short pulse at a specific time. The amplitude of the pulse is equivalent to the amplitude of the signal at the time of the sample. The width of the pulse is designated tp, and the time between pulses (1/Sr) is Tr. The shapes of the pulses themselves come in two forms. One is called **NATURAL SAMPLING,** in which the peak of the pulse follows the signal's actual shape. The second pulse form is a **FLATTOP** shape in which the peak amplitude is held flat by the sample and hold circuit, rather than being allowed to follow the actual signal amplitude change exactly. These shapes are shown in Figure 7.18, which illustrates several different flattop formats based on the type of pulse sampling performed.

The original signal is shown in Figure 7.18a. Sampling pulses set at Sr are drawn in Figure 7.18b and underneath are the natural sampling (Figure 7.18c) and flattop (Figure 7.18d) pulse trains. For flattop sampling, the reconstructed signal

a) Input Signal Sin(2πfs)

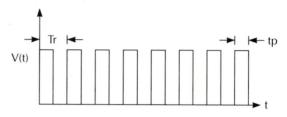

b) Sampling Pulses at Rate Sr. (Pulse Exaggerated for Clarity)

$$Tr = \frac{1}{Sr}$$

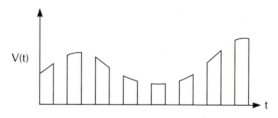

c) Natural Sampling

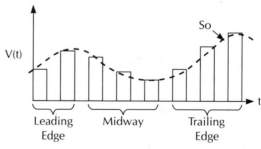

Leading   Midway   Trailing
Edge   Edge

Recovered Signal $So = \frac{tp}{Tr} \times \sin(2\pi fs)$

d) Flattop Sampling

Figure 7.18 Sampling Types

(So), shown as dotted lines in Figure 7.18d, for a given signal f(s) is represented by the relationship:

$$So = \frac{tp}{Tr} [\sin(2\pi fs)] \qquad (7.2)$$

where tp is the time period for the sampling pulse and Tr is the reciprocal of sampling rate (Sr). Since tp/Tr is the duty cycle of the sampling signal, the relationship of So to $\sin(2\pi fs)$ is a direct factor of that duty cycle.

Example 7.4 What is the value of an output signal derived from an input signal f(s) equal to $4\sin(2\pi f)$ sampled at a rate of 40 kHz with a sample pulse width of 1 microsecond?

Solution: Entering the values into equation 7.2:

$$So = 1\mu \times 4\sin(2\pi f) / (1/40 \text{ kHz}) = 0.16\sin(2\pi f)$$

Observe that the amplitude of the reconstructed signal is not the same as that of the original signal. However, the signal itself is correct (that is, it is $\sin(2\pi f)$). The amplitude is recoverable by using an amplifier to boost So back to its prior value.

● **7.13 SAMPLE AND HOLD**

Notice that there is a period of inactivity between sample pulses in Figure 7.18. During the reconstruction process, a method to abridge this gap is required to more accurately reproduce the original signal. This method is called **SAMPLE AND HOLD.** A sample pulse's amplitude is detected and that value retained until the occurrence of the next sample pulse (Figure 7.19). For this method to be effective, the hold time between samples (TH) is relatively small compared with the time period of the original signal.

The most common method used for sample and hold circuits is to employ a capacitance at the output of a buffer amplifier. The capacitor is charged to the sample pulse value. When the amplitude falls to zero between pulses, the capacitor remains charged to the pulse value. The next sample pulse causes the capacitor to charge or discharge to that value. Again the value is held until the next pulse arrives. Figure 7.20 is an example of a sample and hold circuit.

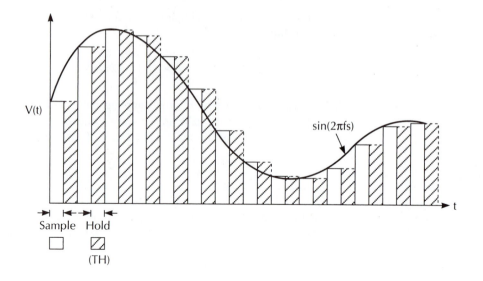

Figure 7.19 Sample and Hold Waveform

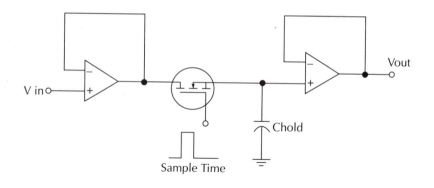

Figure 7.20 Sample and Hold Circuit

● 7.14 QUANTIZATION

QUANTIZATION is the process of approximating sample levels into their closest fixed value. The values are preselected and, since they are fixed, they are easy to encode. The quantized waveform experiences either quantum changes in amplitude or no change in amplitude. Given a signal, fs, with peak voltage points of Vh and Vl, the size (S) of a quantum step is determined by this relationship:

$$S = (Vh - Vl)/n \qquad (7.3)$$

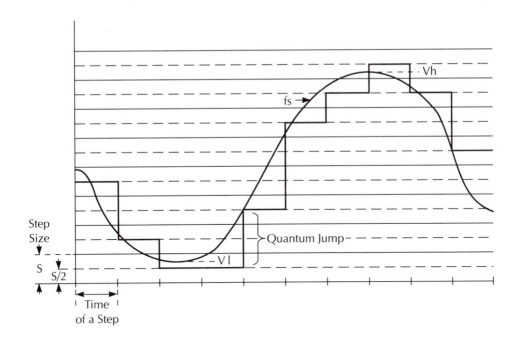

Figure 7.21 Quantized Signal

where n is the number of steps between Vh and Vl. Figure 7.21 illustrates the relationship between fs and a quantized example. Note that the quantized waveform appears similar to a sample and hold approximation of the same fs. The difference is that the normal sample and hold result has voltage levels equivalent to the voltage level of fs at the point the sample was taken. The quantized levels are those fixed levels that are the nearest to fs at the point the sample is taken. The difference between the sample and hold and the quantized results is called the **QUANTIZA-TION ERROR** or **QUANTIZATION NOISE.**

Essentially, the process of sampling and quantizing a signal is a form of pulse amplitude modulation (PAM) where the samples produce pulses of varying amplitudes. When these amplitudes are restricted to discrete quantized values and assigned specific binary codes which are to be transmitted, then a technique called pulse coded modulation is being used.

● 7.15 PULSE CODE MODULATION

Figure 7.22 summarizes the theoretical process of developing a **PULSE CODE MODULATION (PCM)** effect. The range of voltages for signal fs is divided into discrete quantized steps (S). The signal is sampled at each step, with

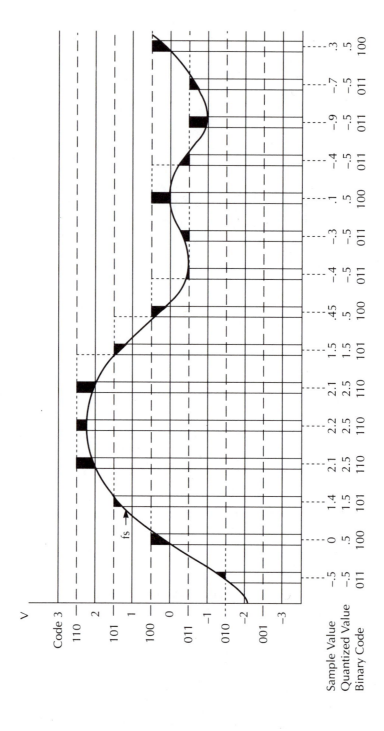

Figure 7.22 Coding a Quantized Signal

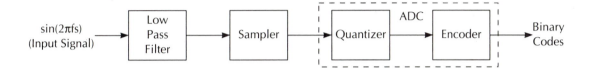

Figure 7.23 PCM Transmit Blocks

the resulting amplitude of the samples coded into binary values. The binary equivalents are actually associated with analog values midway between step amplitudes to minimize errors. These binary codes are shown at the bottom of the figure. The original waveform is transmitted as a serial stream of binary bits representing the quantized levels of each of the samples. At the receive station the binary bits are decoded into the quantized samples and the original signal is reproduced from the resulting samples. Quantizing errors resulting from original levels that were not exactly the same as the quantized step value appear in the replicated waveform.

A typical PCM transmitting system (Figure 7.23) consists of a low-pass filter and sampling circuit followed by a quantizing and encoding unit. These last two functional blocks are combined into a single analog to digital converter (ADC). The most commonly used analog to digital converter for this purpose is a successive approximations converter. A simplified schematic for such a circuit is shown in Figure 7.24.

Initially, the digital to analog converter (DAC) flip-flops are set to 1s. This produces the highest possible analog level from the DAC. This level is compared with the sample's voltage level. If the level is lower than the digital to analog converter analog voltage, all the flip-flops are reset. Next, all flip-flops except the most significant bit's flip-flop are set, and the comparison is made again. This process is repeated until the digital to analog converter output is lower than the sample voltage. The largest value bit of the digital to analog converter remains on while the process of comparing the digital to analog converter output to the input voltage level is repeated to determine the states of the lower bits of the analog to digital converter. This continues until the digital to analog converter's analog output has the closest value below the sample's value. The digital code in the flip-flops then represents that level. In essence, the analog to digital converter does the encoding and quantizing in the same process. It is rare that the digital to analog converter's output exactly matches the sample's level. Instead the actual level is the fixed (quantized) value nearest, but below, the sample's actual value.

The PCM decoder (Figure 7.25) reverses the process. The digital data are fed serially into the decoder. Each one of the data bits is reshaped to remove distortions caused by the transfer along the interconnecting medium used. After shaping, the data bits are fed into a digital to analog converter to produce the quantized samples they represent. These samples are held and filtered to recreate the original signal, fs. The differences between the original signal and the recreated one result from

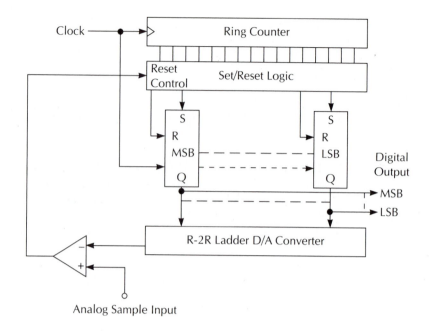

Figure 7.24 Successive Approximations A/D Converter

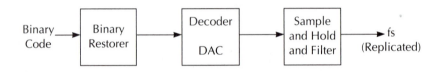

Figure 7.25 PCM Decoder

quantization error and any possible bit errors that might occur in the transmission. The later errors can be detected and/or corrected before the data stream is decoded.

● 7.16 DELTA MODULATION

Pulse code modulation codes each quantized sample into a binary code that is sent and decoded at the receiver. Another form of coded modulation is called **DELTA MODULATION.** The purpose behind this form of modulation is to minimize the effects of noise without increasing the number of bits being sent. This

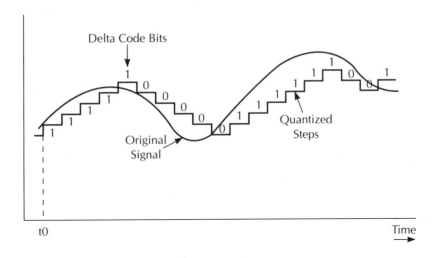

Figure 7.26 Delta Encoding of Quantized Waveform

increases the signal-to-noise ratio, improving system performance. The idea behind delta modulation is to take samples close enough to each other so that each sample's amplitude does not vary by more than a single step size. Then instead of sending a binary code representing the step size, a single bit is sent, signifying whether the sample size has increased or decreased by a single step. This process is illustrated in Figure 7.26. The original signal is first sampled and quantized as with pulse code modulation. If the sample currently being coded is above the previous sample, then a binary bit is set to a logic 1. If the sample is lower than the previous sample, then the bit is set low.

Example 7.5   What binary code is sent for the original signal in Figure 7.26, using delta modulation?

Solution: As shown in Figure 7.26, the original signal is quantized at a given sample rate. A sample whose amplitude is one step size above the previous sample generates a logic 1 bit. One that is lower generates a 0 logic level. The bit pattern generated by the signal in Figure 7.26, starting with the first sample on the left, is:

1111100000111111001

To aid in illustrating this result, the 1 and 0 logic levels are shown in Figure 7.26 above each step.

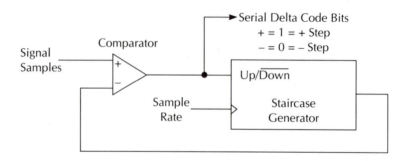

Figure 7.27 Delta Modulator

The functional block diagram for a delta modulator is shown in Figure 7.27. Basically, samples from the original signal are compared with the output of a staircase generator. If the results of that comparison show the original signal to be larger than the staircase voltage, the comparator is set high. This is sent out as a logic 1 and causes the staircase generator to increase by a step. If the comparator indicates that the staircase voltage is greater than the original signal, then the comparator goes low and causes the staircase generator to decrease by one step.

The delta demodulator (Figure 7.28) can be as basic as a staircase generator to replicate the staircase equivalent of the original waveform. As the serial coded delta data are fed into the demodulator, a staircase voltage is generated out. For each logic 1 received, the staircase output is increased by one step and for each 0 received, the output level is decreased by one step. The staircase voltage is then filtered and smoothed out to more closely represent the original signal.

Delta modulation systems are most useful with signals that change at a slow rate. A problem occurs as the rate of the original signal, fs, increases (Figure 7.29). At the slower slope changes, the replicated staircase tracks fairly close to the original fs. A rapid increase in slope of fs causes a widening gap between the fs form and that of the reconstructed waveform at the receiver. This difference is known as **SLOPE OVERLOAD.**

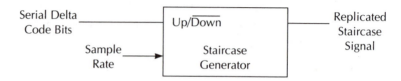

Figure 7.28 Delta Demodulator

DIGITAL & DATA COMMUNICATIONS

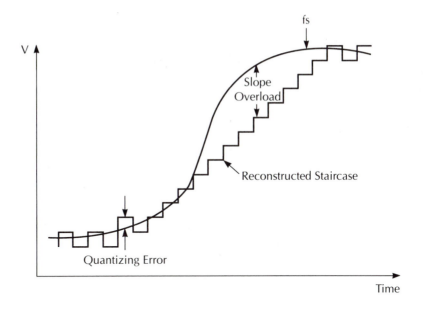

Figure 7.29 Delta Modulation Slope Overload

● **7.17 ADAPTIVE DELTA MODULATION**

Delta modulation works well as long as the original signal does not have large variations in frequency, which cause slope overload. However, analog signals that are derived from audio or video information vary in frequency. For a delta modulation system using fixed step sizes, this presents a problem. The original signal is quantized and sampled as shown in Figure 7.30 in the solid staircase. Note that signal changes are not accurately represented by the staircase signal, particularly on the rising section of the waveform. The problem can be reduced by increasing the sampling rate, except that would cause more digital bits to be transmitted. Another solution is to vary the step sizes as required to more closely resemble the original signal. This is the concept behind **ADAPTIVE DELTA MODULATION (ADM).** The dotted staircase of Figure 7.30 shows the objective of adaptive delta modulation. To make the staircase voltage track the original signal more accurately, the step size is increased on the leading edge and decreased on the trailing edge.

When using adaptive delta modulation, a long string of 1s or 0s indicates that the staircase generator is continually trying to play catch-up with the original signal. That is, it is trying to continually add or continually subtract step voltages from the output. The more frequently the digital bits change from 0 to 1, the more closely the

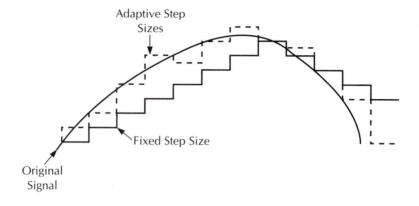

Figure 7.30 Fixed and Adaptive Step Sizing

staircase voltage is tracking the original waveform. An adaptive modulator causes the step size to increase whenever it detects many consecutive 1s or 0s and causes the step size to decrease when it detects alterations in the logic level of the digital data stream. The result of varying the step size is shown in Figure 7.31. The rapidly changing leading edge begins to generate a set of consecutive 1s. Immediately, an adaptive delta modulator increases the step size. As the peak is reached, alternating 1s and 0s are experienced and the modulator reduces the step size. It is increased on the down slope after a string of 0s are detected. The second slope, rising slowly, begins to cause alternating 1s and 0s to appear because of the large step size from the previous down slope. The adaptive delta modulator decreases the step size, causing the staircase output to once again follow the signal more closely.

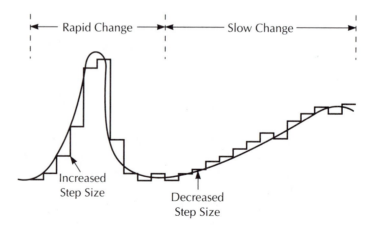

Figure 7.31 Example of Adaptive Delta Modulation

One method of increasing the volume of traffic on a given line is to reduce the number of digital bits that represent a signal to be transmitted without corrupting the signal that is replicated at the receiving end. It is desirable to compress the digital codes at the transmitter and then expand them back to their original form at the receiver. The acronym for this procedure is **COMPANDING (COMPRESSION/ EXPANSION).** Companding can be done in analog or digital form. In analog companding the signal to be digitized is companded first and then sampled and digitized. Digital companding reduces the digital code created after the signal has been digitized. Companding is also used to increase the signal-to-noise ratio of the coded signal.

Analog compression follows one of two algorithms, depending on whether the system that does the companding was developed in North America or in Europe and Japan. The North American μ or 255 law is a system that divides the analog signal range into fifteen segments, each eventually encoded into an 8-bit digital value. The actual voltage value (Y) for each segment is determined by the following formula:

$$Y = \frac{V\ln(1 + \mu v/V)}{\ln(1 + \mu)} \tag{7.4}$$

where Y is the compressed value out of input voltage v for a signal whose maximum peak value (and compressed value) is V. A graph comparing compressed voltage (Y) with input voltage (v) for peak values of $+/-1V$ is shown in Figure 7.32.

Segments are coded as shown in Figure 7.33. The first bit is the sign bit, a logic 1 representing positive values and a logic 0, negative ones. The next 3 bits designates the segment number 000 to 111, or 0 to 7. The fifteenth segment is the 0 segment, which is half positive and half negative. The last 4 bits (at times referred to as bits ABCD) determine the actual level within the range of a particular segment. Table 7.4 lists an example of the voltage levels represented by a compressed signal using μ law. Note that the step size doubles with each segment and that the levels within a segment are divided into sixteen steps (the last 4 bits of the segment code).

The μ law can be applied to digital compression of the signal. The analog signal is digitized and coded using 12 bits. The format for the 12-bit codes is shown in Figure 7.34. Using 12 bits to represent an analog signal improves the digital resolution when compared with 8 bits due to an increased number of quantized levels and reduced step sizes. This improvement in resolution increases the accuracy of the replicated signal. The first bit is the sign bit, with the remaining 11 bits reflecting the magnitude of the sample. The form in the illustration shows eight codes for the 12 bits. Bits ABCD contain any one of sixteen values from 0000 to 1111. The x values have actual digital values, which get discarded during the compression process.

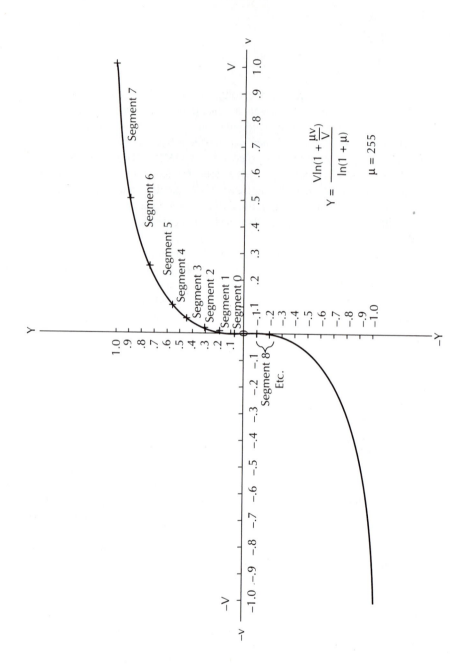

Figure 7.32 Input Sample Voltage (V) versus Compressed Voltage (Y): μ Law Using Unit Peak Value (V = 1.0)

$$Y = \frac{V \ln(1 + \frac{\mu v}{V})}{\ln(1 + \mu)}$$

$\mu = 255$

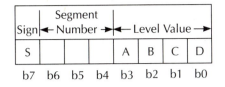

Figure 7.33 PCM Code Example

Digitally, the 12-bit values are encoded into an 8-bit compressed code as follows:

1. Retain the sign bit as the first bit of the 8-bit code.
2. Count the number of 0s until the occurrence of the first 1 bit. Subtract the zero count from 7. This is the segment number.
3. The first occurrence of a 1 is assumed during the expanding process, so it is set aside during compression.
4. Copy the next 4 bits (ABCD) into the 8-bit compressed code.

The compressed code consists of a sign bit, segment number, and the value of the ABCD bits.

Table 7.4 Step Size and Maximum Voltage per Segment for an Example PCM

Segment	Step (mV)	Maximum Voltage (mV)
0	0.5	$16 \times .5 = 8$
1	1.0	$8 + 16 \times 1 = 24$
2	2.0	$24 + 16 \times 2 = 56$
3	4.0	$56 + 16 \times 4 = 120$
4	8.0	$120 + 16 \times 8 = 248$
5	16.0	$248 + 16 \times 16 = 504$
6	32.0	$504 + 16 \times 32 = 1,016$
7	64.0	$1,016 + 16 \times 64 = 2,040$

b11 ◄———————————————————► b0

Segment	S	Data Code Bits
0	S	0 0 0 0 0 0 0 A B C D
1	S	0 0 0 0 0 0 1 A B C D
2	S	0 0 0 0 0 1 A B C D x
3	S	0 0 0 0 1 A B C D x x
4	S	0 0 0 1 A B C D x x x
5	S	0 0 1 A B C D x x x x
6	S	0 1 A B C D x x x x x
7	S	1 A B C D x x x x x x

x–Don't Care Bits

ABCD–Any Digital Value from 0000-1111
Segment Value Code

S–Sign Bit   0 = +   1 = −

Figure 7.34 12-Bit Quantize Sample Codes

Example 7.6   Code the 12-bit code 100001011010 into an 8-bit compressed law code.

Solution: Retain the sign bit (1). Count the 0s until the first 1 occurs (4). Subtract from 7 ($7 - 4 = 3 = 011$). Discard that first 1 and get ABCD = 0110. Place them into the 8-bit format 10110110.

Expanding back digitally reverses the process:

1. Retain the sign bit.
2. Take the segment number, subtract from 7, and add that many 0s.
3. Make the next bit a 1.
4. Next bits are ABCD values.
5. Add a 1 and sufficient 0s to complete the 12-bit value.

The addition of the 1 and 0 bits following the ABCD bits sets a value in the middle between the actual values the remaining bits could have had. While this introduces some error, it is minimal since it occurs in the least significant bits of the 12-bit code.

Example 7.7   Expand the compressed code of example 7.6.

Solution: Retain the sign bit (1). Subtract the segment number from 7 and insert that number of 0s. 7 − 3 = 4 = 0000. Replace the discarded 1. Append the ABCD bits followed by 10 (a 1 and the necessary 0 to complete 12 bits):

100001011010

Note that the last 2 bits of the original 12-bit code from example 7.6 could have been any one of four combinations—00, 01, 10, or 11 (0, 1, 2, or 3). This means that a maximum potential error of 2 could exist. The maximum percentage of error for 2 out of 1011000 = 2/88 = 2.3%. In the example, since the original value of the two lowest bits was 10, there is no error in example 7.7.

In Europe the CCITT standard for companding is the A law, which uses two equations to compute compressed value, Y, depending on the peak value of V in relation to constant A:

$$Y = \frac{Av}{V(1 + \ln A)}; \qquad |v| < V/A \qquad (7.5)$$

and:

$$Y = \frac{1 + \ln(Av/V)}{1 + \ln A}; \qquad V/A < |v| < V \qquad (7.6)$$

These formulas are used with systems that employ signals divided into thirteen segment sections, each segment encoded using an 8-bit converter. The main significance in the difference between the use of the two codes is to facilitate the different systems used in North America and those used in Europe, Asia, and the rest of the world. It starts with the difference in T1 carriers, including the channel capacity of each, and ends with the size of the digital codes used to digitize the analog signals. The repercussion of these differences is to make interconnections between North American and Euro-Asian systems difficult.

● 7.19 CODECS

A **CODER/DECODER (CODEC)** is a device designed to convert analog signals, such as voice communications, into PCM compressed samples to be sent onto digital carriers and to reverse the process, replicating the original analog signal, at the receiver. The term CODEC is an acronym for CODer/DECoder, signifying the pulse coding/decoding function of the device. At first CODEC functions were managed by several separate devices, each performing the tasks necessary for PCM communication. These functions are sampling, quantizing, analog to digital and digital to analog conversions, filtering, and companding. Today all of these functions are available on a single CODEC IC chip such as Intel's 2913.

The 2913 CODEC operates in one of two modes, fixed data rate or variable data rate. In the fixed data rate modes, the chip is capable of operating with one of three fixed clock frequencies: T1's 1.544 MHz, CCITT's 2.048 MHz, and also 1.536 MHz. In the variable data rate, the clock frequency can be set anywhere from 64 kHz to 4.096 MHz. The 2913 allows selection of either the μ law or the A law for the companding function. On the transmit side of the 2913 CODEC (Figure 7.35), the analog signal is fed into an operational amplifier, whose gain is set by external resistor components shown as R1 and R2 in Figure 7.35. If it is undesirable to use the amplifier, the signal can be entered, single ended via the GSx pin and analog ground (GNDA). The gain (Av) for the op amp conforms to the standard op amp noninverting gain formula:

$$Av = 1 + \frac{R2}{R1} \qquad (7.7)$$

The output of the op amp is fed through an antialiasing filter to a sample and hold circuit. The low-pass antialiasing filter attenuates the sampling frequency −35dB. The passband filter following the antialiasing filter is designed to yield a flat response from 300 Hz to 3kHz as illustrated in Figure 7.36. An additional high-pass filter rejects power line frequencies (50 Hz and 60 Hz) and European railroads and ringing frequencies (17 Hz) and their harmonics.

Samples are held for encoding and μ law or A law compression in an internal sample and hold circuit. One additional unit, an auto-zero circuit, is used to correct for any DC offsets that were created by the process to this point. The compressed PCM serialized data are sent out through the Dx transmit data line.

On the receive side of the CODEC (Figure 7.37), the serial data stream on the data receive pin (DR) is shifted into the input register and output in parallel to the digital to analog converter. The resulting analog signal is held in a second internal sample and hold circuit until the next coded sample is shifted into the input register. The decoded samples are then shaped and filtered before being sent out through a power amplifier. The filter is a passband filter with a sharp roll-off at 4,000 Hz. It has a flat response, like the transmit passband filter between 300 Hz and 3 kHz.

The maximum output power amplifier can be used as a single-ended circuit with the signal sensed from either PWRO− or PWRO+ (power amplifier out + or −) and analog ground (GNDA) or as a differential buffer by sensing the output between PWRO− and PWRO+. The maximum voltage level from the power amplifier is 1.50 Vrms. The actual output (Vo) between the PWRO outputs is a function of the gain (Ao) of the op amp circuit and can be set anywhere from the maximum of 1.5 Vrms to a minimum of 0.375 Vrms. The output level is set by the voltage level applied to gain set pin GSr. Maximum output is achieved by connecting GSr to PWRO−, while minimum voltage is achieved by connecting GSr

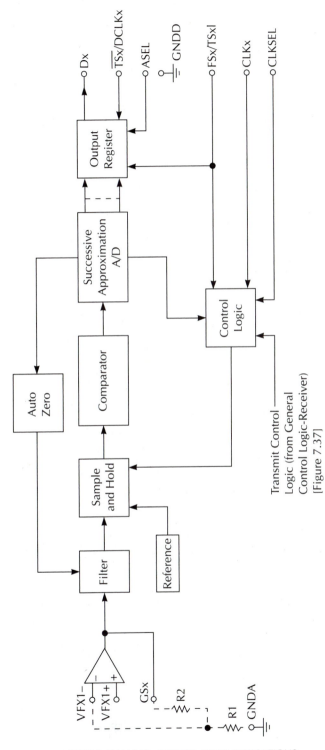

Figure 7.35 CODEC Transmit Blocks

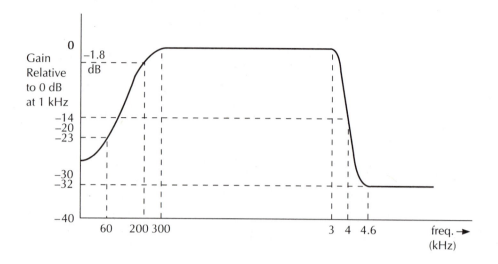

Figure 7.36 CODEC Transmit Filter Characteristics

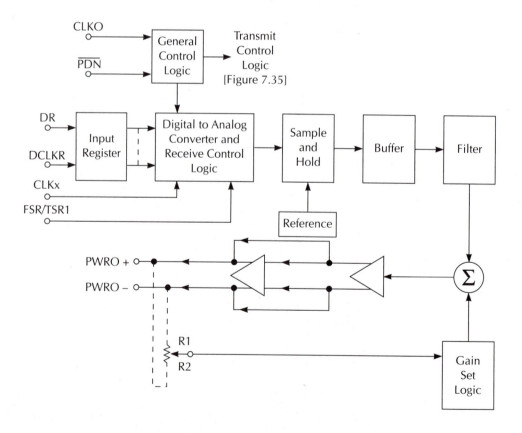

Figure 7.37 CODEC Receive Blocks

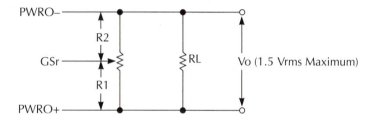

Figure 7.38 Gain-Setting Resistors for Receive Power Amplifier

to PWRO+. Voltage levels between these extremes can be accomplished with a potentiometer connected from PWRO+ to PWRO− whose wiper is connected to GSr as shown in Figure 7.38.

The gain of the power amplifier is calculated by:

$$Ao = \frac{R1 + R2}{R1 + 4R2} \qquad (7.8)$$

By reshuffling the formula, a ratio of resistor values based on the gain can be shown:

$$\frac{R1}{R2} = \frac{4Ao - 1}{1 - Ao} \qquad (7.9)$$

The significance of the ratio of R1 to R2 is that the gain of the amplifier is limited to values that produce a positive ratio of R1 to R2.

Example 7.8　Using the relationship of equation 7.9, determine the ratio for R1/R2 for maximum, minimum, and midrange gains of the power amplifier.

Solution:　A maximum gain of 1 brings the denominator of equation 7.9 to 0, making the ratio of R1 to R2 infinite. This gain is achieved by connecting GSr to PWRO− and disconnecting the resistor network. A midrange gain of 0.5 develops a ratio of 2 for R1/R2. That is, R1 is twice the value of R2. Lastly, the minimum gain occurs when the numerator of equation 7.9 is brought to 0. This occurs at a gain of 0.25. Here the ratio of R1 to R2 is 0. Once again, this cannot be achieved using the resistors. They are removed, and GSr is connected to PWRO+. For any gain between ¼ and 1, the ratio of R1 to R2 can be calculated and used as a guideline for selecting values for the gain network.

This CODEC can operate using either the μ law or A law for companding. The selection is made utilizing the ASEL (A law select) input. Connecting this input to

Vbb sets the A law mode. Connecting it to Vcc or GNDD selects the μ law and causes the least significant bit of the transmit signaling frame to take the state of ASEL.

SUMMARY

Multichannel transmissions allow numerous communications links to be established simultaneously. Time division multiplexing assigns channels to time slots in a transmission stream, while frequency division multiplexed systems fit each channel into a portion of the bandwidth of the transmitted signal.

Reducing noise problems, increasing data capacity, and/or increasing transmission speeds are among the considerations that led to the development of pulse code modulation techniques. These techniques involve sampling, quantization, and companding principles to achieve those goals.

GLOSSARY

Baseband—Single-channel communications system in which that channel is the only one occupying the system's bandwidth.

Bipolar violation (BPV)—Occurs when the voltage level of two successive 1s is the same when using alternate mark inversion data format.

Broadband—Multichannel communications system in which the system bandwidth is shared by the channels using it.

Channel—Single line of communication.

Coder/decoder (CODEC)—Used to convert voice signals to digital codes.

Companding—Compression/expansion of signals to reduce signal-to-noise effects of smaller signals.

Delta modulation—Sends single bit to indicate increase or decrease in sample amplitudes.

Data under voice (DUV)—Method that utilizes normally unused band areas for supplemental voice transmissions.

Flattop sample—Samples which have flat amplitudes rather than following analog amplitude.

Guard band—A portion of unused bandwidth between channels.

Natural sampling—Samples whose amplitude follows original analog level.

Nyquist sampling rate—Minimum sampling rate.

Packet—A section of a message.

Quantization—Process of approximating sample values by using fixed discrete values.

Quantization error or noise—difference between sample level and quantized value.

Slope overload—Difference between original and reconstructed analog signals.

T1 carrier—Communication line that carries 24 digitized voice channels.

QUESTIONS

7.1. Why is full-duplex frequency shift keying an example of frequency division multiplexing?

7.2. What is a guard band?

7.3. How are FDM groups, supergroups, and so on, created?

7.4. What is the purpose of carrier pilots in FDM?

7.5. What are the two largest terminal FDM groupings? How do they differ?

7.6. Which band of frequencies is occupied by a data under voice signal when sent with a mastergroup?

7.7. Give four characteristics of a T1 carrier.

7.8. What type of multiplexing is used on T1 carriers?

7.9. Generally, what type of function does the 193rd bit perform for a basic T1 frame?

7.10. Describe what use is made of the 193rd bits for superframes and extended superframes?

7.11. How many channels make up a superframe and an extended superframe?

7.12. What is the purpose for using B8ZS encoding?

7.13. What is meant by a bipolar violation?

7.14. Define slip. What causes slip to occur?

7.15. What is the difference between jitter and wander?

7.16. List at least five characteristics, data, or error indications that the TM1 T1 Carrier monitor is used for.

7.17. How are multiple channels sent on a single line using TDM?

7.18. What is the essential difference between STDM and STAT MUX TDM?

7.19. What is meant by pulse amplitude modulation?

7.20. What is the difference between natural and flattop sampling?

7.21. What is the purpose of using sample and hold techniques after sampling is done?

7.22. What is the advantage of quantizing a signal's samples?

7.23. What happens to quantized samples in a PCM system?

7.24. Which type of analog to digital converter is commonly used for PCM encoding?

7.25. How does the sampling rate for delta modulation differ in concept from standard PCM sampling?

7.26. What is the basic idea behind delta modulation?

7.27. What is the difference between an adaptive delta modulator and a regular delta modulator?

7.28. When are step sizes changed when using adaptive delta modulation?

7.29. Why is companding used?

7.30. What is a CODEC? How does it differ from a modem?

7.31. What is the data rate range for the 2913 CODEC using variable data rate mode?

7.32. What is the bandwidth of the 2913 CODEC's transmit filter? What is the bandwidth of its receive filter?

PROBLEMS

7.1. What is the frequency range for FDM group 4?

7.2. Draw the RZI form for the data stream 11000111011. The least significant bit is on the left, and the first logic 1 is at a +V level.

7.3. An integrated service requires 768 kbps data to be sent on a fractional T1 carrier. Assuming that the first channel used to hold real data is channel 1, which channels will contain real data and which filler data?

7.4. What is the Nyquist sampling rate when sampling a signal whose highest frequency is 1,630 Hz and lowest frequency is 250 Hz?

7.5. A 3.6cosf signal is naturally sampled at a rate of 56 kHz using 1.25-microsecond sampling pulses. What is the value of the reconstructed output signal?

7.6. What is the step size for a quantized signal whose peak voltages are +16.8V and −22.4V? There are 16 bits in the PCM code for each sample.

7.7. Given a maximum voltage of 3.6V, what is the compressed value for an input voltage of 1.26V using μ law? Using A law? ($\mu = 255$ and $A = 3.0$.)

7.8. Digitally compress the 12-bit code 100101101101.

7.9. Expand the result from question 7.16. How much error is introduced? What is the percentage of error?

7.10. What is the ratio of R1 to R2 to achieve a gain of 2.5 for the input op amp of the transmit side of the 2913 CODEC?

7.11. What is the ratio of R1 to R2 to set the output power amplifier gain to .75?

LAB PROJECTS

7.1. Design, construct, and verify a three-channel FDM system. Each channel is limited to a 0- to 4-kHz bandwidth. Verify the circuit works by setting channel 1's signal frequency to 1 kHz, channel 2's to 1.8 kHz, and channel 3's to 2.5 kHz. This project can be separated into two parts, with one group building the transmit side and one group, the receive side. When the systems are completed, the three channel signals should appear at the received output.

7.2. Devise a TDM system to time multiplex three channels. Verify that the system works by setting channel 1's frequency to 1 kHz, channel 2's to 1.8 kHz, and channel 3's to 2.5 kHz. This project can be done in two parts, transmitter and receiver. The operation of the system can be verified by reproducing the three signals at the receive output.

7.3. Design and construct a sampling circuit that creates sufficient samples of an input sine wave.

7.4. Design and construct a circuit that replicates samples into the original sine wave signal.

7.5. Design and construct a simple PCM coder that converts samples into a digital code.

7.6. Design and construct a PCM encoder that replicates the samples from a PCM-type code.

7.7. Combine the projects from 7.3 to 7.6 into a system.

7.8. Design a system built around the 2913 or 2914 CODEC chip. Verify that a signal driving the transmit side can be replicated at the output of the receive side.

Answers to Odd-Numbered Questions

7.1. Full-duplex originate and answer mark and space signals share the telephone line bandwidth.

7.3. By mixing each channel or group with a different carrier frequency

7.5. Mastergroup mux—2 to 6 mastergroups (1,200–3,600 channels); jumbo group mux—10,800 channels

7.7. Twenty-four channels/frame; 64 kbps/channel; 1.544 Mbps/frame; 8,000-Hz sampling rate; 600 feet between repeaters; 5 to 50 mile distances; uses TDM and PCM; uses RZI

7.9. Framing

7.11. Superframe: 288; extended superframe: 576

7.13. Two consecutive logic 1s with the same voltage polarity

7.15. Wander is low-frequency ($<$10 Hz) jitter.

7.17. Each channel's message is assigned a time slot in the transmission

7.19. PAM—changing of a signal into pulses whose amplitudes vary with the signal's amplitude

7.21. To hold the sampled level until the next sample is taken

7.23. Their amplitudes are coded into binary codes.

7.25. Delta modulation sample rates are high enough so amplitudes of consecutive samples do not vary by more than one step size.

7.27. Step size is fixed using regular delta modulation and varies with adaptive delta modulation as needed.

7.29. Reduce the number of bits transmitted for each sample.

7.31. 64 kbps to 4,096 Mbps

Solutions to Odd-Numbered Problems

7.1. 92 to 96 kHz

7.3. Actual data in channels 1, 3, 5, 7, 9, 11, 13, 15, 17, 19, 21, 23; filler in all other channels

7.5. 0.252cosf

7.7. μ law: 1.023V; A law: 0.955V

7.9. 100101101000; error of 5 from original 365 = 1.37%

7.11. R1 = 8R2

Lab Project Guides

The ability to complete the lab projects will depend on the facilities, availability of components, and time allocated to lab work. These projects are representative of the theory discussed in the chapter.

Eight • Open Systems Network Models

OBJECTIVES

After studying this chapter, the reader will have an understanding of two leading open systems models for communications networking, the open systems interconnection and Systems Network Architecture models, and the protocol standards associated with them. Open systems architectures are flexible structures set into fixed frameworks. Fundamental activities and requirements are detailed at each level of an open systems network. How these activities and requirements are achieved is left open to individual applications. Specific topics include the following:

1. Data system topologies
2. Data switching methods
3. The open systems interconnection (OSI) model
4. IBM's Systems Network Architecture (SNA)

ACRONYMS

HDLC—High-Level Data Link Control
IEEE—Institute of Electrical and Electronics Engineers
ISO—International Standards Organization
LAN—Local Area Network
LU—Logical Unit
MAN—Metropolitan Area Network
NAU—Network Addressable Unit

NETBIOS—**NET**work **B**asic **I**nput **O**utput **S**ervices Interface

OSI—**O**pen **S**ystems **I**nterconnection

PCN—**P**ath **C**ontrol **N**etwork

PU—**P**hysical **U**nit

SDLC—**S**ynchronous **D**ata **L**ink **C**ontrol

SNA—**S**ystems Network Architecture

WAN—**W**ide **A**rea Network

● 8.1 INTRODUCTION

Interconnecting several stations into a viable switch network produces several advantages, such as the following:

1. Resources, such as bulk storage and communications lines, are shared among the stations on the network.

2. High-speed data rates, which result in fast transfers of information, are used on these networks.

3. It is possible to send smaller portions of messages between stations, allowing different stations to communicate with each other in a near-real-time environment unaffected by delays caused by stations waiting for a complete message to be transmitted and received.

These switched networks are local, distant, or a combination of both. In order to have them perform as required, decisions have to be made on how to handle the transmission of data and to establish a means for stations within the network to access the network. In addition, message protocols and station addressing, as well as message integration required to reassemble the portions in their proper sequence, are required to be specified for the network in use. Lastly, methods of error detection and/or correction are incorporated into the network software.

● 8.2 DATA TOPOLOGIES

A communications network supporting more than two stations requires some method to establish a unique link between any two stations within the network. The format of the network is defined by its topology. In chapter 3, point-to-point and multipoint concepts were introduced. The former involves only two stations and does not require further distinction. For the multipoint systems, several topologies exist. In the most direct form, a primary station serves as the controlling station with

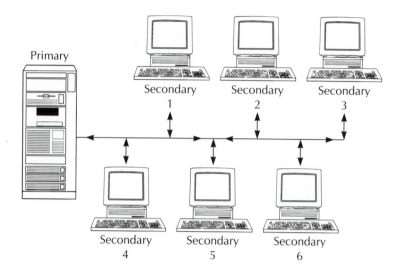

Figure 8.1 Bus Topology

a number of secondaries connected to a common bus. Figure 8.1 illustrates this bus topology. The primary or hub station establishes a communication link with a specific secondary by addressing that secondary through a poll or selection process as detailed by the protocol in use (see chapter 4).

Other direct-connected topologies include ring (Figure 8.2a) and star or hub (Figure 8.2b). Using the ring system, messages are passed from one station to another in one direction until they reach the destination station. A token message is attached to the data being sent around the ring. The token is passed from station to station. As a station receives the token, it can remove messages intended for its use and/or add messages it may want to send on. That station then modifies the token and passes it on to the next station in the ring. Since access to the ring is achieved by passing the token from station to station, a controlling or primary station is not included or required in the ring.

In the star system, the central (primary) station directs traffic from one station to another. The primary essentially acts as the connecting switch, allowing two stations to communicate with each other. A concept referred to as switching provides the means for these stations to operate using one of the topologies mentioned.

● **8.3 DATA SWITCHING**

In any communications system, allowing two stations to send and receive messages requires that they be interconnected. The simplest way to do this is to wire each station to each of the others. Figure 8.3 shows four stations connected in such

DIGITAL & DATA COMMUNICATIONS

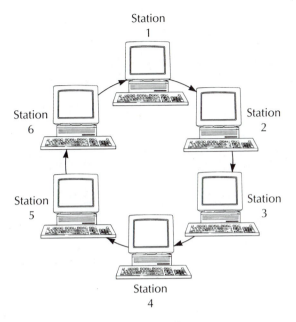

(a) Ring Topology

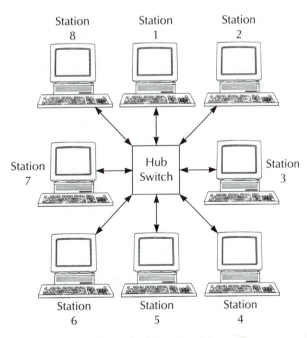

(b) Hub or Star Topology

Figure 8.2 Ring and Star Topologies

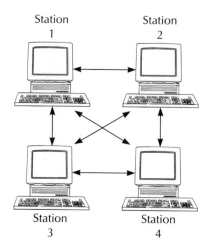

Station 1   Station 2

Station 3   Station 4

Figure 8.3 Four Stations Directly Connected

a fashion. Note that each station has three lines connected to it. Such a system enjoys many advantages. Communications links are always established using the same interconnecting lines each time a message is sent. Information is transferred in real time—that is, as data are sent from one station to another, the only time loss is the propagation time of the cabling; there is no vying (or contention) for a line since each station has an individual connection.

The main disadvantage of a direct-connected system is the number of lines needed to interconnect the stations. As is shown in Figure 8.3, six separate lines are required to interconnect the four stations. While this number does not appear to be too bad, keep in mind that this illustration is for a small number of stations. Now consider increasing the network to 256 stations! To determine the number of interconnecting lines (N) for any number of stations (S) in the system, the following formula is used:

$$N = \frac{(S)(S - 1)}{2} \tag{8.1}$$

Example 8.1   How many lines are required to directly connect 256 stations in a single network?

Solution: Using equation 8.1, the number of lines is:

$$N = \frac{(256)(255)}{2} = 32,640 \text{ lines}$$

One alternative to direct connections is a **CIRCUIT SWITCH,** in which the network stations are connected to a central switching station. This station provides the physical connection between any two stations on the network. The system appears similar to the star topology of Figure 8.2b. The basic telephone network we all use daily is an example of a circuit switch system. Once two stations are connected through the switch, data are sent and received in real time as with direct-connected systems. The actual lines utilized within the many layers (local and long-distance) of the telephone system almost guarantee that a different set of lines is used to interconnect the same two stations at different times. In a sense, this ability to find different routes through the switch network is a benefit. The communications load has to become fairly heavy (as on Mother's Day) before the system fails to interconnect any two stations. The actual path taken to perform the interconnection is TRANSPARENT to the user, who has no idea how the call is routed. The caller's only concern is that the connection is made.

The chief difficulty with a circuit switch occurs when two different stations attempt to establish a data link with the same station. One caller will be successful, while the other hears a busy tone, which is an indication of line CONTENTION. The two callers contended for the connection to the third station. The successful caller's connection was made, while the unsuccessful caller, who receives the busy tone, has to wait until the line is free again before the call can be made and data sent. If both callers attempted to make the call at exactly the same instant, a condition called COLLISION occurs; that is, the call requests collide with each other. The collision must be resolved before one of the calls can be connected. The resolution usually requires both callers to disconnect the line and try making the calling again.

Direct and circuit switch methods transfer data in real time, but both have problems either with the number of interconnecting lines or with line contention. **MESSAGE SWITCHING** is an alternative method designed to overcome these limitations.

In a message switch network, the central switch station accepts the traffic sent to it by the stations connected to the system. The messages are stored in the switch station's buffer memory, and when a line is available, the data are forwarded to the appropriate station. Thus, message switching is referred to as a store and forward method. The benefit with message switching is that when the sending station is finished sending its traffic it can go about its business, content that the messages will eventually reach their destination. There is no waiting for an open line by the sending station.

The problems with message switching, and the main difficulty with implementing systems using message switching, are the need to allocate a sufficiently large data buffer to hold incoming messages and the time it takes for a message to reach its destination. Fairly involved software programs are required to manage the routing and storage of these messages. Additionally, if the system is particularly busy, a message could be delayed a long time before it finally is routed to its destination. What is needed is a method using message switching that allows shorter message ''portions'' to be sent to each station in a continuous sequence. Long messages are broken into smaller units, called **PACKETS.** Each packet contains a header, which includes destination and source identifiers or addresses and a packet number. Packets from several sources are routed to their destinations, where they are reassembled into their original messages. The packets are sent continually as long as there are messages to be sent. They are sent to the various destination stations in a mixed sequence, which results in the receiving stations' continually receiving data until all packets of the messages are sent and received. The receiving stations then utilize the destination addresses to identify packets intended for that station. Next, packet numbers in the headers are used by the receiving stations to assemble the messages into a complete and correct order. This form of network switching, called **PACKET SWITCHING,** is used with many different networks.

● 8.4 TYPES OF NETWORKS

The minute a computer landed on a user's desk and allowed computer processing to be handled by anyone with the time or interest, a new world opened up for office managers. Interoffice communications, record keeping, forecasting, and many other functions came within the reach of several levels of a firm's hierarchy. The problem was to interconnect the personal computers (PCs) within a company to share the work load and data bases. Personal computers, through standard bus structures, also found their place on the manufacturing and quality-testing floors. Again, a need evolved to interconnect these computer WORKSTATIONS to share test programs and data bases held in mainframes within the company. These needs gave birth to the concept of interconnecting numerous communications systems and/or workstations into a single NETWORK. Technologies existed to produce these networks, but concern about compatibility and consistency within networks established a need for standardized guidelines for creating and operating these network systems.

Networks are classified in accordance with the amount of physical territory they encompass. **LOCAL AREA NETWORKS (LANs),** now many years in place, interconnect a high number of access or node points or stations within a confined physical area. An example of the territory covered is a single office building that houses executive, administrative, inventory control, personnel, and payroll offices. All these areas are interconnected using an LAN.

Serving many locations distributed over a large geographical area is the function of a **WIDE AREA NETWORK (WAN).** A system of overnight teller machines used by a banking concern covering the western half of the United States (such as First Interstate Bank) is an example of a WAN. These operate at data rates typically between 14 kbps and 1.5 Mbps.

More localized systems such as teller machines within a single citywide area are interconnected using a **METROPOLITAN AREA NETWORK (MAN).** Data rates of up to 1 Mbps are possible with MANs. A small bank with offices restricted to a city or a single state might well be interconnected through a MAN.

● 8.5 THE OPEN SYSTEMS INTERCONNECTION (OSI) MODEL

As the number of networks increased, the desire to create a standard format and set of specifications to oversee the configuration and use of the networks evolved. A leading author of standards for networks is the **INTERNATIONAL STANDARDS ORGANIZATION (ISO),** which created a seven-layer model guideline for an open systems interconnected communication process (Figure 8.4). The OPEN SYSTEMS INTERCONNECTION (OSI) model refers to a system in which any number of standards or protocols may be used to address one or more of the functions within any of the layers of the model. Overall consistency and application remain as long as the guidelines for each layer are adhered to by the standards and protocols used. The following is a description of each of the layers and, where applicable, some of the standards and protocols that address each of the layers.

● 8.5.1 Physical Layer

The physical layer is the lowest layer of the model. It defines the mechanical, electrical, functional, and procedural aspects of the physical link between stations. RS232C is an example of a standard that addresses the physical layer within the ISO model. Portions of other standards, such as IEEE 802 and CCITT X.25 (both discussed in chapter 9), concern themselves with the physical makeup of a network. This layer defines the electrical and physical requirements of a system. Actual data formats and protocols are defined in the remaining six layers.

● 8.5.2 Data Link Control Layer

The next layer above the physical is the data link layer. This layer—the lowest protocol layer of the OSI model—specifies the data format, sequence, acknowledgment process, and bit error–detection method detailed in a data link protocol used by the network to handle raw digital data. Data link protocols define the rules for node-to-node data transfers. A **NODE** is a point in a network where a connection can be made to a station, network, or another node. Specifically, protocols defining the responses to data link characters within a message fall into this category. Additionally, the establishment of physical addresses for stations on the network is done at the data link layer level. This is accomplished through the use of source and destination address fields within the data link protocol used. Routing of messages

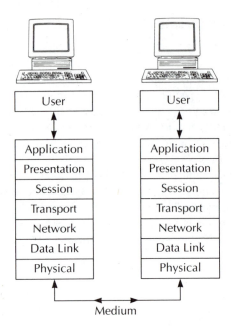

Figure 8.4 OSI Model

and interfacing through one or more networks are *not* included at this layer's level. BISYNC and SYNCHRONOUS DATA LINK CONTROL/HIGH-LEVEL DATA LINK CONTROL (SDLC/HDLC) protocols and portions of IEEE 802 are examples of standards in the data link layer.

● **8.5.3 Network Layer**

How data are routed through a network from source to destination is defined in the network layer. Flow control of packetized information as well as congestion avoidance is a concern of protocols in the network layer. **VIRTUAL CIRCUITS,** which are the logical routes messages can take through a network, are established at this level. Virtual circuits have the advantage of being free of physical restraints. A single virtual circuit may find any number of physical routes to complete the connection. Packets are formalized into frames to include virtual circuit routing and other pertinent data. Protocols at the network level are then responsible for the delivery of these frames in the same sequence they were sent. Specifications for specific networks, such as Ethernet or DECNET, as well as X.25 protocol address this layer.

The interfacing between the applications software and the available hardware is covered in the transport layer. Network connections are created and maintained with considerations toward cost, quality of service, addresses, and error recovery. Processes at the transport level include the following:

1. The mapping of transport addresses onto the network

2. Error detection and recovery to minimize data loss and time lost due to retransmission of bad frames

3. Segmentation or fragmentation of messages to maximize transmission efficiency

4. Flow control between layers below the transport layer (specifically the network layer) and the session layer

In addition, a process called **BLOCKING** is employed when requested by sending stations. Blocking is used to prevent a particular station from receiving specific frames until the sending station is ready for the receiving station to get them.

There are five classes of transport layer protocols defined by the OSI model. They differ in the degree of error recovery and were designed to allow the most flexibility to the network system designer. Class 0, the lowest class, includes minimal error recovery and is used primarily for straight text transmissions. The next class, class 1, increases the amount of error recovery ability, extending use to networks that use the X.25 packet switch at the network level. Class 2 increases error recovery to the point that the network becomes very reliable, requiring very few retransmissions to handle errors. Multiplexing of data connections into a single data link connection for optimizing the network use is part of this class. Class 3 combines the better features of classes 1 and 2. Lastly, the highest level of error recovery is found in class 4. Besides basic error recovery, there are requirements for checking damaged data and lost or out-of-sequence packets. Even though there is a distinct amount of error recovery processes, there is no additional manipulation of data contained in the messages.

● 8.5.5 Session Layer

The remaining layers reach into the higher end of a communication process's structure. The session layer concerns file management and overhead functions. Access availability and system time allocations are included at this layer. Session layer protocols provide a method by which presentation stations can organize, synchronize, and manage the transfers of information between themselves. Included in these processes is the ability to hold messages until the originating station is ready to release them. In this process the session layer protocol is taking advantage of the presentation layer's blocking scheme. Further, the session layer protocol can be directed to discard all data destined for a station without that station being aware that the data were ever sent, if the initiating station requests it through a RESET function. Traffic at the session level can be prioritized if desired by the network manager. Processes at the session level also include:

1. Connection and disconnection from the network and authentication of user access

2. The binding of process names to network addresses

3. Permitting multiple applications to share a virtual circuit

NETWORK BASIC INPUT OUTPUT SERVICES INTERFACE (NETBIOS) is the name of one protocol specification applicable to this layer.

● 8.5.6 Presentation Layer

Higher-level interfacing requirements, including data compression, format conversions, and encryption, are detailed in the presentation layer. These include character set and code translations, format and syntax resolution, and/or data transformation. At this level, devices are treated as virtual devices or stations, being defined as logical rather than physical entities. Three basic forms of protocols address the presentation layer. They are:

1. Virtual terminal protocol, which is used to allow different types of terminals to support different applications

2. Virtual file protocol, which handles code conversions within files, file communication, and file formatting

3. Job transfer and manipulating protocol, which controls the structure of jobs and records

Novell's NETWARE includes software and hardware designed to operate at the presentation and application layers of the OSI model.

● 8.5.7 Application Layer

We finally reach the human interface level, the application layer. This layer defines user interface to lower-level layers and various application processes. This layer is separated into several service elements, each addressing distinct application usage. The common application service element takes care of services from the lower-layer levels independent of the nature of the user application. The application-specific service element deals with handling large amounts of data, including data base access, bulk data transfers, and remote job entry. Lastly, the user-specific service element deals with actual user requirements for the network. Some typical uses are credit checks, inventory, and point of sales.

The identification, authentication, and determination of availability of combined end users is another aspect of the applications layer. This is the highest level of security since it deals with the user entry point into the system. Again, NETWARE is one protocol used in the applications layer area. Digital Equipment Corporation's DECNET is another example of an applications software standard.

● 8.6 SYSTEMS NETWORK ARCHITECTURE

Before development of the OSI model, IBM had created a closed network system based on its own standards and used solely for IBM computers interconnected through a local type of dedicated network. In September 1973 IBM introduced a set of network standards to allow various IBM and non-IBM systems to be interconnected into a common LAN network. This specification, called the **SYSTEMS NETWORK ARCHITECTURE (SNA),** is a multiple-layer model similar to the OSI model discussed in section 8.5. To some degree the concepts of the SNA model provided the source work for the development of the OSI model. Protocols and specifications for each SNA layer provide a means for data generated at one location to be successfully transmitted to any other location on the network regardless of the originating and terminating data format. The actual transformations and physical routing to complete the communications are entirely transparent to the users involved. By 1980 over 2,500 SNA networks were in use. Partially because IBM PCs as well as other devices and mainframe computers can be interfaced into the SNA network, the number of SNA networks continues to increase.

● 8.7 SNA LAYERS

Figure 8.5 is a general diagram illustrating the similarities, in a broad sense, between the SNA and OSI models. It is important to state at this point that even though similarities between OSI and SNA functions are drawn, there are many detailed differences between the specifications of each model. Drawing parallels between the two is done mainly as a point of reference and should not be taken literally. On the lowest scale is the physical layer. The designers behind SNA chose to leave the physical specifications to existing standards such as RS232C and to

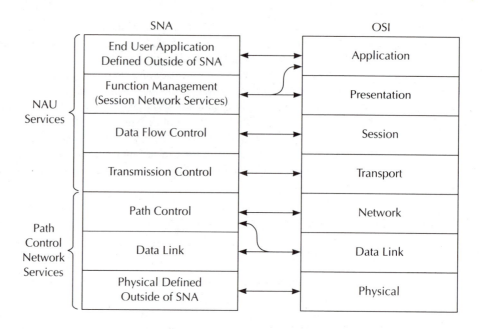

SNA | OSI

NAU Services	OSI
End User Application Defined Outside of SNA	Application
Function Management (Session Network Services)	Presentation
Data Flow Control	Session
Transmission Control	Transport

Path Control Network Services	OSI
Path Control	Network
Data Link	Data Link
Physical Defined Outside of SNA	Physical

Figure 8.5 Comparison of SNA and OSI Layers

concentrate instead on making physical connections to the SNA network compatible to the network. Physical access to the network is provided through specialized nodes, which are discussed later in the chapter. OSI's data link layer finds equivalence in portions of SNA data link and path control layers. SNA utilizes the Synchronous Data Link Control (SDLC) protocol (see chapter 4) to manage DATA LINK CONTROL functions. The **PATH CONTROL LAYER** specifies the virtual circuit that a message takes through the system to reach its destination. The use of virtual circuits means that messages will find the most direct route through the network. No physical route is preestablished, avoiding contention and unnecessary delays in message routing. However, the SNA standard does allow use of permanent virtual circuits, which are dedicated paths through the system for frequently used data links within the network. In a sense they are similar to the leased lines used by the telephone company in that the same path is used by two endpoints each time a data link is established between them. Actual physical paths taken to complete virtual circuits, even permanent virtual circuits, may differ each time the same link is established. Message segmenting (similar to the formation of packets) and sequencing are defined at the path control layer. The setting of transmission priority levels and blocking functions are also found at the path control layer.

Other parts of the path control, as well as the transmission control layer, have equivalent functions in the transport layer of the OSI model. The **TRANSMIS-SION CONTROL LAYER** is responsible for keeping track of the status of the sessions between users, assuring well-paced and sequenced data flow and possible coding and decoding processes. Message headers which define request-message and response-message type data are defined in this layer as well.

The fourth SNA layer, the **DATA FLOW CONTROL LAYER,** deals with responsibilities similar to those of the OSI sessions layer. Identifying the type and mode of data messages and how they are grouped by frame number and address, as well as handling data flow disruption, is included in this layer. Additional functions involve grouping of related messages, called **BRACKETING,** and determining the type of response made to an incoming message. The **FUNCTION MAN-AGEMENT LAYER,** like its OSI counterparts, the applications and presentations layers, supplies the user interface to the network. Details on configuration services, analyzing failures, operator services, interfacing to lower layers, and session services are found in this layer. The conversion of network names supplied by users to network addresses used throughout the system is performed at this level.

Function management services are divided into two sublayers, the **FUNC-TION MANAGEMENT DATA SERVICES,** which coordinate the interface between the user and the network, and the **NETWORK ADDRESSABLE UNIT (NAU) SERVICES MANAGER,** which is responsible for data flow control and transmission control of the lower layers. The services of this layer are further categorized as **END USER** and **SESSION NETWORK SERVICES.** End user services are further subdivided into **SESSION PRESENTATION SERVICES** and **APPLICATIONS-to-APPLICATIONS SERVICES.** The former include establishing a common format between end users and service control between dissimilar devices. Applications-to-applications services include specialized applications functions such as data base accesses and protocol updating.

Session network services encompass three subgroupings, **NETWORK OP-ERATOR SERVICES, CONFIGURATION SERVICES,** and **SESSION SER-VICES.** Network operation services facilitate communication among network operators and **SYSTEM SERVICE CONTROL POINTS (SSCP).** System service control points are programs that manage the network and establish and control interconnections that allow users to communicate within the network. Configuration services are responsible for establishing and disconnecting data links, delivering applications programs to SNA nodes, and maintaining network names, addresses, and status. Lastly, session services perform tasks similar to OSI transport layer services. Capital among these is the translation of logical names into network addresses, which is performed by a system service control point called a **NETWORK DIRECTORY SERVICE.**

A sixth layer called the **APPLICATION LAYER** is not actually specified but is instead left to outside standards and protocols to perform the final user interface into the system. These could amount to the formats and protocols used by the user before that are translated into the formats required by the SNA network. Access to the network is facilitated through the use of a general node point called a **LOGICAL UNIT (LU),** which assures that the incoming data to the network are presented in the correct format.

Logical units are programs that assist in providing translations between the data used by peripheral devices and the message segment formats required for data handling on the network. There are seven types of logical units (LUs) specified by IBM in the SNA model. Some of the more commonly used logical units are:

LU 1—For units that interface directly with SNA character streams. Very little translation is required.

LU 2 and 3—Used to handle IBM 3270 mainframe data streams.

LU 6.2—Used for interfacing units using a general data stream. This data stream does not include headers or trailers, which allows more data to be handled during the same time period as other logical units. The software programs for the LU 6.2 describe the formats and protocols for communication among distributed transaction programs in an SNA network, providing for cooperative processing among distributed dissimilar systems connected to the network.

The physical counterpart to the logical unit is the **PHYSICAL UNIT (PU).** Physical units, in their simplest form, include the hardware to interconnect a single peripheral or terminal to the network. Other physical units form the interface for a host computer or a front-end processor that could be used to control the network. A more complex physical unit manages the interfacing of many peripheral devices to the network through a unit known as a **CLUSTER CONTROLLER.** Logical units and physical units gain access to the network as software and hardware nodes.

There are three basic forms of nodes—**HOST, COMMUNICATIONS CONTROL,** and **PERIPHERAL NODES.** Their relationship to the network is illustrated in Figure 8.6, which is a block diagram of a basic SNA system **SUBAREA.** The host or type 5 node provides control over the other nodes within its subarea through systems service control point software. Often the host node contains a peripheral device directly connected to it to allow direct human interface to the node.

Communications controllers (type 4 nodes) are a subarea node that does contain a system service control point. They maintain control over the peripheral nodes connected to them. Generally, they are interfaced to a host node, providing indirect connections between host and user peripherals.

At the lowest level of node control are the peripheral controllers. These fall into two main categories, the peripheral node (type 1 node), which interfaces to a single peripheral device, and the cluster controller (type 2 node), which manages connections to many peripherals. An additional type 3 node is set aside but not yet defined by the SNA standard. The system may be expanded into several subareas as shown in Figure 8.7. Here the host node has direct control of a cluster controller within its subarea and over communications controllers into two other subareas.

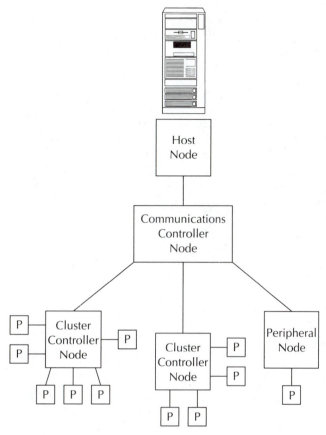

P-Peripheral or Terminal Device

Figure 8.6 Basic SNA Network

Within the subareas additional communications controllers are again interfaced to peripheral control nodes of varying sizes. It is the host node that has direct access to the network path.

An additional unit classification detailed in the SNA standard is the network addressable unit (NAU). NAUs are functional areas that encompass a designated set of logical units, physical units, and system service control points to provide services necessary to move information through the network. Each NAU area has an address that is used to identify itself to other NAUs.

Lastly, a **PATH CONTROL NETWORK (PCN)** is a low-level component that controls the actual physical routing and flow of data through the system. Here explicit routes are assigned to virtual routes as the need arises (Figure 8.8) to assure rapid and accurate data transmission between nodes. Note that there are several possible physical or explicit routes for a given virtual route.

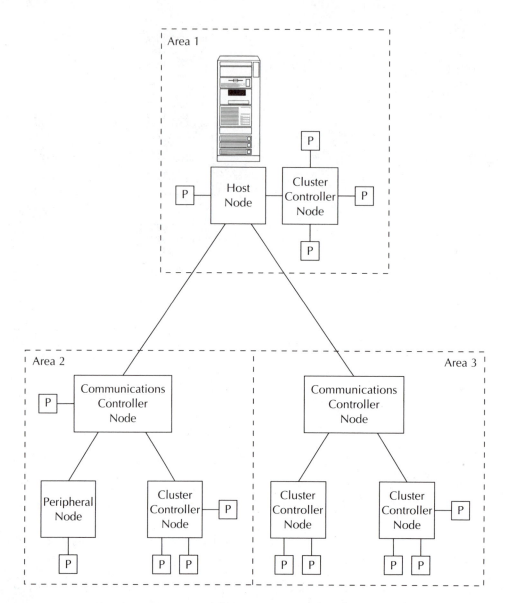

Figure 8.7 Three-Area SNA Network

● 8.9 SNA OPERATING SESSIONS

Each action throughout the system, from the time data entry is made via a
physical unit until it leaves the network through another physical unit, is defined as
an SNA operating **SESSION.** A session is a logical condition connecting two

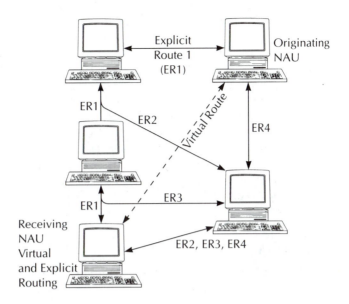

Figure 8.8 Virtual and Explicit Routing

entities to facilitate a succession of data transmissions between them. For instance, logical unit to logical unit sessions provide the actual means for end users to communicate with one another. These sessions are established on a "need to use" basis to minimize unnecessary action on the network.

System service control point to system service control point sessions are established on a multiple domain (subarea) network at the time the network is engaged, and they remain active until the system is shut down. They are used to allow the necessary control service information to be passed between system service control point units, providing cross-information transfers between subarea domains. A domain defines a set of SNA resources that are known to and managed by a system service control point.

Within a given domain, a session between the system service control point and its logical units is required before the logical unit (and hence the user) can access the network. Keeping in mind that system service control points establish the services that manage and control the interfaces to the network, the need for a session between it and the logical units within its domain becomes apparent. There are also sessions between the system service control point and any physical units directly connected to the controlling unit. These physical units provide direct human interface for the controlling processes through peripheral devices connected to the physical unit. There are no specific sessions between physical units. However, communication between adjacent physical units in any subarea may be required to transfer applications programs or to provide initialization or system shutdown commands.

Once the system is active and the necessary sessions are established, data in any format and from any source can be placed on the network through a physical unit directed and interpreted by the logical unit's software. Within the logical unit, system service control point programs manage the translation of the entered data to the format usable by the SNA network. Once the translation is completed, the data leave the logical unit and enter the system. Once on the network, data are dynamically routed to their destination via the most direct route available. This route exhibits the least propagation delay between communicating users. At the receiving logical unit, the data are translated, if necessary, into the form usable by the receiving end user station. The transformed data are then sent to the user terminal through a physical unit. Devices regularly serviced by an SNA network include IBM personal computers and mainframes, workstations, line printers and display terminals, and data communications devices that provide interconnection to other networks.

● 8.10 SNA MESSAGE FORMATS

Data messages on the SNA network are divided into segments similar in form to the packets used for an OSI network. The major difference is that SNA message segments require other detailed routing information beyond the source and destination addresses used in OSI packets. The general format for an SNA segment is shown in Figure 8.9.

The DATA LINK HEADER (DH) consists of a synchronous data link control (SDLC) frame start flag (7E sequence discussed in chapter 4), station address, and control field. The first area of the data field contains the TRANSMISSION HEADER (TH), which is composed of a subarea and element address. The subarea defines a particular node in the network, and the element address points to a specific NAU within that subarea. The REQUEST/RESPONSE HEADER (RH) follows the transmission header and contains codes defining the format and translation

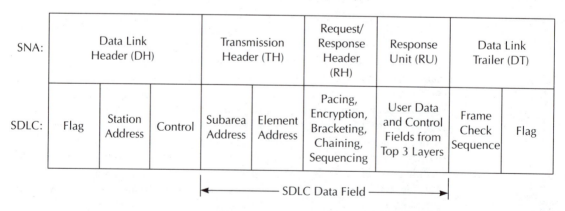

SNA:	Data Link Header (DH)			Transmission Header (TH)		Request/ Response Header (RH)	Response Unit (RU)	Data Link Trailer (DT)	
SDLC:	Flag	Station Address	Control	Subarea Address	Element Address	Pacing, Encryption, Bracketing, Chaining, Sequencing	User Data and Control Fields from Top 3 Layers	Frame Check Sequence	Flag

←————— SDLC Data Field —————→

Figure 8.9 SNA Message Frame

necessary to accommodate the source and destination message formats. These include pacing, bracketing, encryption (or encoding), framing, and sequencing codes. The remaining area in the data field, the RESPONSE UNIT (RU), contains the actual message data as well as control information from any upper network layer. The DATA LINK TRAILER (DT) field holds the last two SDLC fields, the frame check sequence and the ending 7E flag. The system software examines the headers of the message to determine the possible route the message is to take to reach its destination.

SUMMARY

The design, creation, and operation of various networks have been facilitated by adherence to the concepts and standards outlined in open systems architectures. Without some form of standard, the intercommunications required in today's networks between corporations, governments, and consumers would be extremely difficult. Large companies, such as IBM, which have extensive networks in place designed on their network model, must provide the means to interconnect to other networks based on the OSI model. Open systems modeling allows for considerable flexibility in actual equipment and protocol usage. However, by standardizing network access requirements and layer-by-layer specifications, interconnecting at various points within two different OSI networks is easily accomplished.

GLOSSARY

Access right—Security measure which determines a user's ability to access a network based on application needs.

Blocking—Keep a station from receiving messages not addressed to it.

Bridge—Circuit and/or software which allows similar networks to be interconnected.

Circuit switch—A network switching method that physically connects two stations to facilitate communications.

Cluster controller—The hardware and software which manages several stations and/or nodes.

Gateway—Circuit and/or software that allows dissimilar networks to be interconnected.

International Standards Organization (ISO)—Data communications standardizing committee.

Logical unit—Apparent node that services user terminals.

Message switching—Network switching achieved by storing incoming messages and sending them out as lines become available.

NetWare—Software package used in networking to address network and transport layer protocol needs.

Network management—Software used to manage a network. This includes configuration, security, fault recovery, and overall data flow management services for the network.

Node—Network access point.

Open systems interconnection (OSI)—Network model developed by ISO.

Packet—A portion of a message.

Packet switch—A network switching method that interleaves smaller message units, called packets, and sends them together to destination stations.

PAD (Packet Assembler/Disassembler)—Software that performs the conversion between dissimilar network protocols.

PBX (Private Branch Exchange)—Hardware and software that performs on-sight circuit and/or message switching.

Physical unit—Actual hardware that interfaces to user terminals.

Server—Software that facilitates localized interfacing between network functions. These are tailored for files, records, printer and terminal interfaces, and others.

Session—Functional activity between nodes.

SNA (Systems Network Architecture)—IBM's open systems layered architecture.

Virtual circuit—Logical route between two points.

QUESTIONS

8.1. Briefly describe each of the main network topologies: bus, star, and ring.

8.2. Give an advantage and a disadvantage of circuit, message, and packet switching.

8.3. Which switching method allows real-time data transfers?

8.4. Define line contention and collision.

8.5. Which switching method experiences the problem of line contention?

8.6. Why is message switching referred to as "store and forward"?

8.7. What is a packet?

8.8. Explain why packet switching is preferred on local area networks.

8.9. Explain the essential difference between LANs, MANs, and WANs. Which one is used to network a company's varied operations from coast to coast?

8.10. What is a node as it applies to a network?

8.11. For each term below, give the OSI model layer where that term would apply.
   a. RS449   b. BISYNC protocol   c. Modem
   d. SDLC   e. USART   f. Telephone lines

8.12. Which OSI layer provides the human user interface?

8.13. What is the advantage of an open systems environment?

8.14. How does a virtual circuit differ from a physical circuit? What advantage would a virtual circuit provide?

8.15. Give one process or application for each of the upper four layers of the OSI.

8.16. What is blocking, and why is it used?

8.17. Which two SNA layers address functions similar to the transport layer of the OSI model?

8.18. Give two responsibilities for SNA session network services.

8.19. Which OSI equivalent layer is not directly addressed by the SNA model? Applications for this level are left to existing protocols and procedures.

8.20. What are the general names given to nodes in the SNA model that are used as interfaces to user terminals? What function does each type provide to the user interface?

8.21. In general terms, what is the main purpose of logic unit LU 6.2?

8.22. What is the difference between peripheral nodes and cluster controllers?

8.23. Which general data link protocol is used for SNA message formats?

8.24. What are the main functions of network addressable units (NAUs) and path control networks (PCNs)?

8.25. Name the parts of an SNA message and give their function.

PROBLEMS

8.1. How many connections are required to interconnect 1,024 stations directly? Using a circuit switch?

8.2. How many connections are required to directly connect 2,500 nodes? How many are needed to connect those 2,500 nodes using packet switching?

8.3. Using a frequency shift keying (FSK) two-point data link (chapter 5), relate each part of that link to the OSI and SNA models.

RESEARCH ASSIGNMENT

8.1. Many organizations are in the process of writing data communication standards. Only a few, however, such as the International Standard Organization (ISO), which developed the OSI model, have international impact and acceptance. Prepare a report on one of these organizations, including the standards generated and already in use plus standards currently being considered.

8.2. Select one of the OSI layers and research the protocols and standards that apply to that layer. Include the use and authors of those protocols and standards. Discuss how each protocol and standard aids in meeting compliance for that OSI layer.

8.3. Research the current application of IBM's SNA network. Discuss what IBM has done to improve upon that network model.

Answers to Odd-Numbered Questions

8.1: Bus—Single channel onto which all stations are attached; Star—Central switching station with single-channel connections to each individual station; Ring—Each station is connected to a succeeding station in a cascaded fashion. The last station is connected back to the first to complete the ring.

8.3. Circuit switch

8.5. Circuit switch

8.7. A packet is a portion of a message.

8.9. The essential difference between the networks is the size of the area that they service. LANs service a local, usually one-building, area; MANs service a city or regional area; WANs service large regions, countries, or international spheres. A WAN is used for coast-to-coast networking.

8.11. Physical layer: a, c, e, f; data link layer: b, d.

8.13. An open systems environment allows for the flexible use of many different protocols to achieve the same ends—a successful communications linkage.

8.15. Transport—Flow control between lower three layers and the session layer, error recovery, message segmentation

Session—File management, system time allocation, connection and disconnection from the network, bind names to addresses, NETBIOS

Presentation—Data compression, format conversions, encryption, virtual terminal, NETWARE

Application—Human interface, data file access, user requirements, identification, authorization, DECNET

8.17. Network and transport

8.19. Physical

8.21. Interfacing data streams from and to dissimilar systems

8.23. SDLC

8.25. Data link header—Contains starting flag, station address, and control field

Transmission header—Holds the subarea and specific element addresses

Request/response header—Holds information about pacing, encryption, bracketing, chaining, and sequencing.

Response unit—The data and control fields from the top three SNA layers

Data link trailer—Contains the frame check and trailing flag

Problem Solutions

8.1. 523,776; 1,024

8.3. Physical layers—DTE, UART, RS232C, modem, medium (telephone lines)

Data link layer—Character code, asynchronous data link, BISYNC, SDLC, or HDLC protocol

Network through applications layers—Telephone company switching network

Research Assignment Guides

8.1. Standards organizations already mentioned—ISO, IEEE, CCITT. These are standards organizations but by no means the only ones. AT&T (the telephone company) and ANSI (American National Standards Institute) are others.

8.2. There are many data communications companies with lots of hardware and software designed to apply to each of the OSI layers. It is suggested that the instructor minimize the number of students reporting on each layer. Otherwise, you are likely to get a lot of physical layer reports! Other layers are more fascinating.

8.3. Reports that SNA is fading as a platform are premature. Many networks using SNA still exist. However, IBM, to keep pace, has produced an improved network model, called the SSA (System Services Architecture). Reports could show the changes as well as the applications.

Nine • Network Architecture and Protocols

OBJECTIVES

This chapter lays the groundwork for exploring specific network standards and protocols. It begins with a packet switch protocol standard, X.25, includes **INSTITUTE OF ELECTRICAL AND ELECTRONICS ENGINEERS (IEEE)** specifications for different network configurations, and finishes with Novell's NetWare software package. Specific topics include the following:

1. X.25 protocol
2. IEEE 802 standard for networking—including specific applications
3. Ethernet, one of the earliest and most widely used local area networks (LANs)
4. Token bus and ring networks
5. Bridges and gateways to interconnect LANs
6. Private branch exchange (PBX)
7. TCP/IP routing protocol
8. Novell's NetWare

ACRONYMS

CCITT—International Consultative Committee for Telegraphy and Telephony
CRC—Cyclic Redundancy Check
CSMA/CD—Carrier Sense Multiple Access with Collision Detection
DECNET—Digital Equipment Corporation NETwork
DISC—DISConnect
DM—Disconnect Mode

289

FATS—File Allocation TableS

FCC—Federal Communications Commission

FDDI—Fiber Distributed Data Interface

FRMR——FRaMe Reject

GFI—General Format Identifier

HDLC—High-Level Data Link Control

I—numbered Information frame, *or* Integrated interface

IBM—International Business Machines

IEEE—Institute of Electrical and Electronics Engineers

INTERNET—INTERnational NETwork

LAN—Local Area Network

LANCE—Local Area Network Control for Ethernet

LCGN—Logical Channel Group Number

LCI—Logical Channel Identifier

LCN—Logical Channel Number

LLC—Logical Link Control

MAC—Media Access Control

NETBIOS—NETwork Basic Input Output Services interface

ODI—Open Data link Interface

OSI—Open Systems Interconnection

PAD—Packet Assembler/Disassembler

PBX—Private Branch eXchange

PC—Personal Computer

PVC—Permanent Virtual Circuit

REJ—REJect frame

RNR—Receiver Not Ready frame

RR—supervisory Ready to Receive frame

SABME—Set Asynchronous Balanced Mode Extended

SDLC—Synchronous Data Link Control

SIA—Serial Interface Adapter

SNA—System Network Architecture

SSAP—Source Service Access Point

SSCP—Systems Service Control Points

SVC—Switched Virtual Circuit

SYSCON—SYStem CONfiguration

TCP/IP—Transmission Control Protocol/INTERNET Protocol

UA—Unnumbered Acknowledge

UDP—User Datagram Protocol

UI—Unnumbered Information frame

VGM—Voice Grade Medium

XID—eXchange IDentification frame

● 9.1 INTRODUCTION TO X.25 PACKET SWITCH PROTOCOL

The **INTERNATIONAL CONSULTATIVE COMMITTEE FOR TELEG-RAPHY AND TELEPHONY (CCITT)** group's **X.25** protocol is a commonly used network access protocol that complies with the lower three layers of the **OPEN SYSTEMS INTERCONNECTION (OSI)** model. The X.25 protocol is a packet switch **LOCAL AREA NETWORK (LAN)** protocol that incorporates existing standards to specify network requirements. CCITT V.24, similar to IEEE RS232C, addresses the physical layer of the OSI model within the X.25 framework. The data link protocol specified by X.25 is HDLC. The **HIGH-LEVEL DATA LINK CONTROL (HDLC)** protocol frame used is essentially the same as described in chapter 4. Extended addressing is used for the secondary addresses to accommodate the larger number of stations in a network using X.25. Part of the data field is designated as a packet field. This portion is used to identify the packet type so that it may be correctly interpreted. In all other aspects the HDLC protocol remains unchanged in its application by X.25.

X.25 also specifies the manner in which non-X.25 networks can interface to an X.25 network. This is accomplished using a **PACKET ASSEMBLER/DISAS-SEMBLER (PAD).** Essentially, the PAD protocol reads the nonstandard data and translates them into X.25 packets, which are then routed to their destinations on the X.25 network. The process is reversed when an X.25 station sends data to the nonstandard network.

At the network level of the OSI model, X.25 specifies how packets are formed and exchanged using a **VIRTUAL CIRCUIT.** Virtual circuits are those that are specified logically but are not directly connected physically. There are two classes of virtual circuits, **SWITCHED VIRTUAL CIRCUIT (SVC)** and **PERMANENT VIRTUAL CIRCUIT (PVC).** The differences are as obvious as their names. A permanent circuit is one that is always connected between stations, while a switched circuit uses connections that are switchable between stations. It is necessary to remember that virtual circuits are logical and not physical entities. The actual physical connections (even for permanent virtual circuits) are most likely to be different each time a link is established. The network level packet (Figure 9.1) is formatted in the same manner as a basic HDLC packet. The beginning of the information field contains a packet header, which is used to designate the logical channel and packet identifications. It begins with a **GENERAL FORMAT**

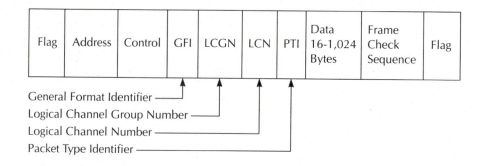

Flag	Address	Control	GFI	LCGN	LCN	PTI	Data 16-1,024 Bytes	Frame Check Sequence	Flag

General Format Identifier
Logical Channel Group Number
Logical Channel Number
Packet Type Identifier

Figure 9.1 X.25 Packet Frame

IDENTIFIER (GFI). Among other information, this field sets the size of the packet frame numbers (NS—the number of the current frame being sent; NR—the number of the next expected frame to be transmitted by the current receiving station).

The general format identifier is followed by the **LOGICAL CHANNEL IDENTIFIER (LCI),** which consists of a **LOGICAL CHANNEL GROUP NUMBER (LCGN)** and **LOGICAL CHANNEL NUMBER (LCN).** The logical channel identifier is capable of identifying a specific channel from among 256 groups of sixteen channels each.

The control fields of the network packet (Figure 9.2) identify the type of packet in use. The types are data packet (same as HDLC information control field), flow control packet (HDLC supervisory packet), and supervisory packet (conceptually the same as the HDLC unnumbered control field). For the data packet, a MORE (M) bit replaces the function of the poll/final (P/F) bit used with HDLC. A 1 logic state of the M bit indicates that there are more data packets to follow.

The flow control frames differ from HDLC supervisory frames by the lack of the P/F bit. Three of four supervisory frames used in HDLC are used in X.25 network flow control packets. They are ready to receive, not ready to receive, and reject, with the one HDLC supervisory frame not used being frame reject.

Supervisory frames for the X.25 network layer use the unnumbered format but are totally different from the unnumbered frames used by HDLC. The X.25 frames include CALL REQUEST, CALL ACCEPTED, CALL CONNECTED, INCOMING CALL, CLEAR INDICATION, CLEAR CONFIRMATION, INTERRUPT, INTERRUPT CONFIRMATION, RESET REQUEST, RESET CONFIRMATION, RESTART REQUEST, and RESTART CONFIRMATION. These satisfy the needs of the X.25 protocol in a manner in which the HDLC unnumbered frames could not. This layer is used to form control and data packets and to establish the connections on a virtual circuit. Also, keep in mind that most of these local area networks use peer stations throughout. That is, there is no controlling primary station in the network.

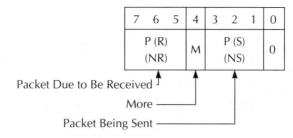

Packet Due to Be Received
More
Packet Being Sent

a) Data Packet

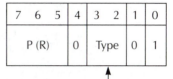

0   0   Ready to Receive

0   1   Not Ready to Receive

1   0   Reject

b) Flow Control

7 6 5 4 3 2	1	0
— Type —	1	1

c) Supervisory

Figure 9.2 X.25 Control Fields

● 9.2 IEEE 802 NETWORK STANDARDS

Another set of standards that follow the OSI model were written by the IEEE network group. This set of standards is designated 802.x and includes:

802.1 — Architecture and internetworking — defines relationship between other 802 specifications

802.2—**LOGICAL LINK CONTROL (LLC)**

802.3—CSMA/CD bus access (the Ethernet network specification closely conforms to the 802.3 specification)

802.4—Token-passing bus access

802.5—Token-passing ring access

802.6—Metropolitan area network (MAN)

802.8—**FIBER DISTRIBUTED DATA INTERFACE (FDDI)**

● **9.2.1 IEEE 802.2: Logical Link Control**

The base protocol used for the 802.2 specification is, once again, a modified form of HDLC. The modifications are based upon the type and modes of operation specified in the standard. Operations under 802.2 are one of two types. Type 1 operation refers to exchanges that do not require a previously established data link connection. Type 1 operations lack flow control and error recovery processes by omitting the use of numbered frames. Instead, type 1 communication relies heavily on using unnumbered frame formats. In contrast, type 2 operations require that a data link be established. The main forms of data sent and received use information and supervisor frames. Unnumbered frames are reserved for establishing and disconnecting data links and frame reject analysis.

The use of frame numbers based on the HDLC protocol facilitates flow control. Frame numbering also aids in error recovery (as described under the **SYNCHRONOUS DATA LINK CONTROL (SDLC)** and HDLC protocols in chapter 4) by specifying which frame is initially detected as possessing an error.

Further application of the HDLC protocol to 802.2 is determined by the class of operation being performed. Class I networks employ type 1 operation only. In this class only unnumbered frames are utilized. They include **UNNUMBERED INFORMATION (UI), EXCHANGE IDENTIFICATION (XID),** and **TEST.** UI frames for X.25 are similar to SDLC/HDLC UI frames discussed in chapter 4. The maximum size of the information field is set by network requirements.

A receiving station is required to respond to an XID frame from another station. The purpose in using this command is to verify that a station in the network is on-line and capable of exchanging traffic. The TEST command informs the receiving stations that a sender is going to perform a loop-back test to check the system. The data the sender transmits following the TEST command are returned or looped back to the sender and not used by the receiving station.

Class II networks allow both type 1 and type 2 operations. Type 1 frames are the same as detailed in class I. Type 2 frames include **INFORMATION (I), SUPERVISORY READY TO RECEIVE (RR), RECEIVER NOT READY (RNR),** and **REJECT (REJ),** as well as the following unnumbered frames: **SET ASYNCHRONOUS BALANCED MODE EXTENDED (SABME), DISCON-**

NECT (DISC), XID, TEST, UNNUMBERED ACKNOWLEDGE (UA), DISCONNECT MODE (DM), and **FRAME REJECT (FRMR).**

SABME is similar in use to the set normal response mode (SNRM) of HDLC. Its purpose is to cause a disconnected station to be placed on-line. Whenever used, this command causes the NS (frame number sent) and NR (frame number expected) counters to be reset to zero. The extended reference of the SABME refers to the extended frame numbers (0–127) used by HDLC (compared with 0–7 of SDLC). The remaining frame types besides XID and TEST are the same as the ones described for the SDLC/HDLC protocol in chapter 4.

● 9.2.2 IEEE 802.2 Frame Format

Figure 9.3 shows the basic format of an 802.2 frame. The least significant bit of an 802 field is designated b1 instead of the more familiar b0. The first field, the **DESTINATION SERVICE ACCESS POINT (DSAP)** (destination address) is a single byte. Bit b1 distinguishes between individual or group addresses. Bit b1 low designates the address as an individual or unique address. Group addresses require bit b1 to be high. The hex broadcast address of FF is a special case of the group address. All lows in the address field is a null address and is used only for system-testing purposes.

SOURCE SERVICE ACCESS POINT (SSAP), or source address, is always unique. As such, bit b1 is used to convey different information. A low in bit b1

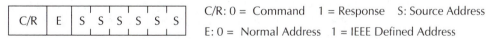

| DSAP 8 | SSAP 8 | Control Field 8: Unnumbered 16: Numbered | Information Field 8m | CRC–32 |

DSAP–Destination Service Access Point

| I/G | E | D | D | D | D | D |

I/G: 0 = Individual   1 = Group   D: Destination Address
E: 0 = Normal Address   1 = IEEE Defined Address

SSAP–Source Service Access Point

| C/R | E | S | S | S | S | S |

C/R: 0 = Command   1 = Response   S: Source Address
E: 0 = Normal Address   1 = IEEE Defined Address

Information Field: m is an Integer

CRC–32: 32 Bit Cyclic Redundancy Check

Figure 9.3 IEEE 802.2 Frame Format

identifies this frame as a command frame; a high indicates a response frame. Command frames are unnumbered commands or originator information frames. A logic low in bit b2 of the SSAP sets aside half of the source addresses, which are defined as a specific 802 application. Examples of these types of addresses are hex 40 and C0, which are individual and group addresses for LLC sublayer management functions. A high logic level in bit b2 signifies that the message is of general traffic application. The remaining bits in the SSAP field are the actual station addresses.

The control field size depends on the type of frame being sent. It is either 8 bits for unnumbered frames or 16 bits for numbered information or supervisory frames. Bit b4 of any frame is the poll/final (P/F) flag as used in the SDLC/HDLC protocol. Unnumbered frames have the following hex codes (one for poll/final frames, one for non-poll/final frames):

1. SABME 6F, 7F
2. DISC 43, 53
3. UA 63, 73
4. DM 0F, 1F
5. FRMR 87, 97
6. UI 03, 13
7. XID AF, BF
8. TEST E3, F3

Control fields for the numbered information and supervisor fields are shown in Figure 9.4. They conform to the HDLC extended form, with NS and NR values limited to 7 bits (0–127) each. Unused bits in the supervisory control field are set low.

Figure 9.5 is the XID information frame format. As with the information frame, there is a preamble, destination, source, and control (or type) field preceding the information field of the XID frame. The first 8 bits of this field describe the format of the XID information field. The next 5 bits identify the logical link control (LLC) type of operation. 10000 binary is class I, and 11000 binary is class II. These bits are followed by four zero bit states. The remaining bits denote the size of the receive window; this is the maximum value the NS number can exceed the NR value. In essence, the window sets the maximum number of numbered frames that can be sent before an acknowledgment is required. The XID control field can have a hexadecimal value of AF or BF, depending on the condition of the P/F flag.

The frame reject (FRMR) field (Figure 9.6) is similar to the frame reject (FRMR) field of SDLC. The first 2 bytes are a copy of the control field of the rejected frame. Bit b17 (recall that 802 specifies the least significant bit as bit b1, not bit b0) is always a low state. The next 14 bits hold the NS and NR number of the rejected frame. W, X, Y, and Z bits designate the reason for the frame reject, which are similar to the SDLC protocol:

1. W—Invalid control field (invalid command)

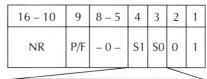

16 – 10	9	8 – 2	1	bit
NR	P/F	NS	0	Use

a) Information Control Field

16 – 10	9	8 – 5	4	3	2	1
NR	P/F	– 0 –	S1	S0	0	1

S1 S0	Function
0 0	RR – Ready to Receive
0 1	RNR – Receiver Not Ready
1 0	Rej – Reject

b) Supervisory Control Field

Figure 9.4 IEEE 802.2 Control Fields

X:– XID Format Identifier (81: IEEE Basic Format)
Y:–LLC Type (10000 Class I 11000 Class II)
W:–Receive Window (Maximum NS Can Exceed NR)

Figure 9.5 XID Information Field (802.2)

2. X—Invalid data field (data sent with supervisory frame or unnumbered frames that disallow data)
3. Y—Information field too large (exceeded specified number of bits)
4. Z—Invalid NR

Added to the frame reject codes is bit V for invalid NS number. This occurs when a receiver detects an incorrect sequence of NS numbers. For instance, the reception of a frame with an NS number of 07 following a frame with an NS number of 10 indicates something is amiss. This condition would cause the receiver to send

1 – 16	17	18 – 24	25	26 – 32	33	34	35	36	37	38 – 40
Control Field of Rejected Frame	0	NS	C/R	NR	W	X	Y	Z	V	0

C/R : 0 = C = Command   I = R = Response
    W: Invalid Control Field
    X: Invalid Data Field
    Y: Information Field too Large
    Z: Invalid NR
    V: Invalid NS

Figure 9.6 IEEE 802.2 Frame Reject (FRMR) Information Field

an FRMR frame with the V bit set. The last 3 bits in the FRMR data field are held low.

● 9.3 STARLAN APPLICATION OF IEEE 802.3

STARLAN is a network developed by AT&T to meet the IEEE 802.3 specification. It uses a star **TOPOLOGY** and **CARRIER SENSE MULTIPLE ACCESS WITH COLLISION DETECTION (CSMA/CD)** access protocol, which is discussed in section 9.4. The center of a STARLAN network is a *HEADER HUB*, which controls the entire network. Connected to this hub on two pairs of twisted wire are **NODE** stations or other *INTERMEDIATE HUBS*. These secondary hubs, which manage a subnetwork, appear as a node to the header hub. Further intermediated hubs can be located in the subnetwork of the intermediate hubs. STARLAN specifies a maximum of five levels of hub activity but does not limit the number of nodes any hub can handle.

A four-wire full-duplex system is used with transmissions applied to one pair and messages received on the other pair. These twisted pairs are limited in length between a node and a hub or between hubs by the STARLAN specifications to 250 meters maximum using 24-gauge wire. Use of different-sized wire affords longer distances that are allowable on a STARLAN system. The maximum distance between nodes attached to a single hub is given as 500 meters (note 250 meters between each node and the bus results in a total of 500 meters between these nodes).

STARLAN uses synchronous data on a baseband, single-channel system. Since synchronous data are used, clocking is recovered from the data stream. To facilitate clock recovery, a form of **MANCHESTER ENCODING** is used as a format for the binary data streams. This type of encoding is illustrated in Figure 9.7.

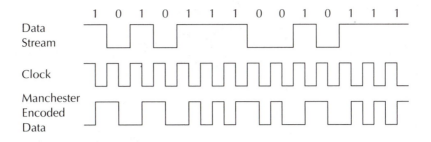

Figure 9.7 Manchester Encoding

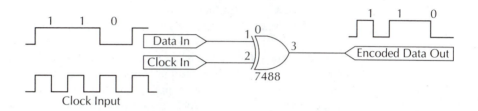

Figure 9.8 Manchester Encoder

The first half of a data bit time is set to the inverse state of the actual data information. This assures that a change of logic state, which aids in clock recovery (see chapter 5), exists at the middle of every data bit sent.

Figure 9.8 shows one method used to generate a Manchester-encoded data stream from a standard binary data stream. At the receiver the numerous changes in state aid in the clock recovery. Once the clock is recovered, the data are decoded from the Manchester form back into original binary information (Figure 9.9).

Access to the network is accomplished through a form of CSMA/CD. In this application the hub monitors the lines for any activity. If the hub senses a carrier from a station, it allows that station access to the network, connecting it to its destination node. Sensing activity from more than one source, the hub declares the existence of a collision condition. It does this by generating a collision-presence signal throughout the network and maintains that signal until all nodes cease transmitting. If, for any reason, a station keeps transmitting after 100 ms from the start of the collision-presence signal, the hub exercises a **JABBER** condition to disable the node.

Stations at the nodes must now wait a random amount of time based on a pseudorandom count sequence known as the truncated binary exponential back-off algorithm, which is discussed in the next section. The first station to time out gains

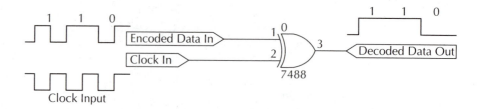

Figure 9.9 Manchester Decoder

access to the network. The remaining nodes sense network activity and wait until the hub is clear to accept their transmissions.

Hewlett Packard produces a STARLAN 10 HUB chip called the HP28663A to meet the STARLAN protocol. It operates at 10 Mbps and handles twelve twisted-pair ports (six full-duplex nodes). It is designed to allow 100-meter connections between the hub and the nodes connected to it. This device is set to handle three levels of cascaded hub connections to allow system expansion. Ethernet networks interface easily to a STARLAN 10 HUB, which makes the device 100% IEEE 802.3 compatible.

A function called autosegmentation is built into the STARLAN 10 HUB to maintain network integrity. Using autosegmentation, a failed node or hub is detected on the network. This faulty segment of the network is disconnected, and an indication is sent so that repairs may be performed on the failed unit. The remaining sections of the network operate normally while the repairs are performed off-line. Once the faulty unit is fixed and sensed by the header hub, it is automatically reconnected to the network.

● 9.4 IEEE 802.3 ETHERNET LOCAL AREA NETWORK

Ethernet is a specification for a local area network. It was developed as a joint effort by Xerox Corporation, Intel Corporation, and Digital Equipment Corporation in 1980. The idea behind its development was to derive a system that would allow numerous computer stations to share resources with each other and a mainframe computer. Information and applications programs are retained at a central location and would be available for use by anyone on the network. The idea is to provide this system to a number of users within a limited physical locality such as an office building or processing plant. The standard details specifications for the network in reference to the first three layers of the OSI model.

● 9.4.1 The Physical Layer Specified by Ethernet

Originally, the Ethernet system was designated to exist within a **BUS** topology using 50-ohm coaxial cabling between stations. The cable would run the needed distances within and between a building's rooms and floors. Today, many Ethernet systems incorporate fiber-optic cable to link stations together, which extends the distances usable by this standard. This section will concentrate on the original coaxial cable application of Ethernet. Fiber optics is discussed in chapter 11 along with an Ethernet application using fiber cables.

Stations at specific locations within the Ethernet network tap into the coaxial cable segments. Specifically, a hundred-station maximum is allowed to tap into an Ethernet segment of 500 meters maximum length. These stations are to be no more than 2.5 meters apart within a second to reduce the possibility of standing wave signals on the cable. Local repeaters are used to interconnect 500-meter segments. The maximum distance, from end to end, between any two communicating stations on the network is specified at 1,500 meters using local repeaters. An additional untapped 1,000 meters can be used between segments with remote repeaters, extending the maximum distance between stations to 2,500 meters. Figure 9.10 shows how these limits are applied. L is used to designate local repeaters, R for remote, and T for a tap.

Example 9.1

What is the maximum number of stations that can be serviced by a basic Ethernet system using local and remote repeaters to maximize the system?

Solution: Three segments can be interconnected, along with an additional 1,000-meter segment as shown in Figure 9.10. Each tappable segment allows 100 stations to be tapped into it. This places a maximum of 300 stations on this basic system.

Ethernet uses a **BASEBAND** exchange at a 10-Mbps data rate. Baseband designates a single-channel system as opposed to a multichannel **BROADBAND** system. All Ethernet stations have equal access to the common single channel. This access is achieved using a system called carrier sense multiple access with collision detection (CSMA/CD). Stations desiring access monitor the line for the presence of a carrier from another station for a minimum period of 9.6 microseconds (μs). The time period assures that no signal is present for a time equivalent to a round-trip propagation between stations 2,500 meters apart. If a carrier is sensed, the station desiring the line must continue to monitor and wait for the line to be free. Once the

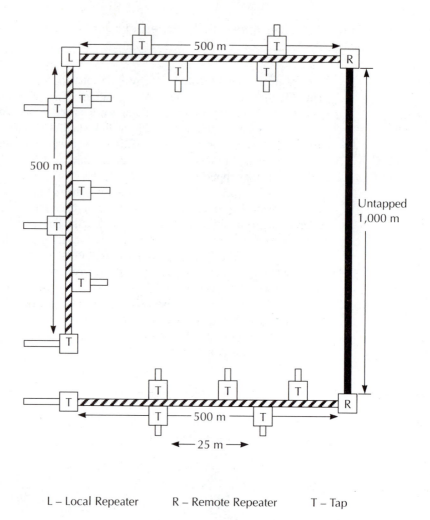

L – Local Repeater   R – Remote Repeater   T – Tap

Figure 9.10 Ethernet Topology Using Coaxial Cable

line is detected as being free of a carrier for the 9.6-μs period, a station can then begin to access the line. This station continues to monitor the line for a period equal to the round-trip propagation time after it accesses the line to detect the possibility of a collision with another station's transmission.

There is the possibility that more than one station has been awaiting use of the system. In that case, two signals may be placed on the line at the same time once these stations sensed the line was clear. The two stations have equal right to the line once it is free. Eventually, these two signals will meet and collide, causing both data streams to become garbled. The sending stations have been monitoring the line for just this possibility. Once the collision is detected, both stations back off and send a jamming signal, composed of 8 bytes of alternating 1s and 0s, onto the line. The jamming signal indicates to all stations that a collision has occurred and that these two stations are going to attempt to regain access to the system. Both stations wait a random amount of time before accessing the line once again. The randomness of this delay very nearly assures that another collision will not occur. One station will access the line well before the other. The amount of the time delay is determined by an algorithm called **TRUNCATED BINARY EXPONENTIAL K BACK-OFF** and is equal to a value from 0 to 2^K time slots. K is the current number of attempts by a station to gain access to the line. This value has a maximum of ten even though a station is allowed sixteen tries before it is disconnected from the network. A time slot is the time it takes for a signal to propagate to the end of the network and back. For Ethernet this period is 51.2 μs. After timing out a station must again monitor the line for 9.6 μs to determine that the line is free of the other station it collided with. The timed-out station can then attempt to seize the line and transmit its data. Since the time delay is random, though, there is still a possibility that both times could finish close enough to cause a second collision. In that situation the process is repeated in the hopes that the randomness of the delay time would not allow a third collision. In some applications the delay can be less random to allow for a priority to be established. In that event, after the collision is detected and the colliding stations have backed off, the station with the higher priority will time out long before the other station. The higher-priority station then is guaranteed access to the network.

● 9.4.2 Ethernet and the Data Link Layer

The OSI-applicable layers for CSMA/CD include both physical and data link layers. CSMA/CD applies to both layers or as a connection between them. For the data link layer, Ethernet uses a form of the HDLC protocol presented in chapter 4. Like the HDLC format, the Ethernet frame (Figure 9.11) has six fields; they are preamble, destination address, source address, type field, data field, and frame check sequence.

There are 8 bytes (64 bits) of preamble used to establish clock recovery and bit synchronization. This fixed preamble contains the following hexadecimal sequence: AAAAAAAAAAAAAAAB. The A hexadecimal character is nothing more than alternating 1s and 0s, providing maximum data transitions to aid in clock recovery. The preamble ends with two consecutive 1s, hence the last B character. The preamble is followed by two address fields. Recall that when using SDLC/HDLC only a secondary's address was specified in the address field. This was because a single primary station controlled message flow and data linking procedures. In Ethernet networks (as with most LANs), there are no primary stations. Any station

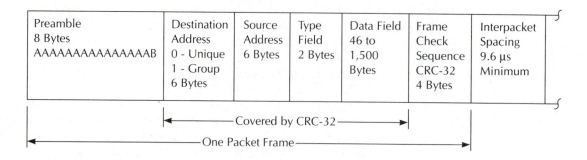

Preamble 8 Bytes AAAAAAAAAAAAAAAB	Destination Address 0 - Unique 1 - Group 6 Bytes	Source Address 6 Bytes	Type Field 2 Bytes	Data Field 46 to 1,500 Bytes	Frame Check Sequence CRC-32 4 Bytes	Interpacket Spacing 9.6 μs Minimum

|←————— Covered by CRC-32 —————→|

|←———————————— One Packet Frame ————————————→|

Figure 9.11 Ethernet Frame

can access the line to talk to any other station using the CSMA/CD method described in section 9.4.1. This requires the sending station as well as the receiving station to be identified. Ethernet provides this in the frame format by including 6-byte destination and source address fields following the preamble.

Destination addresses fall into one of the three types as specified in the Ethernet protocol. They are individual (or unique) addresses, identified by a logic 0 in the first bit position of the address field; group addresses, designated by a logic 1 in the first bit; and broadcast addresses, recognized as a special group address, in hexadecimal, of FFFFFF. The uses of these addresses are the same as in the SDLC/HDLC data link protocol discussed in chapter 4.

The 2-byte type field is an extension of the concepts of the SDLC/HDLC control field. In addition to the functions of identifying frame types (information, supervisor, or unnumbered) and frame numbers, the type field also contains information identifying which one of 247 possible data formats, selected for a specific application, is being used in the data field. The data field itself can contain anywhere from 46 to 1,500 bytes of information, again depending on the application in use. The data field is followed by a 4-byte **CYCLIC REDUNDANCY CONTROL (CRC)** error character in the frame check sequence (see chapter 3 for error-detection methods). The constant used for this CRC-32, in hexadecimal, is 104C11DB6. This frame check sequence checks all the data beginning with the destination address and ending after the last byte in the data field. A minimum 9.6-μs spacing is set aside as a gap between frames to allow the receiving station time to compute and verify the frame check sequence.

● 9.5 ETHERNET INTERFACE DEVICE—SERIAL INTERFACE ADAPTER

Advanced Micro Devices, located in northern California, has developed and marketed a chip set that manages and interfaces devices to an Ethernet network. One unit of that chip set is the **SERIAL INTERFACE ADAPTER (SIA),** which

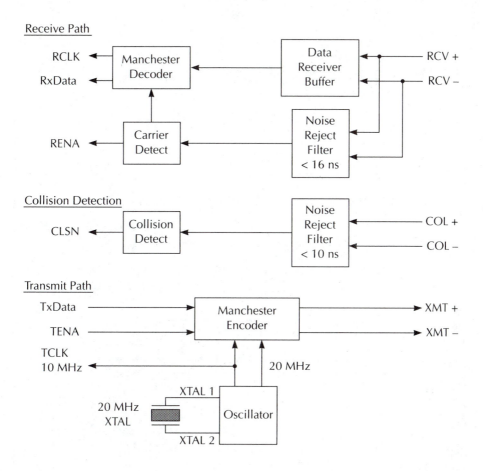

Figure 9.12 SIA Functional Blocks

functions as a Manchester encoder/decoder and collision-detection mechanism. The encoder operates with a crystal-driven 20-MHz oscillator (Figure 9.12). The logic within the oscillator block also divides the 20 MHz by 2 to produce a 10-MHz clock (TCLK). The two clocks create transitions within each data bit to satisfy Manchester coding format. The Manchester encoder combines the transmit data (TX data) with the clocks to produce the Manchester-encoded signal. The encoding process continues as long as transmit enable remains high (usually for the duration of the data stream). Figure 9.13b shows the transmit timing diagram for the serial interface adapter. Data are sent out differentially on transmit (xmt) + and − lines. The transmit select signal determines whether transmit mode 1 or its inverse, transmit mode 2, is being used. In mode 1 the idle time voltage between XMT +/− is 0V, while in mode 2 it is a high (logic 1) level.

The receive side of the serial interface adapter includes carrier detect and Manchester decoder circuits. Incoming data on the differential receiver (RCV)

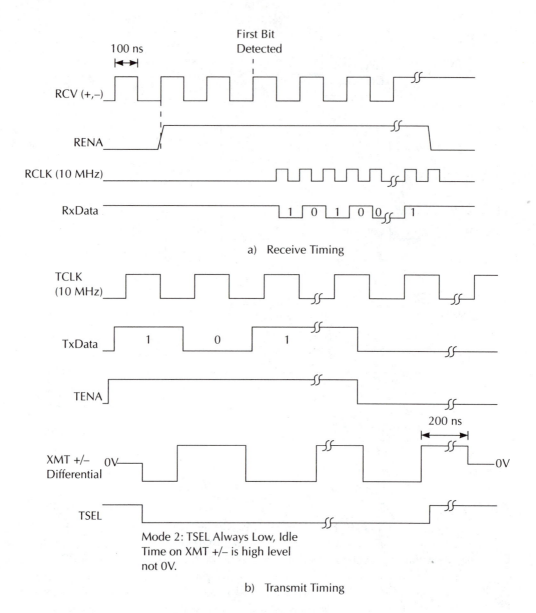

a) Receive Timing

b) Transmit Timing

Mode 2: TSEL Always Low, Idle
Time on XMT +/– is high level
not 0V.

Figure 9.13 SIA Receive and Transmit Timing

+ and − inputs are fed to both circuits. The front of the decoder circuits uses a phase lock loop to capture clock (RCLK) from the preamble within 6 bit times. Sensitivity of the front end is such that a differential signal greater than −300 mV is adequate for detection. As long as the signal does not drop below −175 mV once detection and clock recovery have begun, the receiver will detect the incoming data stream and output the decoded data on RX data. A carrier detect circuit generates a receiver enable (RENA) to indicate that the receiver is detecting the data. Transitions less than 16 ns are rejected by the carrier detect circuit to prevent false starts due to line noise. The relationship between the signals is illustrated on the receive timing diagram of Figure 9.13a.

A third functional block within the serial interface adapter receives a collision indication on COL + and − and generates a collision detection (CLSN) signal to the processor in the system. Again, signals on the differential input must be greater than −300 mV to start and not fall below −175 mV. Transient signals less than 10 ns are rejected by a noise reject filter preceding the collision-detection circuit (Figure 9.12). CLSN remains active as long as the collision jamming signal remains on COL +/−.

Another unit in the Advanced Micro Devices chip set is the actual processor, called a **LOCAL AREA NETWORK CONTROLLER FOR ETHERNET (LANCE),** which creates the Ethernet packet form, computes CRC-32 frame check sequence, and responds to the enable and collision-detection signals from the serial interface adapter. The processor also strips the data from the received string and checks the CRC-32 of the received data stream for errors.

The Ethernet protocol details specifications for the lower two levels of the OSI model. To move up through other layers requires affiliations with other standards and specifications.

● 9.6 IEEE 802.4 TOKEN BUS

A alternate method of access to a network is to employ **TOKEN** passing to control which station has the right to bus access. IEEE 802.4 details the requirements for a token bus network. IEEE 802.5 extends the specification to a **RING** topology. To understand how a token-passing system operates requires familiarity with the concept of a token.

A token is a short message that specifies the station currently using the network and the next station, which gains access to the network after the current station is finished using it. While a station possesses the token it can send messages out on the network and read the messages already present. Tokens may be held for a maximum specified time before they must be passed on to the next succeeding station. The procedure as specified in 802.4 follows this process:

1. Each station knows the *predecessor* and *successor* station address in the token sequence.

2. The token is passed from station to station in descending order.

3. The station currently possessing the token adds its traffic to the data stream. It then modifies the token to include the successor station's address, and the token is passed to it.

4. After sending the token frame, the sending station monitors the line to detect whether the successor station has received the token. This is verified by the successor station sending a valid frame following the token. This frame can be an information frame or a modified token frame. The latter case occurs if the successor station has no messages to add to the existing information frames. In the case the successor station merely modifies the token and passes it along to its successor station.

5. The token is passed to each station in the sequence until it returns to the original station. Once the original sender detects its message returned, it removes that message frame, adds any new ones, reads any messages destined for it, again modifies the token, and passes it along as it did before. This process is repeated as long as the system remains active.

There, of course, is the possibility that the sending station that passed the token does not detect a valid frame following the token frame it sent. In that situation the sending station attempts to retransmit the token to the successor station in the hope that the problem is transient. If the successor station again fails to send out a valid frame, the sending station sends a ''who follows'' frame containing the successor's address. The next station in the sequence following the intended successor station responds with a ''set successor'' frame, which contains its address in the data field. The sending station now modifies the token with this address and passes the token to the new station, bypassing the original successor station.

The ''who follows'' frame is also tried a second time if there is no response to it. Failure to receive a response the second time causes the sending station to try a new tactic by sending a ''solicit successor'' frame. This frame is also sent at regular intervals by each station on the network. It is used to allow a new station to enter the token sequence. Any operational station responding to the solicit successor frame will receive the amended token. It is possible for numerous stations to respond to the solicit successor frame, in which case a resolve contention sequence is initiated. All responding stations wait for given time intervals and respond a second time. This is repeated until there is only one station responding to the solicit successor frame.

Figure 9.14 details the token bus frame formats for **MEDIA ACCESS CONTROL (MAC)** frames, which are specified in IEEE 802.4 and 802.5 to indicate how tokens are formed to allow access to the medium (bus or ring) and how data are included following a token. There are two formats, one for the token and one for the information frame. As with most local area network protocols, these frames are bit-oriented, but these do not follow any established protocol such as HDLC.

The token consists of an access field surrounded by beginning and ending delimiters. Within the access field the first 3 bits establish the priority level of the token. There are eight priority levels (000–111). Stations with equal or higher priority can accept and pass the token, while lower-priority stations must wait until a lower-priority token is passed to them. M, which stands for monitor bit, assures that a frame doesn't continuously circulate in the system. A station receiving its own message with the monitor bit set acknowledges that the frame has been through the

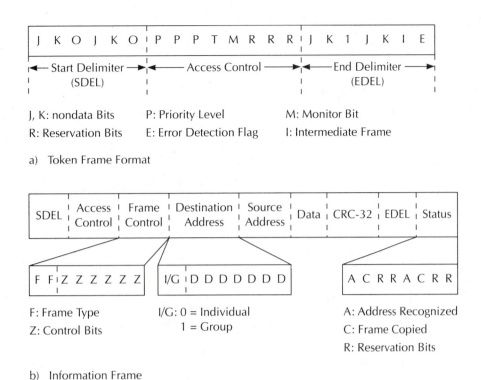

J K O J K O | P P P T M R R R | J K 1 J K I E

← Start Delimiter → | ← Access Control → | ← End Delimiter →
(SDEL) | | (EDEL)

J, K: nondata Bits   P: Priority Level   M: Monitor Bit

R: Reservation Bits   E: Error Detection Flag   I: Intermediate Frame

a) Token Frame Format

SDEL | Access Control | Frame Control | Destination Address | Source Address | Data | CRC-32 | EDEL | Status

F F Z Z Z Z Z Z   I/G D D D D D D D   A C R R A C R R

F: Frame Type   I/G: 0 = Individual   A: Address Recognized

Z: Control Bits   1 = Group   C: Frame Copied

                                                        R: Reservation Bits

b) Information Frame

Figure 9.14 Token Bus Frame Formats

system, and it removes the message from the data stream and resets the monitor bit in the token. Reservation (R) bits are used for requesting token access at the present priority level. The token frame is followed by the information frame, which contains detailed data about the information being sent. It begins with a preamble used to assure clock recovery and bit synchronization. A frame check field follows, containing data about the frame type (Ff bits) and control information (Z bits). The frame types are listed in Table 9.1.

Destination address, which can be individual (first bit low) or group (first bit high), and source address fields are next in the frame sequence. Actual information is next, followed by the frame check sequence, which uses CRC-32 error detection. The last field in the medium access control (MAC) frame is the frame status field. Bit A in this field, called the address recognized bit, is set when the receiving station acknowledges that this frame was meant for it. The frame copied (C) bit indicates that a receiving station made a copy of the frame. These 2 bits tell the originating station that its message was received and copied. The originating station can then remove the frame from the data stream. The R bits are reservation bits used in similar manner to the reservation bits in the token frame.

Discussion of the delimiter fields has been delayed because of the need to mention the encoding format used by the IEEE 802.4 token bus. It is a modified

Table 9.1 Frame Identifier Bits

F	F	Frame Type
0	0	MAC Frame
0	1	LLC Frame
1	0	Not Used
1	1	Not Used

MAC—Media access control
LLC—Logical link control

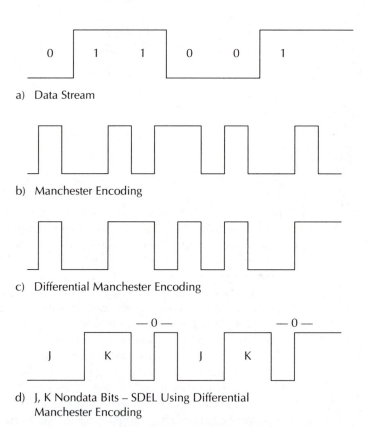

a) Data Stream

b) Manchester Encoding

c) Differential Manchester Encoding

d) J, K Nondata Bits – SDEL Using Differential
   Manchester Encoding

Figure 9.15 Manchester Encoding Forms Used for IEEE 802.x

Manchester encoding scheme called **DIFFERENTIAL MANCHESTER EN-CODING.** Recall from the section 9.4 that Manchester encoding is utilized to assure a transition during each bit period to facilitate clock recovery. This is achieved by inverting the first half of each data bit (Figure 9.15). Differential Manchester encoding still performs that task. The problem with the original Manchester encoding is that the actual data level is detected at the middle of each bit. Differential shifts the detection to the beginning of the bit period. Zero logic levels are detected by the occurrence of a transition of logic state at the beginning of a bit period. A logic 1 is detected by the lack of a change at the beginning of a bit time. The level of the bit is then inverted at the midpoint to assure that a transition occurs for that data bit.

Example 9.2   Compare the encoded data stream for Manchester and differential Manchester encoding for the bit stream: 0 1 1 0 0 1

Solution: This is the data stream illustrated in Figure 9.15. Note that for both Manchester encoding forms, each bit is inverted from the logic level of the first half of the bit period. The differences lie in the data level detection. For regular Manchester the actual data level appears in the second half of the bit periods. For differential you have to observe the beginning of each bit period to see if there is a level change from the previous bit. A level change indicates that the current bit is a 0. No change indicates a logic 1 as shown.

● 9.7 IEEE 802.5 TOKEN RING

Applying the concept of token passing to a ring topology yields a different application than is used with a bus topology. In a ring, tokens and messages are passed around the ring to each station until they reach the originating station. Once back at the originating station, messages that it sent are then removed from the line if they had been received by the destination station. This process is first initiated by a station grabbing a free token from the ring. A **FREE TOKEN** is a short frame that contains a limited amount of information (Figure 9.16a), including starting and ending delimiters, optional priority, monitor, and reservation data. The priority level establishes which stations on the ring may have access to the free token. The monitor bit is used by a node point whose singular concern is to monitor the line for faults and errors. The reservation bits are used to allow higher-priority stations to reserve token use while a lower-priority station has access to the token. The use of reservation bits will become more apparent after the free token has been captured.

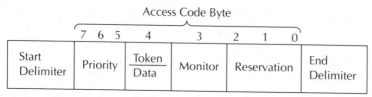

Access Code Byte

| Start Delimiter | Priority | Token Data | Monitor | Reservation | End Delimiter |

a) Free Token (Token Frame)

| Start Delimiter | Access Code | Control Field | Destination Address | Source Address | Data | Frame Check: Sequence | CRC-32 | Address Recognized | Frame Copied |

8 Bit Frame Status

b) Busy Token (Data Field)

| Start Delimiter | End Delimiter |

c) Abort Sequence

Figure 9.16 IEEE 802 Token Ring Formats

A station requiring use of the ring looks for a free token. Upon detecting one on the ring, the station captures the token and immediately amends it to a busy token (Figure 9.16b). The busy token contains the same information as the free token with added fields. Additionally, a bit in the access code is changed to indicate that this is a busy token. Destination and source addresses, control field information (identifying the type of message), data, and frame check sequence (CRC-32) fields are added to the busy token. Following the frame check sequence is a status field, which contains bits that identify the frame as being received (address recognized) and read (frame copied) by the destination station. Once the sending station creates the busy token, it sends it to the next station in the ring. The message is passed from station to station until it returns to the originating station. If the destination address was recognized by the intended receive station and the message was read, the status bits at the end of the token will be correctly set. In this case the originating station, satisfied that its traffic was received, strips the message from the token and amends the busy bit back to free and places the free token back onto the ring.

If during the process of circulating the message a higher-priority station decides it wishes to have access to the line, it must amend the token to indicate its desire. The higher-priority station achieves this by setting the reservation bits to its own priority. When the originating station returns the free token to the ring, the priority level is changed to the higher priority to prevent lower-priority stations from gaining access before the higher-priority level station can capture the free token.

Some token ring protocols maintain a hold timer that limits the time that a station can hold a token before returning it to the ring as a busy or free token. If the originating station receives its message without the address recognized or the frame copied bits set, it will return the busy token back to the ring in an attempt to resend it to the intended destination. If there is trouble on the ring or any reason to abnormally terminate a transmission and free up the token, an **ABORT SEQUENCE** (Figure 9.16c), consisting only of a start and an end delimiter, is sent.

● 9.8 INTERNATIONAL BUSINESS MACHINE'S TOKEN RING

INTERNATIONAL BUSINESS MACHINE (IBM)'s token ring network is implemented through the use of a Texas Instruments chip set designated TMS380 LAN. The network and the chip set actually use a form of **STAR** topology coupled with a logical ring. The logical transfer of data uses token passing and a contiguous ring. Routing and management of the network function under a central hub of a physical ring. Stations connected in the ring are connected through a wire concentrator that acts as a normally closed switch. The hub controls when the switch is on, attaching a node to the ring, and when it is off, disconnecting a node from the network.

The IBM token ring/TMS380 LAN operates at 4 Mbps on twisted-pair wire or fiber-optic cable. With the twisted pair, two pairs of lines are connected to each station. One is a downlink and the other is an uplink to allow data to be passed in a full-duplex, four-wire environment. Token passing is accomplished using the same free token and message frame formats discussed for the token bus. The main difference is that tokens are passed from station to station around the ring rather than in a random sequence afforded by the bus topology. Also, centralized control by the hub station in the star assures orderly, collision-free accesses to the network.

The IBM token ring incorporates a network manager, which is responsible for network security, network configuration changes (node insertion and deletion), data requests and responses, and reconfiguration control. The star hub is an ideal source for the network manager control to emanate from. Access to the network manager functions from a node are available in two forms, an authorized mode and nonauthorized mode. The nonauthorized mode makes all management functions accessible to a node, while authorized mode accesses limit the functions available to a node.

As the size of local area networks grew, the management of the networks became more difficult. Controlling traffic flow, maintaining fault monitoring, and adding and deleting stations from the system required more and more sophisticated system management. An alternate method to handling larger networks is to divide them into smaller networks that are interconnected using hardware/software systems called **BRIDGES** and **GATEWAYS.** Bridges are used to connect two similar networks, while gateways allow systems with differing protocols or message formats to be interconnected.

A bridge is a physical circuit that allows two networks to be interconnected as shown in the IBM token ring application in Figure 9.17. The type of data transferred between the two LANs must have the same format. However, the actual protocols that define the message content can be any usable protocol as long as the source and destination stations use the same protocol. As such, bridges are protocol-independent. For instance, station A on one local area network using X.25 can send a message through a bridge to a station on a second network as long as that station also uses X.25. On the same system, station B can send an Ethernet-type packet to a station on the other local area network through the same bridge.

A bridge works by comparing source and destination addresses within the message packets against an address table. If a destination address is on the same network as the source address, the bridge ignores it, allowing the message to continue on to the destination station. However, if the destination address is on the other local area network connected to the bridge, the bridge copies the message onto the second network so that it can reach its destination. This process is called **FILTERING and FORWARDING,** since messages destined for the same LAN as the source are "filtered" from the bridge, while traffic destined for the other network is forwarded to it.

Address tables are formed by one of two methods. One method is to have the table sent to a bridge by a system manager. Station addresses and associated network numbers are stored directly in the bridge's memory. This method requires the manager to have prior knowledge about the system configuration in order for the table to be formulated. A second method for creating the address table is called **LEARNING.** The bridge sends out a broadcast message asking the various nodes on the two LANs to identify themselves. As each station on the network responds to the broadcast request, the bridge builds the address table. This method requires the software at the bridge to perform the task and also requires time for the table to be built. In either case, once the address table is constructed, the bridge can begin the filtering and forwarding process.

Some bridges have additional filtering capabilities that allow them to filter broadcast or multicast packets so that a situation called a **BROADCAST STORM** does not occur. Essentially, broadcast messages are meant to be received by all stations on the network. Unfiltered systems could experience broadcast packets

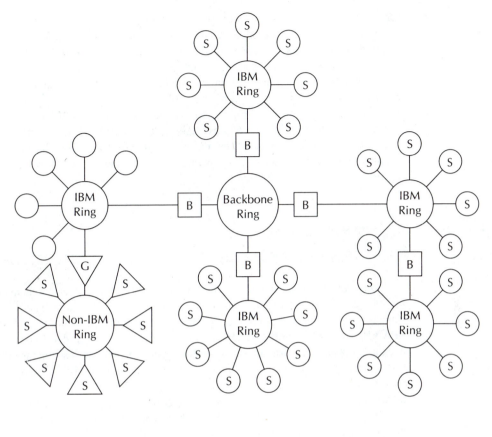

B – Bridge   G – Gateway   S – Station

Figure 9.17 IBM Token Ring Topology

crisscrossing through the bridge until the network completely saturates and quits. A bridge with the capability of sensing a broadcast message would pass it along once between the two LANs so that the broadcast message is made available to all the stations on both networks. Attempts by the broadcast packet to return to the other network through the bridge are filtered out.

● 9.10 GATEWAYS

Gateways are similar in function to bridges in that they connect two networks. The main difference between bridges and gateways is that the latter can interconnect networks with differing message formats. A bridge does not perform any protocol translation in transferring packets between networks. Gateways, on the other hand, are designed to make protocol translations between two dissimilar LANs. For instance, an X.25 station can be made to communicate with an Ethernet station through a gateway. Recall that with a bridge, any protocol can be passed through the bridge, but for communication to be successful the source and destination stations are required to use the same protocol. When using a gateway, the gateway reads the message and determines the source protocol. After comparing the destination address with its address table, the gateway then determines the destination station's protocol. The data from the message are placed into the destination protocol format, and the gateway forwards the translated message to the network that has the destination node.

● 9.11 IEEE 802.1 SPANNING TREE

One concern when using a bridge to interconnect two local area networks involves the loss of network activity in case of a failure of the bridge. The IEEE 802 standard addresses this problem in 802.1, specifying a solution called the **SPANNING TREE PROTOCOL,** which requires a second bridge to be used for backup purposes. The primary bridge used to handle traffic has a secondary parallel bridge connected across it. Under normal operation the primary bridge passes all the messages between the two LANs. If the primary bridge fails, the backup bridge takes over the handling of inter-LAN messages until the primary bridge is repaired.

The spanning tree protocol details the conditions under which the backup bridge takes over and how the takeover is to be accomplished. The main disadvantage of the spanning tree method is the inefficient use of the backup bridge, which effectively idles as long as the primary bridge remains healthy. A process called **LOAD BALANCING** can be applied to a system using the backup bridge. The traffic load is divided between both bridges, which reduces the volume on the primary bridge. In the event that *either* bridge fails, the remaining good bridge takes on the full traffic load until repairs can be made to the failed bridge.

● 9.12 PRIVATE BRANCH EXCHANGE

As the number of telephone connections increased, a need arose for local control of switching services. For instance, calls entering a business office had to be routed to various offices or departments within the office building. The **PRIVATE BRANCH EXCHANGE (PBX)** was developed by the telephone company to facilitate this type of local switching. A PBX station contains the switching system, interface, and control mechanism to relay calls from outside lines to numerous users within a locality. In the simplest form, a PBX switch station consists of a switchboard and an operator who makes the appropriate connections.

By 1963 the switching and control mechanism came under computer control, but the circuits being switched were still analog voice lines. It wasn't until 1971 that the first data lines carrying digital information were installed using PBXs. Voice communications became digitized through the use of analog to digital conversions produced by CODECs operating into a pulse code modulated (PCM) encoded line.

Installation, control and use of PBX systems remained in the domain of the telephone company until 1968, when the **FEDERAL COMMUNICATIONS COMMISSION (FCC)** handed down a ruling permitting outside vendors to interconnect equipment to the existing telephone lines. These vendors were required to adhere to rule 68 of the FCC code, which specified stringent electrical and interfacing requirements for utilizing telephone lines. This ruling opened the PBX networks to integrated data and voice communications.

● **9.13 PBX NETWORKS**

PBX networks are interconnected using node concepts similar to those developed for the systems network architecture (SNA) network. Entry and exit to the network, as well as interconnections within networks, are managed through nodes. Figure 9.18 is an illustration of fully connected PBX nodes. Fully connected systems have a small number of nodes due to the limitation imposed by the number of interconnecting cables used by the network. The number of connections is computed using the same relationship introduced for direct connections in the discussion of switching methods in chapter 8. That is, the total number of connections (C) is:

$$C = \frac{n(n - 1)}{2} \tag{9.1}$$

where n is the number of stations to be interconnected.

Also shown in Figure 9.18 are data terminals (T) and voice phones (V) connected to each node. The disadvantage of a fully connected PBX is the same as for directly connected networks, as discussed in chapter 8. As the number of stations increases, the number of connections rises by a factor close to one-half the square of the number of stations. For larger systems, PBX systems utilize digital switch networks.

The PBX CONTROL MECHANISM manages and coordinates the activities of the digital switch network, manages interfacing to a network, and handles miscellaneous other network resources. This mechanism, a combination of hardware and software, is responsible for accepting input data, translating that data into a form recognized by the PBX network, and supplying output data to the destination station. Notice the functional similarity between the PBX control

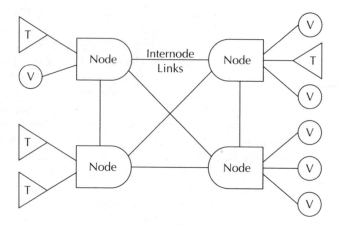

T – Terminal   V – Voice Phone

Figure 9.18 Fully Connected PBX Nodes

mechanism and the systems service control points (SSCPs) and **LOGIC UNITS (LUs)** of the system network architecture (SNA) network model.

A PBX LAN using a digital switch network is shown in Figure 9.19. PBX switching networks handle various types of information switching. Voice entering in analog form connects to the system via an **INTEGRATED INTERFACE (I),** where it is converted into digital form and integrated into a data channel. The voice data are separated from the data at the destination integrated interface. At this point the voice data are returned to analog form. Voice information can also be admitted into the system via an analog interface, which also converts it into digital but does not integrate it onto a data channel. In this instance the digitized voice information is routed separately to a destination node.

Computer terminals supply digital data directly to the system using a T1 data line. Use of this line demands that the data already be placed into the packet form used by the PBX network. Parallel data are converted to serial and translated into the required form through PBX modems or CODECS (DCE—data communication equipment) connected between the network switch data interfaces (D) and the computer terminals.

Totally foreign networks can interface to the PBX network by using a PBX gateway, which is a hardware and software element that translates the foreign network's protocols and data stream into the PBX's format. Translating between differing protocols and packet formats within the gateway is the task of packet assemblers/disassemblers (PADs), which assemble segments of asynchronous data into X.25 formatted packets. Besides setting the data into the correct format, the PAD also translates addresses between PBX extensions and X.25 addresses.

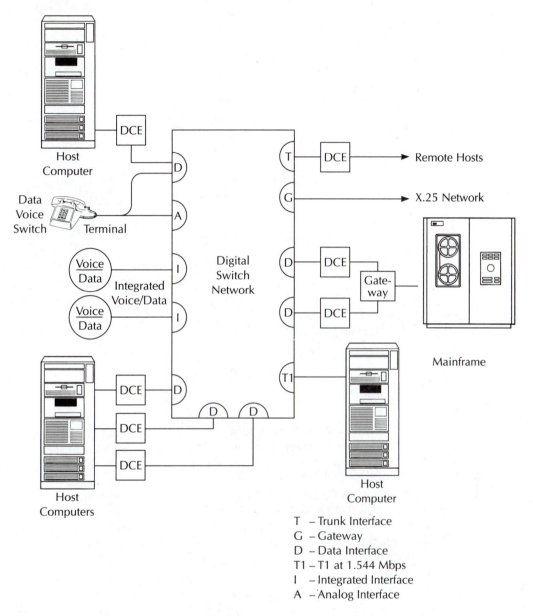

T – Trunk Interface
G – Gateway
D – Data Interface
T1 – T1 at 1.544 Mbps
I – Integrated Interface
A – Analog Interface

Figure 9.19 PBX LAN Example

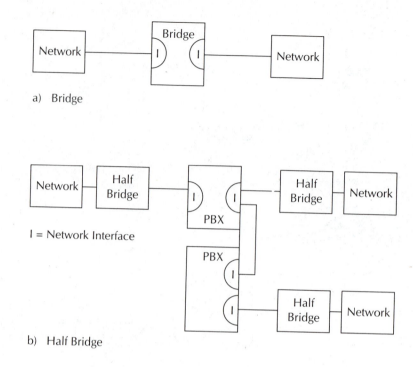

a) Bridge

I = Network Interface

b) Half Bridge

Figure 9.20 PBX Network Bridges

A system called a PBX bridge is employed to interconnect PBX networks. In PBX terminology, a bridge interconnects two adjacent networks, be they PBX or other types, as shown in Figure 9.20a. Half bridges connect a network segment to an intermediate unit. The intermediate unit sends the data to another half bridge, which then routes them to another network segment (Figure 9.20b). It should be kept in mind that the definitions for PBX gateways and bridges differ from the general network definitions. For general use, bridges are used to interconnect similar networks and gateways to interconnect dissimilar networks.

● 9.14 CONNECTIONS TO A PBX NETWORK

Connection paths into the PBX system originate from many sources. They are digitized, if not in digital form already, at various data rates depending on the medium used to convey the data. For instance, an asynchronous analog signal entering a digital PBX arrives on a voice path distinctly different from the physical data path it uses to be transported to a final destination. There is a **VOICE GRADE**

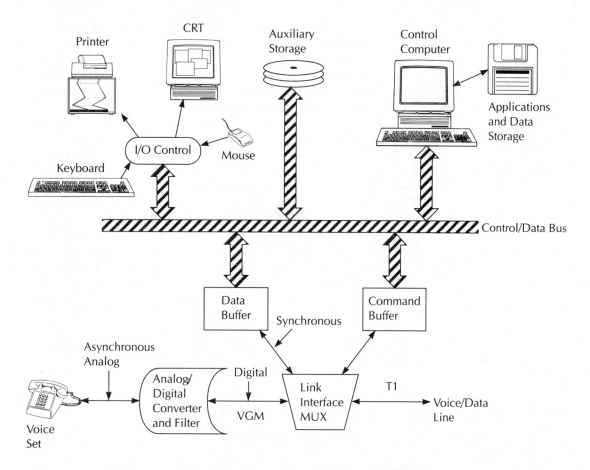

Figure 9.21 PBX Integrated Voice/Data Unit

MEDIUM (VGM) cable that transports the digitized voice signal at 1 Mbps at distances of up to 2,000 feet from source node to destination node.

On the other hand, integrated voice and data are composed of digitized voice, digital data, and control bit streams, which are multiplexed into a single synchronous stream of data (Figure 9.21). This channel is built from a 64-kbps voice channel, one or more 64-kbps data channels, and an 8- to 64-kbps control channel.

A T1 channel that is designed to handle twenty-four channels, full duplex at a total data rate of 1.544 Mbps, is used to interface computer terminals directly into the PBX network using the network's protocols and format. The additional speed is facilitated by the absence of translating software and hardware.

Connections into a PBX network are classed under switch network functions based on a connection table that is updated to include added users and delete nodes no longer connected to the network. Keeping track of stations that are moved from one entry point to another in the network is another function of the connection table. This table also includes the class of service for each node, which includes priority as well as service type (voice, data, etc.).

An example of a communications session on a PBX system begins with a signal from a user node requesting access to the network. This user's address is checked against the connection table for purposes of identification and access authorization, and to determine the type of service the user requires and is entitled to. Once it is determined that the user can have access, it is assigned a time slot that is used by the central control of the PBX switch to acknowledge the user's request. In response to the acknowledgment, the user issues a CALL command identifying the intended destination point. Once again, the connection table is checked, this time to determine the validity and availability of the destination station. If the destination station is in use, a busy signal is returned to the originating user. If the destination station is not currently in use, a time slot is set aside to be used by the destination station. The connection table is upgraded to show that both stations are in use. The destination station uses the assigned slot to issue a READY signal to the PBX controller. The controller then completes the connection to the originating station, and traffic can now flow back and forth between them. At the end of the dialogue between the two stations, one or the other issues a DISCONNECT command, and the connection is broken. The time slots used by the stations are freed for other use, and the connection table is upgraded to reflect the nonbusy condition of the two nodes.

● 9.15 TRANSMISSION CONTROL PROTOCOL/INTERNET PROTOCOL

The **TRANSMISSION CONTROL PROTOCOL/INTERNET PROTO-COL (TCP/IP)** was developed by the Department of Defense as a network-routing protocol. TCP assures a reliable transfer of data in a full-duplex communication. It achieves reliability by including sequence numbers and an error-checking mechanism in its frame format (Figure 9.22). A TCP frame begins with addresses for source and destination nodes (identified as **PORTS**) followed by a sequence and acknowledgment number. These numbers are used to assure that all frames are sent and received. The receiving station can keep track of the frames being sent by checking the sequence numbers. Source ports can monitor the successful reception of acknowledgments to frames it has sent by checking the acknowledgment number. The control field identifies the type of frame being sent, while the value in the window limits the number of consecutive bytes that can be sent before an acknowledgment is to be sent by the destination port. A checksum is used as an error detection for the frame. Urgent pointer information and options can be used to set the routing for the message when **SOURCE ROUTING**—a method whereby the originating station includes routing information within the message rather than leaving routing decisions to the network manager—is employed. Finally, a short buffer PAD precedes the actual message data.

Source Port	Destination Port	Sequence Number	Acknowledge Number	Data Offset		Control	Window	Checksum	Urgent Pointer	Options	PAD	Data

Figure 9.22 TCP Format

Sequence and acknowledgment numbers are used in a similar manner to frame number sent (NS) and frame number expected (NR) numbers in the high-level data link control (HDLC) protocol discussed in chapter 4 in that they are used to keep track of the frame sequences to assure that one does not get lost. They differ in that an acknowledgment number tracks acknowledgments received and not the value of the next expected frame (sequence).

A less reliable form of message transfer is the **DATAGRAM** option of TCP. Datagrams (Figure 9.23) are short messages that do not require acknowledgments. Establishing a connection before sending data is a requirement for TCP that is not required for a Datagram. They are sent out without any method established to verify whether they are ever received by the destination station. The format of the **USER DATAGRAM PROTOCOL (UDP)** frame in Figure 9.23 shows source and destination port addresses, a length field to identify how many bytes of data are in the data field, a checksum for error detection, and the actual data. The frame lacks sequence and acknowledgment numbers. Datagrams are fast and require no preset connection but are not a reliable method of transferring data.

The internet protocol (IP) portion of the combined TCP/IP serves the purpose of routing packets across an integrated system of networks called a **CATANET.** It takes message frames from one network, reassembles them into smaller segments if required, and sends through the complete network to their destinations, where they are reassembled into their original frames. The IP frame shown in Figure 9.24 contains considerably more information in its header. A version number leads the frame to assure that the remaining fields of the frame will be correctly identified according the version in current use. The length of the header is indicated next since it will vary according to the type of frame being sent. Other fields specify the type of service required of the IP frame, an identification field, status flags, a header checksum to validate header information, source and destination addresses, and data. Other fields have specific uses pertaining to the service that may be used by

Source Port	Destination Port	Length	Checksum	Data

Figure 9.23 UDP Format

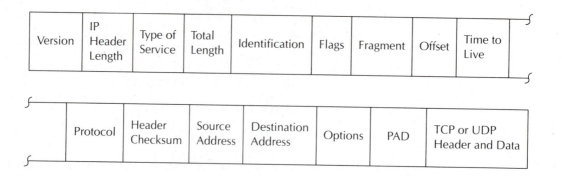

Figure 9.24 IP Format

the frame. These include source routing as described earlier, file transfers, electronic mail service, and terminal emulation services.

● 9.16 OSI UPPER-LEVEL PROTOCOLS

PBXs, SNA networks, as well as other networks are protocol-dependent systems. That is, they can be easily used on networks that use protocols that they are designed for. Devices or networks other than the prime system require some form of translation to the prime network. Numerous other networking companies have developed their own network specifications, including the **DIGITAL EQUIPMENT CORPORATION NETWORK (DECNET)** and Sun Microsystems UNIX-based network. Because of the number of differing network protocols serving these diverse networks, a need arose to develop methods to interface all of them to a common network. In essence, a transparent protocol network was needed into which any network could be connected. Such a network addresses the NETWORK and TRANSPORT layers of the OSI model.

● 9.17 SERVERS

At the network layer, a software mechanism called a **SERVER** is used to perform the local translation between a user's protocol and the common network. These servers (like SNA nodes) come in many variations depending on the service they are required to perform. For example, there are *FILE SERVERS* used to handle data files and *PRINTER* and *MESSAGE SERVERS*, each for interfacing printers and messages to any station on the network. Other types of server software aid in network management while still others assist in developing data routes. Once the data or service is entered into the common network, it must be successfully transported to a destination server, which is responsible for translating the information into the format and protocol recognized by the destination network or station. Realize that the destination system may not necessarily use the same protocols as the source station. In actuality the dissimilarity between the source and destinations is the more common case.

● 9.18 NOVELL'S NETWARE

Figure 9.25 is an overall block diagram of the protocol-independent open systems architecture of Novell's **NetWare** interface. To the prime network, connected through the drivers and **OPEN DATA LINK INTERFACE (ODI)** protocol, the various network transport protocols shown appear identical. The network protocols currently supported by NetWare are XENIX's XNS, IEEE 802.3 (Ethernet through a Transmission Control Protocol/Internet Protocol (TCP/IP)), Macintosh and OS/2 workstations, OSI model, and IBM's SNA model. These networks are fed from host computers or workstations via NetWare server and service application protocols, which are directed and defined within the NetWare **STREAMS** software. **PERSONAL COMPUTERS (PCs)** are used as host

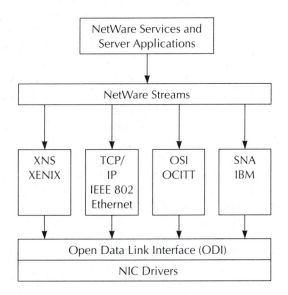

Figure 9.25 Protocol-Independent Architecture

computers to provide intelligent terminal functions available to the user. This means that the user can call up resources and perform control and processing functions at a computer connected directly to the network. NetWare uses distributed data management to control, monitor, and recover from faults throughout the prime network. It is this *NETWORK MANAGEMENT* that has the overall control of the entire system.

● 9.19 NETWORK MANAGEMENT

The Network Management software of the NetWare package is responsible for configuring management control, tailoring it for the specific needs of the system. Included under this management are user access rights and other system security measures like encoded passwords, fault management, and recovery and accounting services that determine the amount a user is charged for using the system.

To configure a system, the associated servers are set to recognize access rights for the users it serves. These servers must be able to read the data fed to them by the user and convert the form and protocol to fit into the common network patterns. Network Management by Novell uses software routines called **SYSCON** for **SYSTEM CONFIGURATION** and **CLIENT SERVER PROTOCOL** (Figure 9.26) for user interfacing. SYSCON software can configure any file server on the network from any workstation on the network, thus providing a method for easily adding workstations to the system. The client server protocol is protocol-specific in that it is used to match the protocol of the stations (as shown in Figure 9.26), interfacing to the network to the network itself.

Fault tolerance, the detection of errors and system recovery from a fault situation, is managed through several facilities within NetWare. Duplicate directories and **FILE ALLOCATION TABLES (FATS)** for all data and process software handled in the system are maintained. This is coupled with a disk mirroring and duplexing system that reduces recovery time in the case of disk failure anywhere within the network. Duplicate or mirror copies of disks holding data or programs are updated as frequently as the master disks they are mirroring. If the master disk fails, the backup disk is put on-line once the failure is detected. This reduces fault recovery time. The faulty disk will eventually need to be replaced or updated, but the immediate information is not lost because of the failure.

In addition to the fault tolerance recovery scheme that is managed directly by Network Management, a second software package called **NetWare CARE** is used to supply system diagnostics to determine what and where the fault is. NetWare CARE maintains a network map and can be used to perform point-to-point testing anywhere in the network. Control of NetWare CARE is manageable from any workstation or host computer on the system. In addition to diagnostics and testing, NetWare CARE creates performance evaluations that may be viewed at any station and at any time provided that station has access rights to those evaluations.

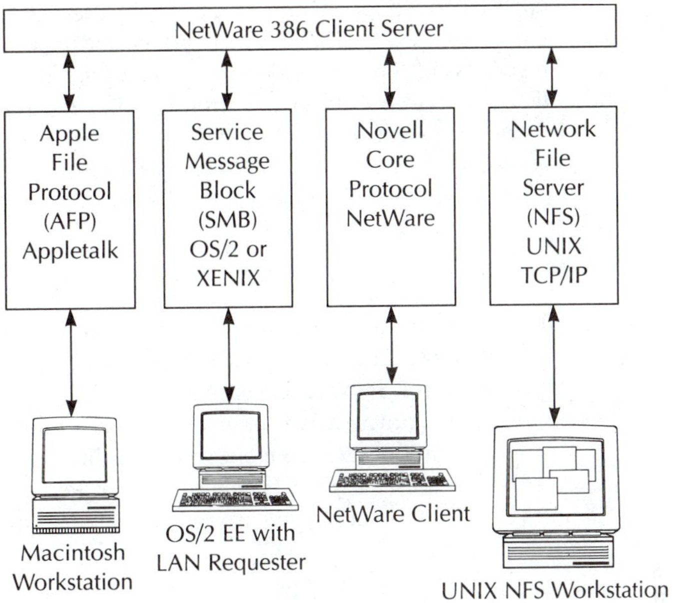
NetWare 386 Client Server
Apple File Protocol (AFP) Appletalk
Service Message Block (SMB) OS/2 or XENIX
Novell Core Protocol NetWare
Network File Server (NFS) UNIX TCP/IP
Macintosh Workstation
OS/2 EE with LAN Requester
NetWare Client
UNIX NFS Workstation

Figure 9.26 Client Server Support

• 9.20 NETWARE 386—DISTRIBUTED NETWORK MANAGEMENT

NetWare 386 is the latest version of Novell's Network Management System. It is based on distributed management as opposed to a centralized management. Centralized management puts all management and server applications at a centralized point in the network. This provides easy access of the manager to the system and localizes all control. If something goes wrong with this section of the network, however, the entire network would cease to operate. Distributed management is more difficult to develop initially, but failures are localized, allowing portions of the system to remain active despite failures in other sections of the network.

Within distributed processing, the management tasks are delegated, in part, to the servers on the system. Besides the file and printer servers mentioned earlier, additional servers offer a wide range of services. There are plotter, mail, and archive servers, records management, accounting, and encryption servers, and many others. There is also a Network Management Server that oversees the activity of the other managers. NetWare's Print Server can manage multiple print servers, each capable of dealing with up to sixteen printers per server. In addition to the printers directed under server control, printers can also be maintained at the workstation site. These printers are controlled directly by the workstation as opposed to server control. Printers are managed under one of four operating modes—*QUEUE ONLY*, which prints documents according to highest priority regardless of the type of document to be printed; *FORMS ONLY*, which prints all the same types of documents from all files regardless of priority before printing other document types; *QUEUE BEFORE FORM*, which prints documents according to priority first and then by type; and *FORM BEFORE QUEUE*, which sets up print order by type and then by priority.

Example 9.3

Ten documents are to be printed. A documents have the highest priority, followed by B and then C documents. The types of documents are payroll, inventory, and quality control. In what sequence are the following documents, received by a print server in the order shown, printed using each print server mode?

    A—Payroll-execs, inventory-shoes, quality control-shoes
    B—Payroll-managers, inventory-cars, payroll-clerks
    C—Quality control-cars, payroll-production, inventory-bikes, quality control-bikes

Solution: **For queue only,** everything under A is printed first, then B and then C. There is no regard to document type. **For forms only:** payroll documents first, then inventory, then quality control. Priority is disregarded. The documents would be printed in the order they were received according to type. **For queue before forms:** A-payroll is followed by B-payroll and then C-payroll (execs, managers, clerks and production). This sequence is repeated for inventory, then for quality control. **For forms before queue:** payroll-execs from A is printed first, then B-payroll, payroll-managers, payroll-clerks, followed by C payroll-production. The next type, inventory, would be printed in likely manner with all inventory documents in A printed first, followed by those in B and finally C. This holds true for the quality-control documents as well.

As a means of establishing security on the prime network, users are assigned access rights to the network, under NetWare, based upon on their needs for use of the network. Those access rights are summarized as follows:

1. READ—User can open and read files.
2. WRITE—User can open and write to files.
3. CREATE—User can create files or subdirectories within existing files.
4. ERASE—User can delete files and subdirectories.
5. ACCESS CONTROL—User can modify other user access rights.
6. FILE SCAN—User can see files and subdirectories when scanning the directory.

7. MODIFY—User can change the name and attributes of files or subdirectories.
8. SUPERVISORY—User has all the rights to this directory and any subdirectories.

These access rights are assigned to specific users through the servers on the network. Only applicable access rights are delegated. As such, most users will have a partial set of access rights. Specifically, access rights like access control and supervisory are restricted to manager consoles and limited users. Additional security is assured through encrypted passwords that prevent unauthorized access to the network.

● 9.21 NETWARE IN SUMMARY

The NetWare system has been developed to manage different networks onto a single common prime network. The OSI levels of management and application are the network and transport levels. It is presumed that the host networks are operating within their confines according to the specifications of their protocols. That is, the physical and data link layers of the OSI model are accountable by the host networks. Servers, each with their specific functions, handle the translation of the host to the prime network, while other software, such as STREAMS, accounts for transporting data about the prime network. The entire system is managed through distributed data management, which gives more control of management to the servers on the system. Fault management is handled through several means, including duplicate or mirror disks, system monitoring, and disk duplexing. Lastly, security is maintained through a structure of access rights that are established at the server level when the system is configured.

● 9.22 SESSION LAYER PROTOCOL—NETBIOS

In 1984 a company called SYTEK developed a protocol for maintaining session layer applications called NET/20. This protocol included DATAGRAM services, peer-to-peer communications and server services at the session layer level, and workstation function applications. NET/20 has since been enhanced and made applicable to a wide range of personal computer–based networks. It is known today as **NETBIOS (NETWORK BASIC INPUT OUTPUT SERVICES INTERFACE)** and has become the de facto standard for session layer specifications of PC-based networks. The NETBIOS protocol fits into the network model as shown in Figure 9.27. A user interfaces directly through an applications layer operation (called simply an applications) into the NETBIOS service. The applications enters

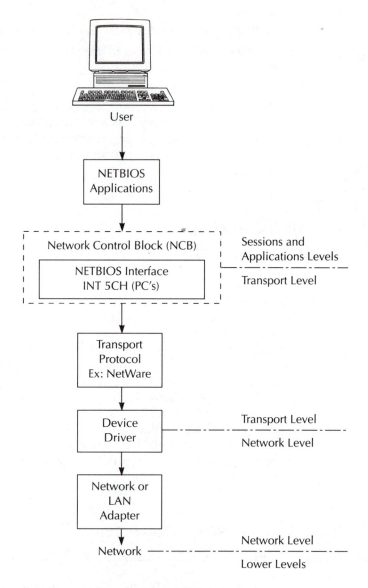

Figure 9.27 NETBIOS Interface in the OSI Model

the network session through the NETBIOS interface, which controls and manages
adapter nodes for this purpose. Various NETBIOS commands are used to establish
sessions and assure network operation. These commands are generated as a result
of applications issuing them through **NETWORK CONTROL BLOCKS,** which
are discussed later. Session layer activity is then integrated through transport layer

protocols and device drivers to the **NETWORK** or **LAN ADAPTER** and onto the network.

IBM PCs can access network facilities through a node called an **LAN ADAPTER,** which is hardware and software independent of the network it is interfacing to. That is, NETBIOS through the LAN adapter can be configured to interface with most of the popular LAN networks (Ethernet, token ring, STARLAN, etc.). LAN adapters maintain access rights to networks through the use of names and name tables. Names are identifying codes for adapter nodes and can each be one of three types—*UNIQUE, GROUP,* or *BROADCAST.* Each adapter maintains a name table of all the names applicable to that adapter. A second table holds the names of the rest of the adapters on the system. Names can be added or deleted using NETBIOS name support commands *ADD NAME, ADD GROUP NAME,* and *DELETE NAME.*

To add a name to an adapter, the application using that adapter first sends a broadcast message containing a **NAME CLAIM** format. If the name being claimed does not exist in any of the adapters in the system, then it is all right for the issuing adapter to use the name. This is accomplished by issuing an ADD NAME command with the intended name. The NETBIOS service program places the name in its own tables and issues a success report to the adapter along with a unique 1-byte name number. This number is the code used to reference the name in future communications with the adapter. Names may be deleted at any time by issuing a DELETE NAME command with the name number. NETBIOS service then removes the name from its tables and makes the name number available for future use.

The first name in an adapter's table is its permanent node name, which uniquely identifies that adapter. It is made up of 10 bytes of 0s (to assure that there are no errors caused by data preceding the number) followed by a 6-byte name number. When the NETBIOS sends a reset command to initialize the adapters, all names in the name table except the first permanent node name are erased. The RESET command also cancels any current active sessions and sets up buffers to hold the maximum number of allowable sessions and outstanding commands as specified for a particular application. That is, each application determines the maximum number of sessions that can be active at a given time and how many message frames can be sent before an acknowledgment is issued. These maximums are set through software based on system use and needs.

Additional general-purpose NETBIOS commands include *CANCEL, ADAPTER STATUS, UNLINK, FIND NAME,* and *TRACE.* The CANCEL command cancels any pending command. This is helpful if a command is issued incorrectly or if an adapter, for some reason, cannot respond to a command. By issuing an ADAPTER STATUS command, the NETBIOS can gather status information about a particular adapter. This includes the names the adapter is using, whether there are any active sessions involving the adapter, and whether any errors have occurred. The *UNLINK* command terminates any boot (initializing) operations initiated from a remote disk source. The FIND NAME command is used to identify which adapter is using a particular name. NETBIOS service uses the TRACE command as a diagnostic tool to activate a trace of all commands issued to the NETBIOS interface.

Two basic types of messages are supported by NETBIOS—DATAGRAM and SESSION. Datagrams as explained earlier, are unreliable, unacknowledged, short, fast messages. Since they are not acknowledged, the originating station has no way of knowing whether they were ever received. Two types of DATAGRAMs are used—unique and broadcast. Unique datagrams are destined for a particular node, while broadcast datagrams are intended for all nodes set to receive them. An adapter sets itself to receive DATAGRAMs at any time by first issuing a **RECEIVE DATAGRAM** or **RECEIVE BROADCAST DATAGRAM** command. Once the appropriate command is issued, any DATAGRAM destined for that adapter node can be received by that adapter. Again, the originating station has no way of confirming whether the datagram was received at all or received without errors. A unique use of the RECEIVE DATAGRAM command occurs when the hexadecimal name number FF is used with the command. This allows an adapter to receive datagrams destined for any adapter on the system.

Originating nodes send DATAGRAMs through **SEND DATAGRAM** or **SEND BROADCAST DATAGRAM** commands. DATAGRAM messages can contain anywhere from 0 to 512 bytes of data and are sent using a simplex communication (one-way communication), since acknowledgments are not required.

Session messages, while more reliable, require additional overhead to assure the reliability. Headers include sequence numbers and acknowledgment status, along with source and destination name numbers. Session messages allow up to 64k bytes of information and operate in the full-duplex mode. More than one session can be run at a single time, and each adapter is also capable of operating with multiple sessions. This is facilitated by the use of session numbers, which keep track of each separate session in process. Sequence numbers maintain the order of each session's block, as well as verifying that each frame of a session is sent and received. Sessions are initiated by the issue of a *LISTEN* command, which includes the adapter names that an adapter is willing to have a session with. An * in the name number of a LISTEN command designates that adapter as one that is willing to open a session with any adapter in the network.

The session is established by a second application issuing a CALL command from its adapter. The CALL command contains the name number of the adapter it is calling. NETBIOS service then issues a **LOCAL SESSION NUMBER** to both adapters, which may or may not be the same number for both applications. Each can use either local session number while the session is in progress. Messages are sent and received through the use of *SEND* and *RECEIVE* commands. Messages must originate from a contiguous block of memory for the SEND command and cannot exceed 64k bytes. Another form of the SEND command, the *CHAIN SEND* command, allows messages located in two separate blocks of contiguous memory to be concatenated into a single message. This command allows messages to be expanded to 128k bytes in length. At the end of the session, either application terminates the session by issuing a *HANG UP* command through its adapter.

Two additional commands are available for session support. They are *RECEIVE ANY,* which allows an application to receive a message from any open active session, and *SESSION STATUS,* which allows the NETBIOS service to get data about any active session using a unique name or about all active sessions (ALL NAMES).

Commands are issued through a mechanism called a **NETWORK CONTROL BLOCK.** The command-issuing agent forms the network control block, which contains information, such as any applicable name numbers, sets a pointer to memory where the block is stored, and issues an INT 5CH, which is a microprocessor interrupt command. This command redirects program processing to the NETWORK service program so that the command can be issued.

SUMMARY

The development of large communications networks interconnecting many different terminals and stations gave rise to the need to create a standard operating model for those networks. The open systems interconnection (OSI) model developed by the International Standards Organization (ISO) describes a seven-layer network model. The concept of an open system model allows sufficient flexibility so that system components produced by different manufacturers can be adapted into a network based on the OSI model. Additionally, numerous software companies, such as Novell, have produced programs for the protocols required to operate on open system networks. As networks will be required to carry larger volumes of traffic, faster, and to more locations, these systems and software will have to be expanded to accommodate this growth.

The size of the networks and the need to interconnect several separate networks into a unified system led to the development of special circuits and programs called bridges and gateways. Bridges are used to interconnect similar networks, while gateways interconnect differing networks.

The IEEE 802 standard specifies the requirements for standard networks. These specifications can be used in the planning of local area networks, metropolitan area networks, wide area networks, and networks based on bus or ring topologies. These standards, out of necessity, address one or more layers of the OSI model, making both the OSI model and the IEEE 802 standards compatible.

GLOSSARY

Abort sequence—Frame which contains only a start and end delimiter used to abort token message sequence.

Baseband—Systems that use a single communication channel.

Bracketing—Grouping related messages.

Bridge—Device and software to interconnect two similar networks.

Broadband—System that uses multiple channels on a single communication line.

Broadcast storm—Overload of traffic on a network due to the issuing of too many broadcast messages.

Bus—System topology that uses a common set of lines which each station taps into.

Catanet—Integrated system of networks.

Client server protocol—Novell's NetWare user interface protocol.

Cluster controller—Multiple peripheral interface controller.

Carrier Sense Multiple Access with Collision Detection (CSMA/CD)—A method to resolve multipoint system accesses by contention.

Datagrams—Unacknowledged TCP messages.

Ethernet—First practical LAN specification that uses baseband bus topology and CSMA/CD accessing.

Filtering and forwarding—System which determines into which network portion a message is to be sent.

Free token—Token without a message which can be captured by any station.

Gateway—Device and software used to interconnect dissimilar LANs into a common network.

IEEE 802—Network standard authored by IEEE organization.

Integrated interface—PBX unit that digitizes analog voice and integrates it onto the PBX network.

Jabber—Condition used by a hub to disable a node.

Learning—A process used to establish a network address table.

Load Balancing—Sharing traffic between a bridge and a back up bridge to reduce bridge overloading.

Logical unit (LU)—SNA entry access node function.

Name Claim—broadcast message used to add a station's name to a LAN adapter.

NetWare—Novell's open systems protocol.

NetWare CARE—Novell's diagnostics package.

Network or LAN adapter—Interface between drivers and the network.

Node—Local area network access point.

Open data link interface (ODI)—Novell protocol that interfaces external networks onto a prime network.

Packet assembler/disassembler (PAD)—Translates between data formats.

Ports—TCP source or destination node.

Private branch exchange (PBX)—Local switching network.

Ring—Network topology that connects all stations into a ring. Messages are passed from one station to the next through the ring.

Server—Protocol translation between elements on a network.

Serial interface adapter (SIA)—Manchester encoder/decoder chip.

Source routing—Message routing information is maintained within the message instead of managed by an external controller.

Spanning Tree Protocol—Bridge protocol which manages a bridge and a backup bridge to maintain network integrity.

Star—Network topology that uses a central station. Each peripheral station is connected via a "spoke" to the central station.

STARLAN—AT&T star topology network that satisfies IEEE 802.3 CSMA/CD protocol.

Streams—Novell's collection of service applications protocols.

SYSCON—Novell's system configuration software.

Transmission Control Protocol/INTERNET Protocol (TCP/IP)—Department of Defense routing protocol.

Token—Short message that precedes transmission on a token bus or ring network.

Topology—Physical configuration of a network; typical forms are bus, star, and ring.

Virtual circuit—Logical route between two points.

X.25—Packet switch network protocol.

QUESTIONS

9.1. Name a standard and a protocol that apply to X.25 physical and data link layers.

9.2. Which OSI layer does an X.25 PAD comply with?

9.3. How many total channels are available to be selected from using the LCI field?

9.4. What are the differences between X.25 and HDLC information frames?

9.5. What are the differences between X.25 flow control and HDLC supervisor frames?

9.6. Which IEEE 802 specification is similar to the Ethernet standard?

9.7. What is the difference between 802.2 type 1 and type 2 operations?

9.8. Which HDLC supervisor frame is not used with either 802.2 or X.25 formats?

9.9. List five Ethernet specifications that comply with the OSI model.

9.10. Which protocol discussed in chapter 4 supplies the basis for Ethernet data link processes?

9.11. How does CSMA/CD resolve the problem of line contention?

9.12. What topology is used by Ethernet networks?

9.13. Indicate whether the following statement is true or false: All stations on an Ethernet network are peer stations.

9.14. What is the purpose of the preamble preceding Ethernet packets?

9.15. What is the difference between Ethernet address fields and HDLC address fields?

9.16. What is the maximum number of 802.2 numbered frames that can be sent sequentially without an acknowledgment?

9.17. Which frame reject cause is added using the 802.2 FRMR information field? Give an example of what this reject cause is designed to detect.

9.18. Which 802 specification addresses the following network environments?
a. Bus topology   b. MAN   c. Token ring   d. Token bus

9.19. What is used in the token specification to assure that delimiter sequences are never detected within data fields?

9.20. What is sent on a token bus by an originating station if it does not detect a valid frame response after passing the token a second time to the successor station?

9.21. Referring to question 9.20, what does the originating station send if it does not receive a valid response after repeating the sequence required because of the failure of the successor station to issue a valid frame response to the token message?

9.22. What is the prime difference between a token bus and a token ring?

9.23. Describe the main difference between the functions of a bridge and a gateway.

9.24. What is the main use of the parallel bridge used in the spanning tree protocol?

9.25. What is the purpose of balance loading in the spanning tree application?

9.26. Which type of switching—circuit, message, or packet—were the original PBX systems designed to perform?

9.27. What data form, digital or analog, are the data on present-day networks connected through PBXs?

9.28. What is the difference between gateways and bridges as it applies to the PBX network?

9.29. Describe the difference between PBX bridges and half bridges.

9.30. What is the main advantage and disadvantage of a DATAGRAM message?

9.31. How are routes for messages determined using source routing?

9.32. What main functions is Network Management responsible for?

9.33. What is the purpose of disk mirroring and duplexing?

9.34. What is the prime advantage of centralized network management?

9.35. What is the prime disadvantage of centralized network management?

PROBLEMS

9.1. Code the following data using first Manchester encoding and then differential Manchester encoding. Assume the beginning level for both is at a logic 0.

011010111010010001

9.2. The following sets of files were received at a print server in the sequence shown:

Station	File
1	Data Base I
3	Spread Sheet I
1	Word Proc II
2	Spread Sheet V
2	Data Base III
3	Word Proc III
1	Spread Sheet III

In which order are they printed, given that station 1 has higher priority than station 2, which in turn has higher priority than station 3?

9.3. A user has the following access rights on a network: read, write, file scan, and create. Which activities listed below can be performed by this user?
   a. Save updated information to an inventory file.
   b. Remove middle management payroll files from the system.
   c. Change the name of parts inventory file to parts needed.
   d. Open a new employee list file.
   e. Find the hire date of employee 375 from an existing employee list file.
   f. Change the access rights of an account executive.
   g. Check the files listed under the payroll subdirectory.

Answers to Odd-Numbered Questions

9.1. Physical: V.24, RS232C, X.21bis (not discussed); data link: HDLC, PAD
9.3. 4,096 channels
9.5. X.25 flow control frames do not use the poll/final flag. Frame reject format is not used with X.25 flow control.
9.7. Type 1 uses unnumbered frames only; type 2 uses all frame formats. Type 1 does not require previous data linking; type 2 does.
9.9. Coaxial cable, 500-meter segment length, 1,000-meter extensions, 10 Mbps data rates, use of repeaters, 1,500 meters maximum distance between stations, 100 taps per segment, baseband channel all address physical layer considerations.
9.11. Stations monitor the line for one round-trip propagation delay time before attempting to access the line. If more than one station accesses the line at the same time, a collision detection and back-off process is performed to resolve the contention between the two stations.
9.13. True
9.15. Ethernet address fields include both source and destination addresses, while HDLC includes only secondary station addresses.
9.17. Invalid NS number designated as bit V. Any example that shows a gap in NS numbers causes this error to occur. Example NS sequence of 24, 25, and 33 creates the condition because frame 33 does not follow frame 25, frame 26 does.
9.19. Use of J and K nondata bits
9.21. Solicit successor frame
9.23. Both bridges and gateways interface two networks together. The prime difference is that bridges are used for similar networks and gateways for dissimilar networks.
9.25. Balance loading makes use of the secondary backup bridge to handle some of the traffic load instead of idling while waiting for failure of the primary bridge.
9.27. Digital
9.29. A PBX bridge interconnects networks together. Half bridges connect networks to PBXs.

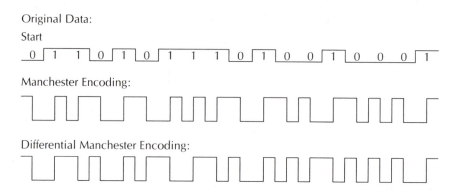

Original Data:

Start

0 1 1 0 1 0 1 1 1 0 1 0 0 1 0 0 0 1

Manchester Encoding:

Differential Manchester Encoding:

Figure 9.28 Answer for Problem 9.1

9.31. The routing information is included in the message header of the source station.

9.33. A mirror disk contains the same information as a main disk. It is multiplexed into use when the main disk fails.

9.35. If the management node fails, the entire network is down.

Solutions to Odd-Numbered Problems

9.1. See Figure 9.28.

9.3. Can do: a (write), d (write), e (read), and g (scan); cannot do: b (erase), c (modify), or f (access control)

Ten • Integrated Services Digital Network

OBJECTIVES

The discussion in this chapter centers on Integrated Services Digital Network (ISDN). The concepts and uses for ISDN are explored first, followed by applications and users of ISDN. An additional section covers ISDN applied to a fiber-optic network for broadband usage.

A section is included on protocol analyzers to test, monitor, simulate, and/or emulate many of the networks and communications links discussed in this and preceding chapters.

ACRONYMS

ANI—**A**utomatic **N**umber **I**dentification

ATM—**A**synchronous **T**ransfer **M**ode

BDT—**B**roadband **D**istant **T**erminal

BISDN—**B**roadband **I**ntegrated **S**ervices **D**igital **N**etwork

BNT—**B**roadband **N**etwork **T**ermination

BRI—**B**asic **R**ate Interface

BTI—**B**roadband **T**erminal **I**nterface

CCITT—International **C**onsultative **C**ommittee for **T**elegraphy and **T**elephony

C/R—**C**ommand/**R**esponse

CRC—**C**yclic **R**edundancy **C**heck

DCE—**D**igital **C**ommunications or **D**ata **C**ircuit **T**erminating **E**quipment

DTE—Data Terminal Equipment

ET—Exchange Termination

HDLC—High-Level Data Link Control

ISDN—Integrated Services Digital Network

LAN—Local Area Network

LAPB—Link Access Protocol for B channel

LAPD—Link Access Protocol for D channel

LRC—Longitudinal Redundancy Check

LT—Line Termination

MAN—Metropolitan Area Network

NT—Network Termination frame

NT1—Network Termination 1 unit

OSI—Open Systems Interconnection Model

PBX—Private Branch EXchange

PRI—Primary Rate Interface

SAP—Service Access Point

SDLC—Synchronous Data Link Control

SNA—System Network Architecture

SPN—Subscriber's Premises Network

TA—Terminal Adapter

TE, TE1, TE2—Terminal Equipment, 1 and 2

TEP—Terminal End Point

XID—transfer ID

WAN—Wide Area Network

● 10.1 INTRODUCTION

The data communications field is in a continual state of flux. New standards and communication network facilities are constantly being introduced and implemented. Many of the older standards are being updated to accommodate current technology. A comparatively new integrated public service is among the latest network systems currently in the process of implementation by the telephone company. This service, called **INTEGRATED SERVICES DIGITAL NET-WORK (ISDN),** incorporates voice, data, video, and facsimile information on the same common carrier. The system utilizes existing public networks to bring these various services to subscribers. Additionally, ISDN uses numerous established **LOCAL AREA NETWORKS (LANs), WIDE AREA NETWORKS (WANs), METROPOLITAN AREA NETWORKS (MANs), PRIVATE BRANCH EXCHANGES (PBXs),** and so on, interconnecting them via common public carriers. The benefits of using an integrated network such as ISDN include line sharing, high data throughput, easy identification of data packets, and error recovery. The establishment and termination of calls are done the same regardless of the originating station's configuration.

● 10.2 INTEGRATING SERVICES

Numerous types of information, including voice, video, and data, are combined or integrated into a single channel. There are a number of types of ISDN data channels used for this purpose, which are detailed in **INTERNATIONAL CONSULTATIVE COMMITTEE FOR TELEGRAPHY AND TELEPHONY (CCITT)** specifications for ISDN networks. The current type being implemented is the full-duplex transparent **BASIC (B) CHANNEL.** Data from several devices are circuit switched into the B channel for transmission to a receiver at a 64-kbps rate. B channels carry voice, digital, and/or high-speed data transmissions.

A D channel, operating at 8 or 16 kbps, is used for control signaling or low-speed packet-switching transmissions. Specific uses of the D channel include call setup and termination and system maintenance control. Use of packet-switched data on the D channel provides a means for telemetry information to be passed along while integrated data travel on the B channel. Telemetry data include low scan alarm, energy monitoring, and security data.

Two established forms of ISDN service are commonly available — the **BASIC RATE INTERFACE (BRI),** consisting of two B channels and one D channel (2B + D) and a **PRIMARY RATE INTERFACE (PRI).** In North America, Japan, and Korea, the primary rate interface consists of twenty-three B channels and one D channel (23B + D). In Europe and the rest of the world, the primary rate interface uses thirty B channels and one D channel. Basic rate interface systems require bandwidths that can accommodate 144 kbps to handle the two 64-kbps B channels (B1 and B2) and the 16-kbps D channel. Primary rate interface systems, because of the larger number of B channels, need bandwidths of 1.544 Mbps for North American applications and 2.048 Mbps for European systems.

Data, in any form originating at a subscriber (customer or user), are converted into ISDN form by a **TERMINAL ADAPTER (TA),** transported onto the common carrier (telephone system), and delivered to the destination station. On the receiving side, the process is reversed, returning the data to a form usable by the receiving customer. Because of the use of terminal adapters, the sending and receiving customers do not have to use the same type of networking or protocols at their sites.

Essentially, the communications methodology is transparent to the user. What happens to the messages along the way is insignificant to the user as long as the data are successfully communicated.

● 10.3 ISDN CONNECTIONS AND INTERFACES

ISDN units and interfaces throughout the network are clearly defined by function and reference in the network. Users may access the network through one of two categories of entry terminal devices, **TERMINAL EQUIPMENT 1 (TE1)** and **TERMINAL EQUIPMENT 2 (TE2).** Terminal equipment 1 units contain standard ISDN interfaces and require no data protocol translation. Raw data are entered and immediately configured into ISDN protocol format. TE1s are illustrated in Figure 10.1 as a terminal and digital telephone. Non-ISDN terminals are classified as TE2. They are connected to the system via physical interfaces such as RS232C and X.21. Terminal adapters (TA) are required to perform the required translation between the non-ISDN data protocol and the ISDN network protocol formats. TAs translate user data rates into ISDN's 64-kbps B channel or 16-kbps D channel rates. X.25 packets, if used, are converted to ISDN packet formats, and any additional signaling required is added by the terminal adapter. These terminal adapters support analog phones and facsimile, as well, by use of a 3.1-kHz audio bearer service channel. These audio signals are digitized and formatted into ISDN and sent onto the network. The entry point of the TE2 device to the terminal adapter is referenced by the letter R. An S is used to designate entry points of the terminal adapters or TE1 units into the ISDN lines.

S reference points indicate interface lines that are presently in ISDN format. They supply the 2B + D data at 192 kbps for the basic rate to **NETWORK TERMINATION 1 (NT1)** units. These units supply the physical interface between the customer to the common carrier at reference points designated by the letter T. As such, these units are often called **CUSTOMER EXIT/ENTRY NODES.** An additional network termination unit, the NT2, may be used to terminate several S point connections, providing local switching. NT2s may also include protocols for data link and network layers of the **OPEN SYSTEMS INTERCONNECTION (OSI)** model to complete the interface between the common carrier local loop and the central office (U reference point). NT2s also perform two-wire to four-wire conversion for ISDN basic rate access to the common carrier.

The U loop is the media interface point between NT1 and the central office. U loops are terminated at the central office by a **LINE TERMINATION (LT)** unit, which incorporates physical layer interfacing between the central office transceiver and the loop lines. The LT unit is connected at reference point V to an **EXCHANGE TERMINATION (ET)** unit such as a PBX that routes the data, through a local network, to an outgoing channel or to a central office user.

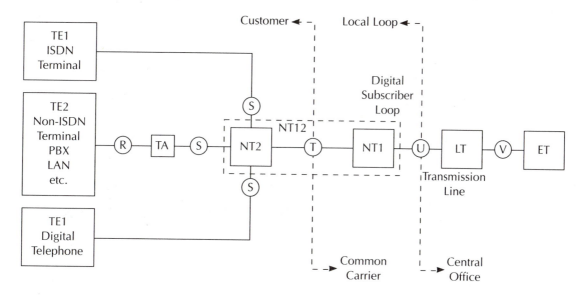

Figure 10.1 ISDN Connections and Reference Points

● 10.4 ADDITIONAL ISDN CHANNEL TYPES

The preceding discussion was limited to two types of transmission channels, B and D. These are the channels in current use by existing ISDN subscribers. Additional channels specified by the ISDN system are presently usable or still in the development stages. They include the following:

A channel—Conventional analog voice

HO channel—Six-channel circuit switch at 64 kbps per channel used for high-speed data or signal imaging

H11 channel—Twenty-four-channel switch at 64 kbps per channel for a total data rate of 1.536 Mbps on the channel

H12 channel—European version of H11 using 30 channels instead of 24 for a total rate of 1.92 Mbps

E channel—Packet switched using 64 kbps similar in function to the D channel

The common T1 channel described in chapter 8 can be implemented on ISDN using various combinations of the channels described earlier. The most direct definition is supplied by the H11 channel, which is designed principally for the twenty-four-channel T1 carrier. However, T1 can also be implemented using 23B + D or 24B channels, 3HO + D or 4HO. The main difference in these is whether or not the communications link requires a signaling (D) channel or not.

ISDN specifications define a network architecture modeled according to the open systems interconnection (OSI) layer network model. The network termination units address the physical layer by describing the interfaces required by users to access the system. Devices included as part of a network termination 1 (NT1) are transceivers, modems, interfaces (such as RS232C), and network media (such as twisted-pair cabling).

Figure 10.2 illustrates the frame format for the basic rate interface (BRI) as it applies to the physical entry and exit (S) points to the network. There are two frame formats, **TERMINAL EQUIPMENT (TE),** defining those sent by terminal equipment to the network (entry), and **NETWORK TERMINATION (NT)** frames, defining frames sent the opposite way (exit). The first two bits of these frames (F: framing; L: DC balancing) are used for synchronization. Each B channel information byte following the F and L bits is followed by D channel bit information. Additional L and F bits are used to maintain framing and synchronization. The A bit in the network termination frame activates or deactivates terminal equipment, putting this equipment on-line or removing it from the network.

Line arbitration for the sending of D channel control information by a terminal is achieved through use of the E bit in the network termination frames, which is a copy of the previous terminal equipment frame D bit information. S bits are undefined at this time. When a terminal has D information to send, it first observes the E bits of the NT frames. A specified consecutive number of E bits must be high before the station can send its D information. The consecutive high states of the E bits indicate that no D control information has been sent in any previous terminal equipment frames and the station can now send a terminal equipment frame with D control information.

Network termination and terminal equipment frames are transmitted at a rate of 4,000 frames per second. Each frame contains 48 bits, thus making the data rate 192 kbps. In contrast, primary rate interfaces transfer data at 1.544 Mbps (North America) and 2.048 Mbps (Europe), which are the rates specified by the T1 channel specification. The North American version of a primary rate interface (PRI) frame (Figure 10.3) contains a framing bit followed by twenty-three or twenty-four B channel fields and a D channel bit.

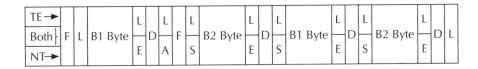

Figure 10.2 Basic Rate Interface (BRI) Frames

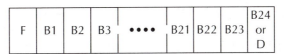

1.544 Mbps U.S. PRI Frame

2.048 Mbps European PRI Frame

Figure 10.3 Primary Rate Interface (PRI) Frames

The European primary rate interface frame also begins with a framing bit but is followed by sixteen B channel fields, a D channel bit, and fifteen more B channels. Frames are transmitted at a rate of 8,000 frames per second, resulting in the different data rates for U.S. and European primary rate interface frame transmissions.

● 10.6 ISDN DATA LINK LAYER

The protocol used by ISDN at the data link layer is closely related to the **HIGH-LEVEL DATA LINK CONTROL (HDLC)** protocol and is called **LINK ACCESS PROTOCOL FOR D CHANNEL (LAPD)** and **LAPB** for B channels. Details of the protocol specifications are found in CCITT Q920 and Q921. The basic LAPD frame (Figure 10.4) is identical to the HDLC frame. It starts with the 7E flag, followed by an address field. The use of this field differs from the HDLC basic form

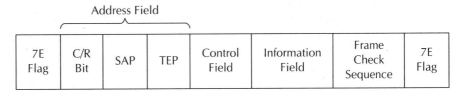

C/R-Command/Response
SAP-Service Access Point
TEP-Terminal End Point

Figure 10.4 LAPD Frame

by including a **SERVICE ACCESS POINT (SAP)** address and a **TERMINAL END POINT (TEP)** address. An additional bit in the address field, called the **COMMAND/RESPONSE (C/R) BIT,** is used to identify the frame as a command (0 for traffic from terminals, 1 for traffic from the network) or a response (opposite states) message. The control fields identify the type of frame and keep track of the frame sequence as described in chapter 4 for SDLC and HDLC protocols. Information, supervisory, and unnumbered control fields are the same as their HDLC counterparts with the following exceptions. The set normal response unnumbered frame is replaced with set asynchronous balanced mode extended. It functions similarly to the set normal response mode by establishing a data link for acknowledged data transfers (information frames). An additional unnumbered frame, **TRANSFER ID (XID),** is included to allow stations to identify themselves for line management purposes.

LAPB frames are similar to those of LAPD. XID and unnumbered information (UI) frames are not used with LAPB. Also, the frame numbers are restricted to modulo 8 (3-bit frame numbers) while LAPD uses modulo 128 (7-bit frame numbers).

Network layer specifications are included in the LAPD protocol. Call setup, flow control, call teardown, routing, error detection and recovery, and other network functions are described in CCITT Q930 and Q931 for the LAPD protocol.

• 10.7 ISDN APPLICATIONS

In 1984 the hamburger chain giant, McDonald's Corporation, headquartered in Oak Brook, Illinois, decided that there was a requirement to integrate their numerous network and communications needs into a more efficient and cost-effective system. Contracts between McDonald's and Illinois Bell Telephone and AT&T were signed for the telephone companies to provide an integrated network based on the ISDN standard as specified by CCITT/ISDN standards. Locally, McDonald's was divided between the home office (known as the Plaza), the Lodge, and Hamburger University. These facilities were up to 1.5 miles apart, with most interconnecting communications being provided on leased lines or the public telephone system. Networks at the sites and other services included an IBM Systems Network Architecture (SNA) network, office automation (AO), UNIX-based network supporting the product development area, a CENTREX system handling regular telephone and Telex communications, and various leased lines and packet-switch facilities to interconnect the networks.

Illinois Bell offered to establish a centralized system based on ISDN to integrate the functions of the various networks into a single information-transport network. Employees at the home office were equipped with ISDN telephone sets that included a ISDN data module that serves as an NT1 interface directly into the ISDN network. **AUTOMATIC NUMBER IDENTIFICATION (ANI)** is included to display the caller's number before the call is answered. Employees can use this ability to either screen calls or to tap into a caller's data base before actual dialogue is begun with the caller.

To manage the network, AT&T provided a package called **NETPARTNER.** This management software took the responsibility of maintaining the network from McDonald's and placed the maintenance and monitoring tasks onto AT&T. The network can still be monitored from a local terminal, but data are directly available to the AT&T support teams. NETPARTNER tests ISDN lines for integrity and reports any problems that are detected. Translations between ISDN and analog lines are also checked and/or changed as required.

Office automation (OA) users access the OA system by dialing a four-digit access number into either a X.25 packet-switched D channel or a B-channel circuit-switch connection. The McDonald's office automation system makes the following services available to their employees:

1. Word processing
2. Data base management
3. Graphics
4. Spread sheets
5. Electronic mail
6. Time management
7. Access into other ISDN services
8. Printer accesses to all areas to obtain hard copies quickly

Printer interfaces throughout the system are served by a 9.6-kbps B-channel circuit switch connection allowing high-speed (eight to ten pages per minute) laser printer capability. No printer is more than 25 feet from employees. The quantity of printers coupled with the speed of printing results in efficient hard copy service. Other benefits of office automation via ISDN to McDonald's include:

1. Easily accessing multiple hosts and applications
2. Large and quick file transfer
3. Easy implementation of employee situation change—office location change, new hire, termination, and so on.

Use of IBM PC and facsimile services is also incorporated into the McDonald's ISDN network through terminal adapters (TA) that do the necessary conversion between IBM, facsimile, and ISDN formats and protocols.

McDonald's is noteworthy because it was the first large, nationwide business to incorporate the use of ISDN. The integrated nature of ISDN was used to consolidate existing services and to add many of the newer features described earlier. McDonald's is not the only large company using ISDN, but it is was the first large company willing to invest money and time to reap the benefits provided by ISDN.

The Mellon Bank of Pittsburgh, Pennsylvania, uses ISDN to handle many services between its Pittsburgh and Philadelphia banks, using Bell Pennsylvania as a local carrier and MCI for long-distance connections. The main application is for video teleconferencing, transferring digitized video data at a rate of 112 kbps. Additional tests using ISDN with Mellon are being performed in the hopes of raising the rate to 128 kbps without degrading the resolution of the video.

Other uses of ISDN at Mellon are customer-related. Using a system called Global Cash Management Services, Mellon handles cash disbursements, collections, and electronic funds transfers using ISDN as the prime network. Mellon is also testing the use of automatic number identification (ANI) through the use of the D channel. Identification information about the calling party's number arrives on the D channel and is used to call up the customer's files to an agent's terminal before the voice call is connected through to that agent.

An October 1987 article in *Data Communications* entitled "What ISDN Will Mean for Australia" detailed projections by Australia's Telecom service to implement ISDN throughout that country by early 1989. The plans included interconnecting PBXs to the public telephone network once ISDN adapters were put into place. Another large user that has plans to integrate ISDN into its nationwide communications facilities is the West German Bundespost.

Two additional users of ISDN services, Shell Oil and Tenneco Corporation, receive the service from Southwestern Bell. Another major supplier, Southern Bell of Atlanta, provides commercial ISDN services for AT&T, Hayes Microcomputer Products, Contel Corporation, Southwest Banks, and Digital Equipment Corporation (DEC).

A large number of users employ the ISDN network for teleconferencing, telemarketing, and to build onto existing wide area networks (WANs). Eastman Kodak implemented ISDN in 1988 for voice teleconferencing. American Express uses ANI to speed customer assistance, as does American Transtech, a telemarketing firm. The list of ISDN users keeps growing, with each user finding its individual need for integrated and special services.

● 10.8 BROADBAND ISDN

Proposals by the CCITT to expand upon ISDN features to promote broadband communications in the form of a **BROADBAND ISDN (BISDN)** are currently under discussion and development. The new standard is to be based upon the concept of an **ASYNCHRONOUS TRANSFER MODE (ATM),** incorporating fiber cable as the medium for data transfer. The BISDN specification sets a maximum length of 1 km per cable segment but is making provision for repeater-type interface extensions. The data rate expected on the wider-bandwidth fiber-optic cables will be either 100, 155, or 600 Mbps, depending on application and point in the network.

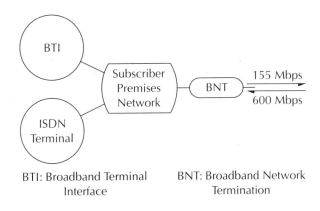

BTI: Broadband Terminal
Interface

BNT: Broadband Network
Termination

Figure 10.5 BISDN Configuration

In composing the new specification, the authors are required to meet existing ISDN interface specifications as well as new broadband network needs. A typical network access arrangement is shown in Figure 10.5. A standard ISDN terminal (or network interface) and a **BROADBAND TERMINAL INTERFACE (BTI)** are serviced by the **SUBSCRIBER'S PREMISES NETWORK (SPN),** which multiplexes the incoming data and transfers them to the broadband node, called a **BROADBAND NETWORK TERMINATION (BNT),** which codes the information into packets used by the BISDN network. Data transfers to and from the BISDN network may be asymmetric. That is, access on and off the BISDN network may be done at different rates depending on system requirements. The illustration shows data entry to the network to be at 155 Mbps and from the network, 600 Mbps.

Subscriber's premises networks can be a private branch exchange (PBX), token ring, or other local area network configuration. The BNT interface with supporting software is to be capable of recognizing the protocols and frame formats of the data required by the interfacing network.

● **10.9 ASYNCHRONOUS TRANSFER MODE**

The asynchronous transfer mode (ATM) is a method by which data enter and exit a network in an asynchronous fashion rather than a time-dependent synchronous arrangement. ATM uses labeled channels transferable at fixed rates, which can be set anywhere from 16 kbps up to a system's maximum rate. Data entering the network are transferred into fixed time slots called **CELLS,** which are identified by a **LABEL** in the cell header (Figure 10.6). The label defines a **VIRTUAL CHANNEL IDENTIFIER,** which indicates the node source and destination of the

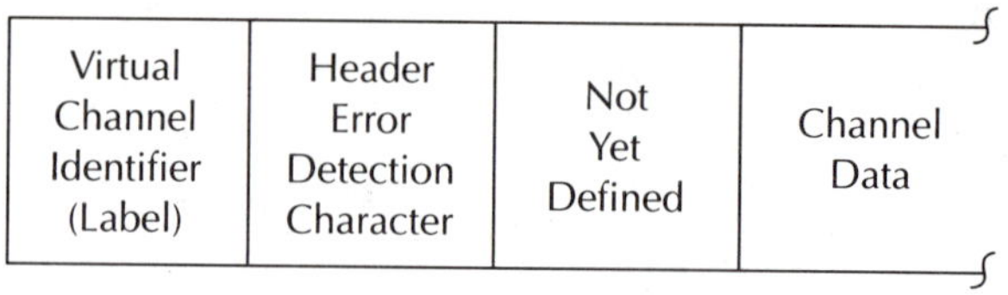

Virtual Channel Identifier (Label)	Header Error Detection Character	Not Yet Defined	Channel Data

Figure 10.6 ATM Header Cell

data packet. The channel is virtual rather than specific, allowing the actual physical routing and network entry and departure times to be determined based on network availability and access rights. Following the header label is the header error-detection character, which could be a form of **CYCLIC REDUNDANCY CHECK (CRC)** or any other error detection format. This error detection is for the header label only. Separate error-detection methods are employed for the data field. A third section in the header has yet to be fully formalized and remains unidentified at this time.

● 10.10 BISDN CONFIGURATION

A possible BISDN configuration (Figure 10.7) begins with data supplied to the system through various peripheral devices or networks. Each is interfaced to the **ACCESS NODE** of a BISDN network through a **BROADBAND DISTANT TERMINAL (BDT),** which is responsible for electrical to optical conversions, multiplexing of peripherals, and maintenance of the local system. Access nodes concentrate several BDTs into a number of high-speed fiber lines directed through a feeder point into a **SERVICE NODE.** The service node manages the majority of control functions for system access. It handles call processing, administrative functions, switching, and maintenance functions. The service node uses distributed management through functional modules within its area of influence. These functional modules are connected in a star configuration (Figure 10.8) and include switching, administrative, gateway, and maintenance modules. The central hub oversees the operation of the modules and acts as the end user interface for control signaling and traffic management.

User terminals close to the central office bypass the use of access nodes and are connected directly to the BISDN network through a service node (Figure 10.8). The larger bandwidth of fiber cables allows for higher data rates and numerous channel-handling capacity for the BISDN system. Once the specifications are completed, applications and use are expected to follow quickly.

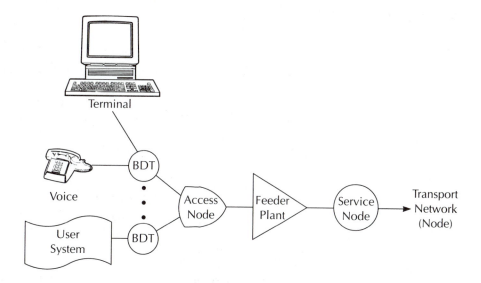

Figure 10.7 BISDN Access

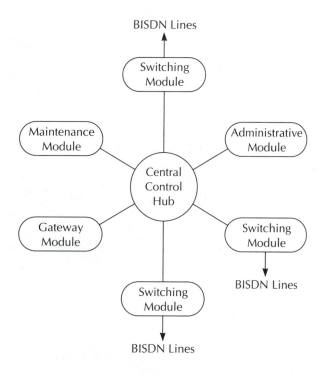

Figure 10.8 BISDN Service Node

One of the preliminary digital and data communications test instruments in use today is the **PROTOCOL ANALYZER.** This necessary test set is used to monitor the performance and help determine problems on a network. Many analyzers have been developed, some to meet specific types of networks and others to be more universal. One such analyzer is the Hewlett Packard HP4954A series Protocol Analyzer (Figure 10.9). The set has a monitor screen, keyboard, internal 20-Mbyte hard drive, and accessible 3½″ floppy drive. Additionally, there are a number of LED indicators on the front panel. Connectors are provided on the back of the set to allow access into various networks. These include connections for RS232C, RS449, ISDN, X.25, and System Network Architecture (SNA) access. An additional RS232C connector is provided for connection to a printer used for making hard copies of analyzer results.

The HP4954A series was developed to be used with X.25, SNA networks and systems that use high-level data link (HDLC), **SYNCHRONOUS DATA LINK CONTROL (SDLC),** BISYNC, or asynchronous protocols. The type of test being performed on the network is controlled through the keyboard using ''soft keys'' defined by previously written or user-developed software. The analyzer incorporates a version of the C programming language called **DataCommC** to allow user-specific applications to be written and executed.

To analyze a system's performance, protocol analyzers use statistical methods to record and display the amount of data being transferred and the number and types of errors experienced during the transmission. These errors include bit errors (parity, cyclic redundancy check [CRC], and longitudinal redundancy check

Figure 10.9 HP4954A Protocol Analyzer
Photo Courtesy of Hewlett Packard

Figure 10.10 With the X.25 and SNA network performance analyzers, easy-to-read graphs let you quickly view and assess such key performance statistics as line utilization, overhead, response time, and network errors. Photo and Caption Courtesy of Hewlett Packard.

[LRC]), framing errors, and loss of carrier. The transmission of data can be captured and displayed on the screen at various selected points. These points are triggered by detection of framing elements, errors, or after a specific number of bits have been transferred. The screen then displays the data, along with interpretation of certain bits or characters. When capturing data based on an error, the position of the error is shown, assisting the user in analyzing and determining the cause of the error.

Protocol analyzers are capable of data and error simulation, which allows the user to inject a specific data pattern with or without errors. Capturing the data on the receiving end, the user can determine what errors occurred, as well as where they occurred. Since the data sent are known, any deviations from the data will be actual errors. Errors can be injected to test the error-detection or error-correction capabilities of the receiver.

Figure 10.10 shows a HP4954A being used to emulate a device entry into a network. The operator can select the operation to be run on the system by keying in the selection through the keyboard or by running emulation programs from a floppy disk set into the drive. The HP4954A series analyzers are capable of emulating data terminal equipment (DTE) or data communications equipment

(DCE) using any number of data link protocols. The same types of networks that can be tested by the analyzer can also be accessed under emulation. The emulation facility aids system developers, who can test the response of the system to a device planned to be added to it. Customized applications can be tested before they are adopted to the network, saving a lot of development and network application test time.

SUMMARY

Integrated services of voice, data, video, teleconferencing, and other applications are receiving wider and wider acceptance within the data and digital communications community. The need to communicate a lot of information and ideas quickly between distant and far-reaching locations increases as the information age continues to move forward. Providing integrated services on the public network is the thrust of the ISDN system.

With the growth of networks and the use of public networks for ISDN and other communications needs, the requirement to test and develop applications for these systems has become a priority concern. Many types of testing and developing equipment have been developed to assist in this endeavor. The protocol analyzer is one of the basic instruments in use today to monitor, test, and emulate network activity and applications.

GLOSSARY

Access node—Entry point into an ISDN network.

Basic (B) channel—A 64-kbps data channel.

Basic rate interface (BRI)—Defines an ISDN network consisting of two B channels and one D channel.

Cell—BISDN time slot.

Control (D) channel—An 8- or 16-kbps channel used for control information.

Customer exit/entry nodes—Terminal Equipment 1 nodes used for user interfaces.

DataCommC—C language version used with the HP4954A protocol analyzer.

Label—Asynchronous transfer mode cell identifier.

Link access protocol (LAP)—ISDN data link protocol closely related to HDLC.

NETPARTNER—AT&T ISDN network management software.

Primary rate interface (PRI)—Defines an ISDN network consisting of twenty-three B channels and one D channel.

Protocol analyzer—Test set used to analize digital traffic.

Service node—Manages ISDN system access control functions.

Terminal equipment (TE)—Physical user network entry equipment.

Virtual channel identifier—Designates asynchronous transfer mode data packet source and destination identifier.

QUESTIONS

10.1. What does the acronym ISDN mean? How does it differ from regular telephone service?

10.2. What type of information is carried by the ISDN B and D channels? At what data rates is this information conveyed for each channel?

10.3. How many channels make up the basic rate interface (BRI) and the North American and European primary rate interfaces (PRIs)?

10.4. Contrast the bandwidth requirements for the North American and European primary rate interfaces. Why are they different?

10.5. What purpose does an ISDN terminal adapter (TA) provide?

10.6. What is the difference between TE1 and TE2 ISDN interfaces?

10.7. Describe each of the ISDN termination reference points: R, S, T, U, and V.

10.8. What ISDN interface category do PBXs fall into?

10.9. Which ISDN unit applies to the OSI physical layer?

10.10. What is the purpose of the F and L bits of the terminal equipment (TE) frame format?

10.11. Which D channel bits are used for line arbitration?

10.12. What is the name for the ISDN data link protocol? Which data link protocol does it closely resemble?

10.13. What is ANI? What is it used for?

10.14. What is the purpose of the XID unnumbered ISDN frame?

10.15. What is the purpose of broadband network termination and the subscribers premises network?

10.16. How are BISDN cells identified? What information is conveyed by the identifying process?

10.17. What element of a BISDN network allows peripherals to interface to the network?

10.18. What is the purpose of a BISDN service node?

10.19. Protocol analyzers are one of the most significant pieces of test equipment available for use on today's networks. What are the principal functions that can be performed, monitored, or tested with a protocol analyzer?

RESEARCH ASSIGNMENT

10.1. You are to establish an ISDN-type network for your office. PBXs, Ethernet, and token ring LANs are already in place. Consider and list the hardware and software requirements to interface the existing networks into one cohesive network that also interfaces with ISDN. You will have to conduct research in the subject areas of previous chapters in addition to this chapter's material to complete your list. Give a reason for each selection on your list.

10.2. Research the current applications for ISDN. Include why the selection to use ISDN was made.

10.3. Research BISDN applications. Include why BISDN was selected over ISDN.

Answers to Odd-Numbered Questions

10.1. Integrated services digital network. It provides voice, video, digital data, and other services on a single line. Regular telephone service can supply only one service at a time.

10.3. Basic rate interface: 2B + 1D; North American primary rate interface: 23B + 1D; European primary rate interface: 30B + 1D

10.5. Terminal adapters provide the translation between non-ISDN and ISDN network formats.

10.7. R: TE2 to TA point; S: TE1 and TA to NT2 point; T: customer to common carrier point; U: local loop to central office point; V: local loop to exchange termination point

10.9. Network termination (NT)

10.11. E bits

10.13. Automatic number identification; it allows receiver to identify the calling party's telephone number.

10.15. Subscriber's premises network multiplexes incoming data and transfers them to the broadband node called a broadband network termination, which places the data into ISDN packets to be used on the network.

10.17. Access node or service node

10.19. Protocol analyzers are used to monitor activity on the network lines and interfaces between DTE and DCE equipment. They may be used to gather statistical information about errors. Data on the line can be captured and analyzed in binary, HEX, ASCII, or protocol codes. Transmit and receive stations can be emulated by a protocol analyzer.

Eleven • Fiber-Optic Communications

OBJECTIVES

The purpose of this chapter is to give the student a firm understanding of the basics of fiber-optic systems. Topics include the following:

1. Light propagation theory
2. Fiber-optic cables
3. Light sources and detectors
4. Core indexing
5. Cable losses and distortions
6. Fiber distributed data interface
7. Network applications

It is not the intent of this chapter to be a thorough course on fiber optics, for that would require a separate text. Rather, the thrust is toward a good understanding of optic basics and their application for network usage.

ACRONYMS

AlGaAs—**Aluminum Gallium Arsenide**
ANSI—**American National Standards Institute**
CRC—**Cyclic Redundancy Check**
EMI—**ElectroMagnetic Interference**
EMP—**ElectroMagnetic Pulse**
FCS—**Frame Check Sequence**
FDDI—**Fiber Distributed Data Interface**
FDDN—**Fiber-Optic Distributed Data Network**
LED—**Light-Emitting Diode**

MAC—**M**edia **A**ccess **C**ontrol Protocol

MFOTS—**M**ilitarized **F**iber-**O**ptic Tran**S**mission Network

NRZ—**N**on-**R**eturn to **Z**ero

NRZI—**N**on-**R**eturn to **Z**ero Mark **I**nversion

OSI—**O**pen **S**ystems **I**nterconnection

PHY—**PHY**sical Protocol

PMD—**P**hysical **M**edia-**D**ependent Protocol

SMT—**S**tation **M**anagemen**T** Protocol

● 11.1 INTRODUCTION

Early data communications owes its existence to the telephone network. Without it, any form of communications would have been difficult. Because the network was in place when data communications came into being, it was not necessary to create an interconnecting network of cables and switch stations. The telephone company uses combinations of twisted pairs of wires and coaxial cables for much of its land-based interconnections. It has been shown that these media restrict the rate of data flow due to limitations impressed by bandwidth specifications. Equally as important is the effect of various impairments and degradations inherent to twisted-pair and coaxial cabling.

With the advent of optical media for the purposes of transferring digital data, an era of expansion in the data communications field was born. The advantages of fiber-optic cables and communications systems include increased bandwidth, less sensitivity to magnetic interference, and increased security. To understand why these advantages exist is to understand the nature of light as a transmitting medium. Light rays are high-frequency electromagnetic waves with a short wavelength in the micron (μm) range. For optical fiber communications, light is expected to travel through a medium of glass. Light in the visible range exhibits wavelengths between 0.4 and 0.7 μm and is highly attenuated by glass. This attenuation does not make for efficient transfers of data through optic cables. Ultraviolet waves, which are attenuated through glass more than visible light, also are not used for fiber-optic communications. On the other hand, light waves with wavelengths between 0.85 and 1.6 μm (infrared range) travel very efficiently through glass.

Since glass is used as the conducting medium and light waves are at such high frequencies, fiber-optic transmissions are immune to electrical interferences. Glass, being an insulator, will not conduct electrical signals or radiations. When using coaxial or twisted-pair cabling, currents can be induced from one metallic wire to another by the magnetic fields created by the current in the wire. Light rays are propagated free of these magnetic effects. This results in a lack of magnetic interference such as cross talk between neighboring lines and effects of static electricity from causes such as electrical storms.

The transmission of light is not affected by electromagnetic effects or reactive elements of capacitance and inductance that occur in coaxial or twisted-pair cables. Instead, light travel through a medium is restricted by the light signal's wavelength (discussed later) and the medium's opaqueness. The important result of this difference is the increased bandwidth available to signals traveling via light versus electrical energy. Fiber-optic cables can manage data rates exceeding 1000 Mbps due to increased bandwidth.

One additional advantage of fiber-optic cables is their small size and light weight. A complete cable measures between 0.1 and 0.15 millimeters in diameter, with core sizes ranging from 5 to 600 μm in diameter. The size does have a negative side, though. Because of the smallness and the need to align cable ends precisely to sending and receiving devices, and to each other (in the case of a splice), couplers and splices used for fiber-optic systems must be precisely aligned, requiring very careful mating to cable ends. This makes fiber-optic cables more difficult to install than coaxial or twisted-pair cables, but this is a minor drawback considering the benefits that are gained. These benefits are (1) more bandwidth capacity; (2) immunity to **ELECTROMAGNETIC INTERFERENCE (EMI)**; (3) immunity to **ELECTROMAGNETIC PULSES (EMPs)** from nuclear explosions; (4) less susceptible to weather conditions (does not corrode or rust); and (5) more difficult to tap into (better security). Another consideration is that the use of optical systems does require that electrical signals be converted to light signals and back again at the receiving end. This is a process that is incorporated when a light source is modulated by the digital data stream and demodulated at the receiver.

● 11.2 BASIC CONCEPTS OF LIGHT PROPAGATION

One theory of light propagation is that light travels from a source to a receptor as a collection of ray elements, simply called **RAYS.** These rays have direction and are of varying strength depending on their energy content. As long as these rays travel through a single material (or medium), they move in a straight line. Once they encounter a second medium, they are bent, either continuing forward at a new angle **(REFRACTION)** or bounced backward going in a direction not quite exactly from where they came **(REFLECTION).** The amount of refraction or reflection is dependent on the density and composition of the two media involved. A **REFRACTIVE INDEX** (n) has been developed to predict the effect on light as it attempts to pass from one medium to another. This index uses air as a standard reference of 1. Refractive indexes for any other material are found by dividing the speed of light through air by the speed of light through the other material. Other common indexes are quartz at 1.46, water (1.33), fused silica (1.46), silicon (3.5), and glass (1.5).

Figure 11.1 illustrates a light ray being reflected as it meets a surface after traveling in air. The initial ray striking the surface is called the **INCIDENT RAY.** An angle measure between the incident ray and a **NORMAL** reference line that is

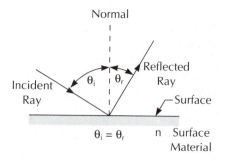

Figure 11.1 Incident and Reflected Rays

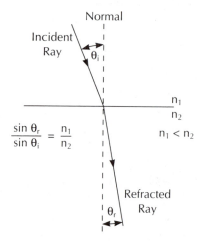

Figure 11.2 Refracted Ray

perpendicular to the surface is called the **INCIDENT ANGLE,** θ_i. The reflected ray and its angle of reflection (θ_r) are also shown. To achieve reflection, the refractive index of the surface material must be greater than the incident medium. If the reverse is true, as in the case of an underwater light source projecting a beam of light into the air, refraction occurs (Figure 11.2). Angles of incidence and refraction are measured compared with a reference that is the perpendicular normal to the meeting line of the two media.

A formula called **SNELL's LAW** is used as a measure of the relationship between the ratio of refractive indexes and the ratio of the incident and reflective or refractive angles. That equation is:

DIGITAL & DATA COMMUNICATIONS

$$\frac{\sin\theta_r}{\sin\theta_i} = \frac{n_1}{n_2} \qquad\qquad (11.1)$$

where n1 is the incident medium and n2 the reflective or refractive medium. θ_i is the angle of incidence, and θ_r is the angle of reflection or refraction.

Two important angular measurements are the critical angle (θ_c) which is measure of the angle of *MAXIMUM REFLECTION* from the normal (a line perpendicular to the reflecting surface), and Brewster's angle (θ_B), which is a measure of the *MINIMUM ANGLE OF REFLECTION*. Light rays reflecting on a surface must meet that surface at an angle between those two values. Their equations are:

$$\sin \theta_c = n_2/n_1 \qquad\qquad (11.2)$$

$$\tan \theta_B = n_2/n_1 \qquad\qquad (11.3)$$

Example 11.1   Given the angle of incidence as 12.5°, what is the reflective/refractive angle of the light ray if it enters from air into water? Is this a reflected or refracted ray?

Solution: The refractive index for air is 1.0 and for water, 1.33. Using equation 11.1, the angle can be calculated:

$$\theta_r = \sin^{-1} [n1 \times \sin \theta_i/n_2] = 11.37°$$

Equations 11.2 and 11.3 determine the range of reflective angles for this example. The critical angle is 48.75°, and the Brewster angle is 36.94°. That means that this ray is refracted into the water rather than reflected, since 11.37° is below the Brewster angle (minimum reflection angle) and does not exceed the critical angle of maximum reflection. Had the results lain between these two values, the light ray would be reflected rather than refracted.

The importance of reflection and refraction cannot be overemphasized. In a fiber-optic cable, light is directed into the core of the cable. The most ideal situation is for the light to enter right at the center line of the core. When this happens, the light does not approach the walls of the core and travels fairly unimpeded through the cable. However, to align a light source so that the light is all directed down the core's center is extremely difficult and cost-prohibitive. As such, quite a bit of the light source is going to enter the cable at angles away from the center. Figure 11.3 shows a basic fiber cable entry point. Light enters the core and is slightly refracted due to the differences in refracted indexes. The rays of light travel in the core until they hit the core wall, which is encased in a differently doped material called the **CLADDING.** By changing the doping (insertion of foreign material into a pure material), the refractive index (n) is changed. The purpose of the cladding is to

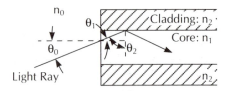

Figure 11.3 Angle of Incidence (θ_1), Refraction (θ_r), and Reflection (θ_2)

reflect most of the light rays to keep the light inside the core. The reflected rays continue traveling down the cable's core, bouncing off the cladding walls as it goes along. Eventually, the light will exit the cable at the receiving end.

● 11.3 PROJECTING LIGHT INTO THE CABLE

Many methods have been developed to direct light into a fiber-optic cable. Each involves the use of a light source and some method to align the light source with the cable entry. Two commonly used sources are **LIGHT-EMITTING DIODES (LEDs)** and **INJECTION LASER DIODES.** The principal differences between these two sources are the amount of light emitted by each and the cone of emitted light projected from their surfaces. Light-emitting diodes emit a much lower level of light (-15 dBm power level) than do the laser diodes (-6 dBm), which concentrate light into a tighter cone pattern. The emitting patterns for both sources are illustrated in Figure 11.4. The light-emitting diode pattern is broader, since the

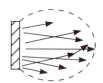

(a) LED (Microwatts)

(b) Injection Laser (Milliwatts)

Figure 11.4 Radiating Patterns for LED and Injection Laser

Cross-Section of Connector

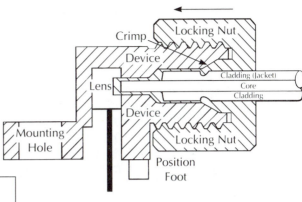

Termination Instructions

1. Cut cable squarely with sharp blade or hot knife.

2. Strip jacket back with 18 gauge wire stripper to expose 0.10–0.18" of bare fiber core.

   Avoid nicking the fiber core.

3. Insert terminated fiber through locking nut and into the connector until the core tip seats against the molded lens inside the device package.

   Screw connector locking nut down to a snug fit, locking the fiber in place.

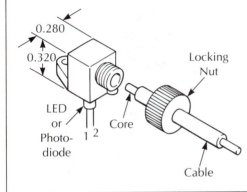

Figure 11.5 Device Connector for Fiber-Optic Cable. Courtesy of Motorola, Inc.

light is emitted from the surface of the diode. Lasers project light from their edge, forming a more intense and narrow cone. Coupling using lasers is more precise than with the light-emitting diode.

To couple light from the source into a fiber cable requires the source to be held firm to the cable. If the coupling shifts at any time, the amount of coupled light is reduced. Devices used to connect light sources to cables vary in shape and size according to the type of light source and the actual diameter of the fiber to be used. An example of a connecting device is shown in Figure 11.5. The light source is mounted inside the connector. The screw end is slipped onto the cable, which is then inserted into the connector. The screw cap is then brought to the connector and screwed on. As the cap is hand tightened onto the connector, it crimps down onto the cable and holds it in place. The end of the fiber cable must be cut flat and clean to assure best coupling.

Figure 11.6 is a cross-section of fiber-optic cable. The core, constructed of silica (silicon dioxide, or SiO_2), occupies the center with a diameter between 5 and 600 microns (μm). A silica cladding with a different refractive index acts as a reflective wall surrounding the core. It brings the cable up to a diameter of between 125 and 750 μm. The cladding is covered with a silicone coating and an inner jacket to provide strength against the bending the inner cable will experience as a result of installation and environmental shifts during its lifetime. An additional strength member and an outer jacket surround the inner jacket to bring the overall cable size to between 0.1 and 0.15 millimeters in diameter. The light signal to be transmitted is directed to the small core.

Application of imaging for fiber optics involves determining where to place a light source and a photo receptor (fiber cable, etc.). Figure 11.7 illustrates the concept of using several light rays and a photodetector as a receptor. The receptor element is limited in size, so some of the light sent through the lens does not strike the detector surface. Note that ray a is bent at such an angle that it misses the photodetector. Ray b is shown as having the maximum angle of incidence that allows it to be detected. This angle is measured in relationship to a normal line, which is in parallel to ray c. Ray c is the ideal situation—a light ray projected along the center line of the lens or cable.

The angle formed by ray b and the normal is called the **ACCEPTANCE ANGLE.** All incident light rays approaching the lens at this angle or less, when grouped together form an **ACCEPTANCE CONE.** Acceptance cones define the range of incident angles that rays can approach and successfully enter the core. The maximum angle of acceptance (θ_A) is found by using the following formula:

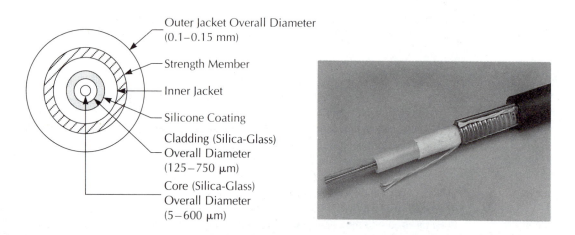

Outer Jacket Overall Diameter
(0.1–0.15 mm)

Strength Member

Inner Jacket

Silicone Coating

Cladding (Silica-Glass)
Overall Diameter
(125–750 μm)

Core (Silica-Glass)
Overall Diameter
(5–600 μm)

Figure 11.6 Fiber Cable Components. Photo Courtesy of Anixter Brothers, Inc.

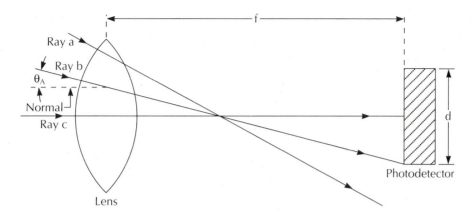

Figure 11.7 Acceptance Angle (θ_A)

$$\theta_A = \tan^{-1} (d/2f) \qquad (11.4)$$

where d is the diameter of the photodetector (or cable core) and f is the focal point of the lens.

The effect of acceptance cone, refraction, and reflection as considerations for a fiber-optic cable is illustrated in Figure 11.8. A single incident ray is shown approaching the cable within the acceptance core. The refractive index differences between n_0 and n_1 cause the incident wave to be refracted slightly. This could cause the ray to miss the focal point as it enters the core. This is not crucial. What is important is that the ray entered within the acceptance cone. Any rays trying to enter outside the cone will be refracted into and completely absorbed by the cladding. The illustrated ray continues on a straight line within the cable core until it strikes the cladding (refractive index n_2). The difference in the refractive indexes causes most of the light to be reflected back into the core. However, the cladding is not completely opaque, so some of the light energy is absorbed into the cladding. This absorbed amount will travel through the cladding until it meets the silicon coating. Here some of this absorbed energy is radiated into the silicon, and some is reflected back into the cladding.

The original ray continues through the fiber core, with most of its energy being reflected against the cladding until it emerges at the distant end. The radiated energy that results from the energy absorbed into the cladding reenters the core at a refracted angle. Most of it is reflected back into the cladding. The level of absorbed energy is sufficiently low that for most applications it can be ignored. The bulk of the light ray's energy travels through the cable. Notice that the closer the ray enters the cable at the normal angle (parallel to the normal or center line), the fewer times it meets the cladding and the less energy will be lost due to absorption. As can be seen, the placement of the light source to the fiber is of considerable importance.

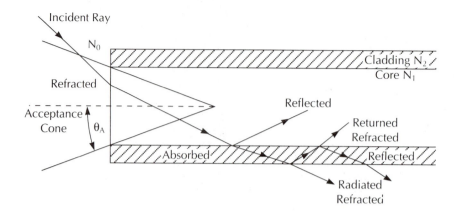

Figure 11.8 Incident Ray within Acceptance Cone

Another aspect that affects the results of data transmissions using fiber optics is the propagation time through the core.

Light that enters the core at the center line travels in an unimpeded straight line through the core (as long as the core itself is straight). Light entering at any other angle will eventually hit the cladding and be ''bounced'' down the cable. These rays travel greater distances than the ray entering at the center of the core. The larger the angle of incidence, the further the rays have to travel before exiting the core. As a result, the rays emerge at different times, resulting in a phenomenon called **PULSE SPREADING,** which causes the replicated electrical information to be distorted by the varying arrival times of the light rays through the cable. Fortunately, the distortion is not great, but it does present a limiting factor to the length of the fiber cable and the data rates that are propagated through it. If the length is too long, then the spreading can cause the loss of digital bits and create data errors.

● 11.5 FIBER-OPTIC CABLE MODES

Fiber-optic cables are constructed to operate in one of two modes, *SINGLE MODE* or *MULTIMODE* (Figure 11.9). The use of a particular mode depends on the requirements for the communications network. Tight coupled systems using laser diodes for sources employ single-mode cables, which are used with a highly directional light input. The single-mode cable concentrates light energy passage to the center of the core, which is very narrow (6 to 12 μm in diameter). Recall from the previous discussions that a light ray concentrated at the core's center moves the quickest through the core with the least distortion and attenuation.

Multimode cores, on the other hand, are designed to accept the entrance of light in a fairly wide acceptance cone. The core diameter is between 50 and 200 μm.

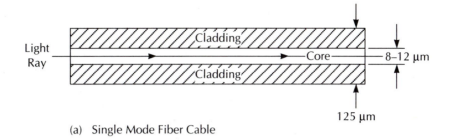

(a) Single Mode Fiber Cable

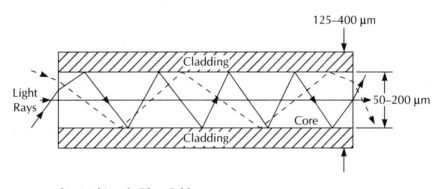

(b) Multimode Fiber Cable

Figure 11.9 Comparison of Single- and Multimode Fiber Cables

Each light ray entering at a specific angle experiences a given amount of attenuation loss and propagation time, which characterize its mode within the cable. While the advantage of the multimode cable is the wider acceptance angle, allowing more light energy to enter the cable, the disadvantage is the distortion of the demodulated signal. This distortion occurs because the different light rays arrive at the end at different times due to the actual distances they must cover through the core. Pulse spreading, similar to that caused by propagation problems in coaxial or twisted-pair cables, results from these propagation differences.

● 11.6 REFRACTIVE INDEXES IN FIBER CORES

In a *STEP INDEX* core used throughout the chapter thus far, the incident ray enters the core, is refracted slightly, and travels through the core as it is reflected from one side of the cladding to the other (Figure 11.10). It has been shown that the cladding absorbs some of the energy each time the ray strikes it. A graded index

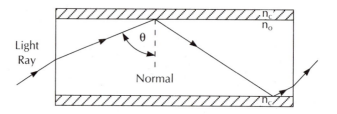

Figure 11.10 Step Index Fiber Cable

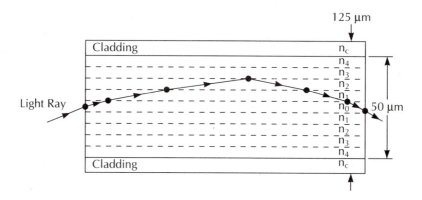

Figure 11.11 Graded Index Fiber Cable

core bends the ray back toward the center, reducing the physical distance it travels. Many of the rays entering the core never reach the cladding, reducing the absorption of the light energy.

The attenuation and propagation problems of a multimode core are reduced by the use of a *GRADED INDEX* core (Figure 11.11). Unlike the step index core in which the refractive index is the same throughout, in the graded index core the refractive index gradually changes from the center of the core out to the cladding. The incident ray enters the cable as with a step index. However, instead of traveling in a straight line until it hits the cladding, it is refracted in small increments as it travels through the core. This refraction bends the ray away from the cladding and back toward the core's center as shown in Figure 11.11. The amount of variation in the refractive index of the graded core is computed using the following formula:

$$n_r = n_0 \sqrt{1 - [(n_0{}^2 - n_c{}^2)/2n_0{}^2]} \tag{11.5}$$

where n_r is the variance of refractive index, n_0 is the refractive index at the center of the core, and n_c is the refractive index of the cladding.

Example 11.2   By what factor should the refractive index of each step in a graded index fiber change if the center index is 1.2 and the cladding index is 1.1?

Solution: Enter the value of 1.2 as n_0 and 1.1 as n_c into equation 11.5, and compute n_r:

$$n_r = 1.2\sqrt{1 - [(1.2)^2 - (1.1)^2]/[2(1.2)^2]} = 1.15$$

● 11.7 LIGHT SOURCES

The two primary sources of light used in fiber-optic systems today are light-emitting diodes and laser diodes. Within these broad areas lie different types for each source. Light-emitting diodes work on the principle that an electrical current flowing through the junction of P and N materials in a semiconductor causes light energy in the form of photons to be released from the semiconductor material. The amount of photon energy that is released is set by the form and amount of impurity doping of the semiconductor material and the amount of current passing through the diode. The more current, the more photon energy is released. Light-emitting diodes come in two basic forms, surface-emitting and edge-emitting. For surface-emitting diodes, the light pattern is almost semispherical, allowing light to be dispersed in a wide area. The light from this type of pattern, illustrated in Figure 11.12, can be easily coupled into an optic fiber cable. Given the large area of dispersion, a certain amount of light will enter the core within the acceptance

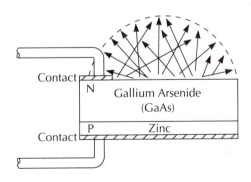

Figure 11.12 Surface-Emitting LED

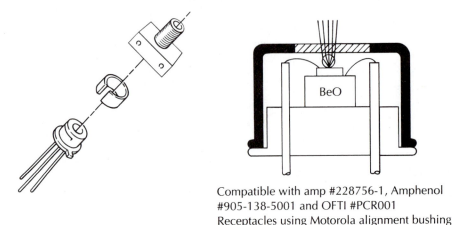

Compatible with amp #228756-1, Amphenol
#905-138-5001 and OFTI #PCR001
Receptacles using Motorola alignment bushing
MFOA06 (included)

Package Cross-Section

Figure 11.13 Photodiode Courtesy of Motorola, Inc.

angle. On the down side, much of the light power emitted from the light-emitting diode is lost outside of the acceptance angle.

An example of a high-speed surface-emitting light-emitting diode (LED) is Motorola's MFOE1201/2 diode. This LED is made of the material **ALUMINUM GALLIUM ARSENIDE (AlGaAs).** The material selected is based on the wavelengths the photodetector is designed to detect. This light-emitting diode is specified to operate at digital rates up to 200 Mbps or analog rates up to 100 MHz. Physically, it is mountable into a fiber-optic connector as shown in Figure 11.13.

Edge-emitting diodes are constructed in a way that forces light to be emitted in a much smaller conical pattern. This pattern, shown in Figure 11.14, allows more of the light power to be directed to and coupled into the fiber core. A tighter pattern also requires tighter alignment of the coupling between the light-emitting diode and the cable. Light-emitting diode sources, because of their lower power and wider dispersion, are suitable for short-distance connections, which minimizes attenuation losses due to the length of the cable. For longer distances, injection laser diodes are used.

Lasers generate highly concentrated light beams by forcing as much light energy produced by them into a tight, highly directional dispersion pattern. The light produced by various forms of lasers has found use in eye surgery and other applications requiring intense and finite use of energy. Injection lasers used for optic fiber communications do not have the intense power required for eye surgery (an

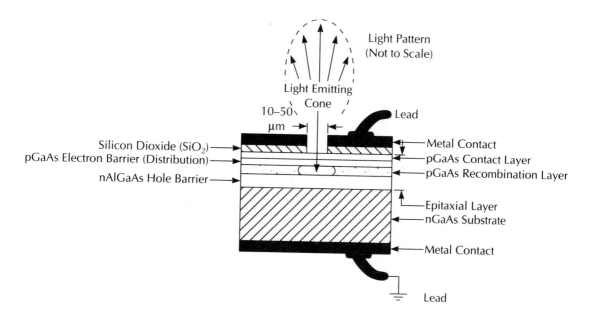

Figure 11.14 Edge-Emitting LED

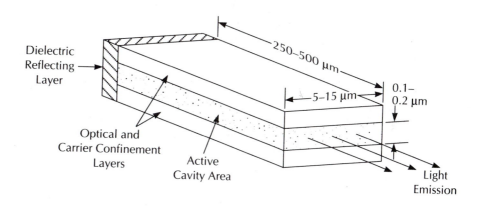

Figure 11.15 Laser Diode Construction

example of the typical configuration of a laser diode is shown in Figure 11.15). They are produced on a much smaller physical scale and require less power to operate. Injection lasers do produce a tight, directional dispersion pattern that can be directed into the fiber, entering at an angle and position close to the core's center. Losses due to attenuation are reduced, increasing the efficiency of energy transfer through the cable. Fiber cable and splicing losses (those incurred because of the connectors used to splice cables together) for lasers are typically 0.5 dB per kilometer (km) of cable, compared with 2 dB per km using light-emitting diodes.

● 11.8 OPTICAL DETECTORS

On the receiving end of the fiber-optic cable is a light-sensitive detector encased in a special adapter (Figure 11.16). The connection must hold the alignment of the end of the cable tightly to the lens of the detector to allow the maximum amount of light to be sensed by the device. A photosensitive device like a photodiode or phototransistor operates on the principle that light striking an exposed section of a PN junction of semiconductor material will cause electron and hole flow activity to begin. The more light applied, the higher the current flow. In essence, as the light increases, the effective resistance of the semiconductor is reduced by the increased activity. Electrical current from an external current source increases in inverse proportion to the reduction of the device's resistance. A typical example of a detector is Motorola's MFOD71/2/3 PIN Photodiode (Figure 11.16), used for short-distance cables. The PIN designation refers to the inclusion of a large amount of intrinsic material between the P and the N material that form the detector. The intrinsic material assures a low leakage current when there is an absence of light and the diode is off. The lack of leakage current also allows for fast responses of the diode to the presence of light, since the diode does not have to overcome this leakage current first. Detector response to light changes is one of the most significant specifications for these devices. The MFOD7X series response time is less than 5 nanoseconds. A 1.0-MHz PIN TTL receiver using the MFOD71 detector is shown in Figure 11.17. The potentiometer, R5, sets the sensitivity of the circuit to the amount of current produced by the photodiode in response to the light it detects. A reference voltage for comparator U1 is set by the voltage divider of R5 and R4. Current changes through the photodiode increase or decrease the bias into the bases of Q1 and Q2. Changes in conduction of Q2, resulting from the bias change, are reflected in the emitter current through R3. U1 acting as a comparator switches from ground (0V) to Vcc (5V) as the voltage across R3 decreases below the reference voltage across R5. Thus, as the light intensity changes between one level representing a logic 1 to another representing a logic 0, the TTL output of the comparator changes between +5V and 0V.

For longer fiber cable applications, a photodarlington transistor such as Motorola's MFOD2302 (Figure 11.18) is used. This particular device detects infrared light sources operating at medium-frequency ranges over fiber cable distances up to 1,000 meters. In comparison, the MFOD7X series is designed for cable systems with maximum lengths of 25 to 32 meters.

```
┌─────────────────────────────┐
│         MFOD71              │
│         MFOD72              │
│         MFOD73              │
└─────────────────────────────┘
```

Cross-Section of Connector

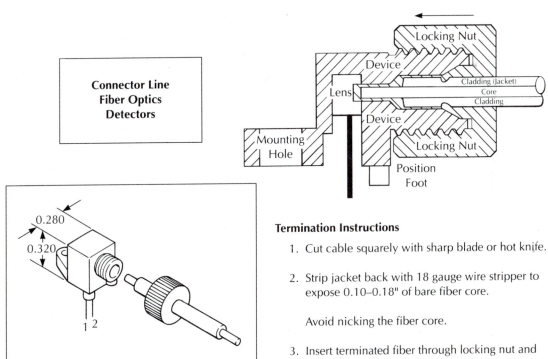

┌─────────────────────────────┐
│   **Connector Line**   │
│   **Fiber Optics**   │
│   **Detectors**   │
└─────────────────────────────┘

Termination Instructions

1. Cut cable squarely with sharp blade or hot knife.

2. Strip jacket back with 18 gauge wire stripper to expose 0.10–0.18" of bare fiber core.

   Avoid nicking the fiber core.

3. Insert terminated fiber through locking nut and into the connector until the core tip seats against the molded lens inside the device package.

   Screw connector locking nut down to a snug fit, locking the fiber in place.

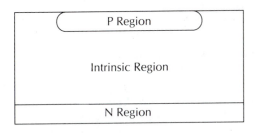

Figure 11.16 PIN Photodetector Courtesy of Motorola, Inc.

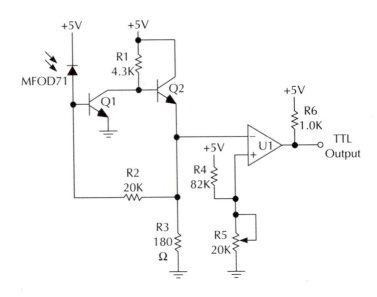

Figure 11.17 1.0-MHz PIN Receiver Courtesy of Motorola, Inc.

● 11.9 LOSSES DUE TO MECHANICAL COUPLING

One impairment to signals in a fiber cable has been mentioned already—absorption attenuation of the cable itself. Larger losses result from the physical connections that bring light sources and detectors into alignment with the cable. Additional losses are experienced with splices, which are connectors that marry two cables together. Even the best alignment of light sources to a cable can still result in some of the light power being lost by occurring outside of the cable's acceptance cone. This type of loss increases as the misalignment of light source to cable increases. The same type of problem occurs at the detector end. Misalignment there causes some of the light power to miss the detector's window. Figure 11.19 illustrates splice-type misalignments, which are the same for misalignments at the source and detector ends as well.

The lateral or axial misalignment is the most common problem when connecting cables together or when connecting sources and detectors to cables. The fiber end stays straight, but the center doesn't match with the other cable. The misalignment illustrated in Figure 11.19a is exaggerated for clarity. Md is the amount of distance of the misalignment. Light rays exiting the left side of the connection enter at a different point in the right-side cable. Some of the rays fail to make the transition, while others are refracted at a new angle.

A second type of physical misalignment, longitudinal or space gap, is shown in Figure 11.19b. Light is forced to exit the cable on the left and reenter the cable

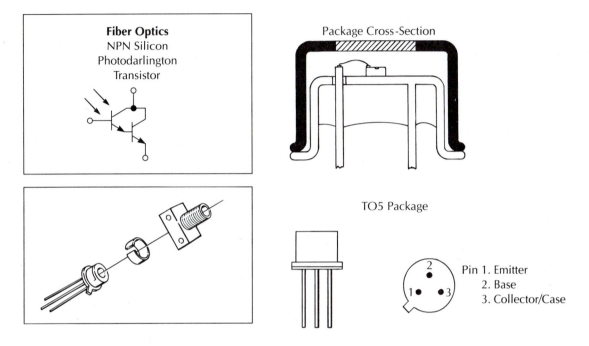

MFOD2302

Fiber Optics
NPN Silicon
Photodarlington
Transistor

Package Cross-Section

TO5 Package

Pin 1. Emitter
2. Base
3. Collector/Case

Figure 11.18 Long-Range Infrared Detector Courtesy of Motorola, Inc.

on the right after transversing the gap distance (Ms). Once again, some of the exiting light reaches the right cable outside of the acceptance cone and is lost. The third type of misalignment (Figure 11.19c) is angular misalignment. Here the cables are not matched in a straight line, with the edges separated at an angle of θ_m. Light power is lost by the combination of rays falling outside of the acceptance cone again and those entering the right-hand cable at a new incidence angle.

These misalignments are reduced by precision-fitting couplings and splices and by carefully following the connecting processes outlined by manufacturers of these devices. Care must be taken when cutting the end of the fiber to assure as straight as possible an edge is created. Once the cable is inserted in the coupler, an outer knurled knob is screwed over the coupling, crimping the cable end in place (Figure 11.20). This aids in alignment and prevents the coupling from shifting as the cable is moved about.

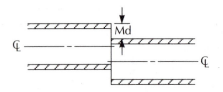

(a)   Lateral or Axial Misalignment

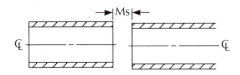

(b)   Longitudinal or Space Gap

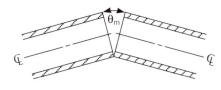

(c)   Angular Misalignment

Figure 11.19 Mechanical Misalignments of Fiber Connections

● 11.10 EFFECTS OF BENDS IN THE CABLE

   Fiber-optic cables are not always laid out in a straight line. More realistically, a cable is expected to undergo many bends as it is physically positioned. A bend in the cable is illustrated in Figure 11.21. Breaks in the cable do not occur if the bending radius remains greater than 150 times the fiber cable diameter. Tighter bends place too great a stress on the glass core and can create microbends or result in a breakdown of the material. A light ray traveling through the core is reflected, as described earlier, as long as the angle of incidence (θ_i) remains greater than the critical angle (θ_c). However, upon approaching the bend, the angle of incidence (θ_2)

Cross-Section of Connector

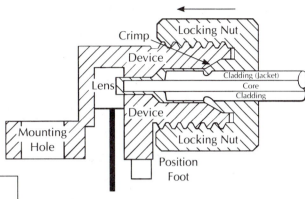

Termination Instructions

1. Cut cable squarely with sharp blade or hot knife.

2. Strip jacket back with 18 gauge wire stripper to expose 0.10–0.18" of bare fiber core.

   Avoid nicking the fiber core.

3. Insert terminated fiber through locking nut and into the connector until the core tip seats against the molded lens inside the device package.

   Screw connector locking nut down to a snug fit, locking the fiber in place.

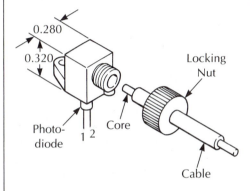

Figure 11.20 Device Connector for Fiber Detector Courtesy of Motorola, Inc.

as the ray strikes the cladding in the bend changes. If this angle becomes less than the critical angle, the ray is refracted and energy is lost into the cladding. Less light is reflected back into the core. As the bend becomes tighter, θ_i becomes smaller and more light is refracted and less reflected. Care must be taken to minimize bends and, when forced to use them, to keep the bend as wide as possible.

● 11.11 ADDITIONAL LOSSES IN FIBER CABLES

An additional, although comparatively small, absorption loss is experienced due to the intrinsic nature of any material. Impurities in the core cause some light

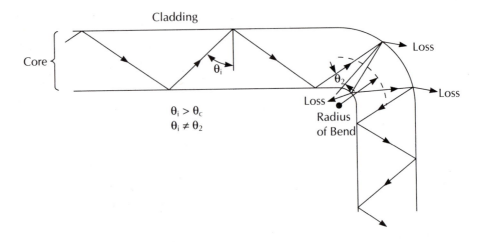

Figure 11.21 Effects on Reflection Angle Due to Bend in Fiber
Cable

energy to be absorbed as the ray transverses the cable. A larger consideration is
given to the effect of scattering in the core. During the manufacturing process,
doping molecules move freely through the hot liquid glass. As the glass is cooled
and formed into the shape of a core, these molecules stop moving and settle
randomly throughout the core material. This creates uneven molecular density
within the core, resulting in localized changes in the refractive index of the core.
Additional pockets are created by small air bubbles trapped in the fiber during the
cooling process. A ray, traveling through the cable, strikes these pockets, and part
of the ray's energy is scattered away from the bulk of the ray, resulting in loss of
light power at the exit end of the cable.

● 11.12 FIBER DISTRIBUTED DATA INTERFACE

The American National Standards Institute (ANSI) has developed ANSIX3T
95, a 100-Mbps network standard for fiber-optic ring networks more commonly
known as **FIBER DISTRIBUTED DATA INTERFACE (FDDI)**. The FDDI
token ring is composed of two counterrotating rings, a primary ring and a secondary
one that passes tokens and messages in the opposite direction to the primary ring.
This secondary ring may be used as a separate data channel or as a backup ring in
case of failure of the primary ring. In contrast to standard ring configurations in
which the token is held by the sending station until it detects the return of its
messages, the FDDI ring provides for release of the token immediately after data

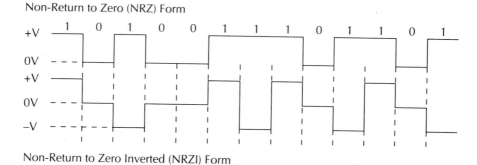

Non-Return to Zero (NRZ) Form

Non-Return to Zero Inverted (NRZI) Form

Figure 11.22 NRZ and NRZI Binary Formats

transmission by the sending station instead of waiting for a forwarding acknowledgment from the next station in the ring. Further, the ring is equipped with optical bypasses to allow stations in the ring to be bypassed in the case of failure.

The maximum frame size specified for FDDI is 4,500 bytes, and the network is managed by four protocols, **PHYSICAL MEDIA-DEPENDENT (PMD), PHYSICAL (PHY), MEDIA ACCESS CONTROL (MAC),** and **STATION MANAGEMENT (SMT).** PMD specifies the physical properties associated with fiber optics, including bandwidth, transmitter and receiver specifications, optical waveform characteristics, and power requirements. Clock synchronization, symbol alignment, and data encoding are detailed in the PHY protocol. The encoding scheme used by FDDI is referred to as NRZ 4B/5B, with NRZ meaning basic **NON-RETURN TO ZERO** encoding (Figure 11.22) and 4B/5B referring to a scheme that converts 4 bits per symbol into 5-bit digital codes. These two protocols address the physical layer of the **OPEN SYSTEMS INTEGRATION (OSI)** model.

Media access control (MAC) is responsible for frame formation and error detection. Error detection is accomplished using CRC-32 as a frame check sequence (FCS). MAC configures the data stream into a timed token protocol, converting the non-return to zero (NRZ) encoded data into a 125-Mbps **NON-RETURN TO ZERO MARK INVERSION** (NRZI) data stream applied to the fiber cable network. NRZI is shown in Figure 11.22 to contrast it with non-return to zero. Signals vary from low to high as in any digital format, but with this method, 0 or space logic is high and 1s or marks are low. MAC is also responsible for maintaining an error-recovery scheme to assure data integrity. MAC applies to the data link layer of the open systems integration model.

Station management (SMT) protocol, which lies outside of the OSI model, oversees the network and the three other protocols. This involves the start and recovery of network ring operation, including station insertion and removal from the network. Initialization, configuration, and statistics involving the network are included in the management protocol. SMT defines error-detection and fault-isolation algorithms, including usage of optical bypasses to circumvent bad stations without losing network operation.

● 11.13 SUPERNET—FDDI APPLICATION

Advanced Micro Devices (AMD) has developed a five-chip application of FDDI. Uses for this network include, but are not limited to, interconnecting mainframes, supermini computers, and high-speed storage devices; backbone networks that tie together different types of networks such as token ring, Ethernet, star or hub, and so on; and front-end networks such as workstation environments. SUPERNET, which includes the hardware and software for this system, is designed to maintain a maximum of 500 stations over a network with a maximum fiber cable length of 200 km. It uses two counterrotating rings passing data at 100 Mbps. The secondary ring application for SUPERNET is to back up the primary ring.

There are two classes of stations connected to an FDDI network, class A and class B. Class A stations are connected to both the primary and secondary rings (Figure 11.23). Class B stations, on the other hand, are interfaced only to the primary ring and require a specialized class A station, called a wiring concentrator, to make that interface. A specified maximum of 2 km between stations limits the 500-station ring to its maximum distance of 200 km around the ring.

Two forms of token passing are performed on an FDDI network. The first, called **UNRESTRICTED TOKEN,** may be captured by any station on the network. The second, **RESTRICTED TOKEN,** is reserved for stations with a specific priority code. Other stations on the network must wait until these stations are finished before capturing a token and adding messages to the network. Access to the network is through a function called **ROTATION TIME.** All the stations on the ring negotiate for rotation time based on their required network bandwidth. The larger the bandwidth requirement by a station, the more required rotation time for the station to access, receive, and send messages. Stations with the shortest rotation time capture tokens first. It is presumed that they will hold the token for the shortest time. Besides rotation and hold time, there is also a **TRANSMISSION VALID TIME,** whose sole purpose is to assure that packets do not exceed the maximum size specified for a particular network.

In the event of a station or temporary ring failure, the station management applications programs respond by making the ring whole again. This may include bypassing the faulty station or dropping it from the ring entirely. In the case where the secondary ring is used as a backup ring and the failure on the primary ring is catastrophic, SMT causes the switch between rings to be accomplished. Once the necessary corrections are made, a station is selected to transmit a **BEACON TOKEN** around the ring. When the originating station detects the return of the beacon, it is assumed that the ring is whole again and normal transmissions can be resumed.

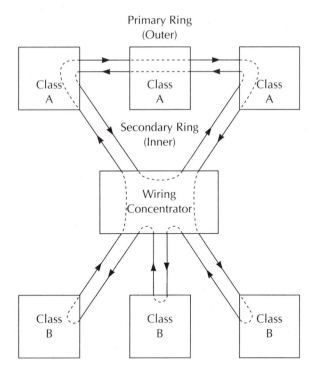

Figure 11.23 FDDI Ring Network

● 11.14 INCORPORATION OF FDDI INTO AN AVIONICS SYSTEM

Northrop Corporation has developed a fiber network utilizing FDDI for the military. It is called **FIBER-OPTIC DATA DISTRIBUTED NETWORK (FDDN)** and is responsible for fast data transfers within an aircraft. FDDN is fault-tolerant, providing high-performance interprocessing of advanced aircraft functions, including guidance and missile control. Fault tolerance refers to a system that is capable of detecting a fault and maintaining system integrity by bypassing the fault so that communications within the system are not lost.

The network, illustrated in Figure 11.24, consists of several operational levels. These include the host interface responsible for interconnecting the host system into the network and the transfer level, which assures the dependable transfer of data from the host to the FDDI interface. Some system error detection and recovery is included in the transfer level. The FDDI interface delivers the data from the host transfer level to the central network through a fiber-optic interface based on the FDDI standard. The resulting data stream is further encoded and formatted as it

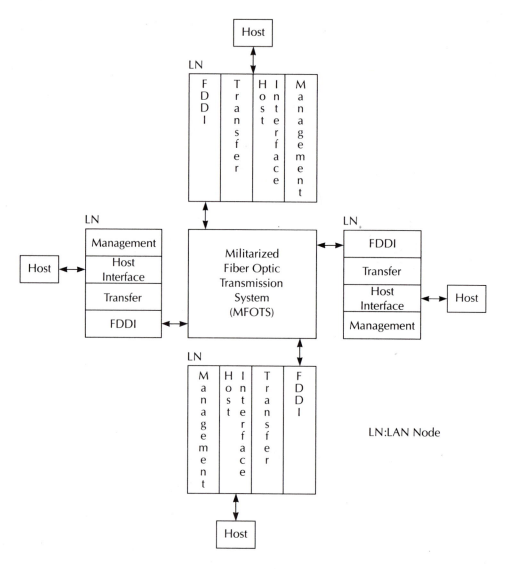

Figure 11.24 FDDN Network

enters and leaves the **MILITARIZED FIBER-OPTIC TRANSMISSION (MFOTS)** network. Within this network, processed data are routed to their proper destination at speeds of up to 10 Mbps. The high volume and speed required by advanced aircraft systems make for a suitable and desirable environment for a fiber-optic network.

SUMMARY

Fiber-optic applications in the digital and data communications field are a continually expanding area. Additionally, older technologies are becoming updated by the replacement of coaxial and twisted-pair cables with fiber-optic cables. For example, the transatlantic phone cables, laid at the turn of the twentieth century, are being replaced by fiber-optic cables. This will increase the number of transatlantic connections by a large factor. Localized networks have been expanded beyond the confines of their original specifications. Characteristics of fiber cables, such as their large bandwidth, allow higher data rates to be transmitted greater distances. Early local area network systems based on Ethernet limitations have been expanded by replacing the coaxial segments with fiber-optic cables.

Use of fiber-optic communications is finding numerous applications in distributed networks such as FDDN and integrated services networks such as ISDN. This is a growing area in data communications in which innovations are expected for many years to come.

GLOSSARY

Acceptance angle—Angle at which light rays can enter a fiber core.

Acceptance cone—Group of all light rays that can enter a fiber core.

Beacon token—Used to check integrity of the main FDDI ring following the repair of that ring.

Cladding—Material surrounding a fiber-optic core which has a refractive index that causes light rays to be reflected back into the core.

Fiber distributed data interface (FDDI)—A fiber-optic network protocol.

Fiber-optic distributed data network (FDDN)—An avionics application using FDDI.

Focal point—The point at which light converges on one side of a lens after passing through that lens.

Incident angle—The angle at which light meets a surface.

Incident ray—Light ray that initially strikes a surface.

Injection laser diode—Semiconductor device which emits a concentrated cone of light.

Media access control (MAC)—One of the four subprotocols of FDDI.

Normal—Line perpendicular to a surface by which all light ray angles are measured.

PHY—Physical protocol of FDDI.

PIN diode—A type of diode that has a large amount of intrinsic material between the P and N material.

PMD—Physical media-dependent protocol of FDDI.

Pulse spreading—Signal distortion caused by different propagation times for each light ray traveling through a cable.

Ray—Unit of light traveling in a line.

Reflection—Return of light in a direction opposite to its original direction.

Refraction—Bending of a light ray so that it continues to travel in the same direction but at a slightly different angle.

Refractive index—Value that determines the amount a light ray will be reflected or refracted by comparing the values of the indexes of two surfaces.

Restricted token—Reserved FDDI tokens for higher priority stations.

Rotation time—Access algorithm for the FDDI network.

Snell's law—Defines the relationship between the ratio of critical and incident angles to the ratio of refractive indexes between two surfaces.

Transmission valid time—Restricts the length of time a FDDI station can send a packet.

Unrestricted token—FDDI free token that can be captured by any station on the ring.

QUESTIONS

11.1. Give some advantages of using light as a means of conveying digital data.

11.2. Compare fiber-optic cable parameters to coaxial or twisted-pair parameters. Why is fiber cable preferred? What are the disadvantages of using fiber cable?

11.3. What is meant by reflection and refraction?

11.4. What is used as the reference for refractive indexes?

11.5. Define incident ray and normal.

11.6. How do the core and cladding differ?

11.7. What are the advantages and disadvantages of light-emitting diodes compared with injection laser diodes as a source of light for a fiber-optic system?

11.8. What are the advantages and disadvantages between edge- and surface-emitting diodes used as a light source for fiber-optic systems?

11.9. How does pulse spreading occur in a step index fiber core? What effect does this pulse spreading have on the received information?

11.10. Compare the advantages and disadvantages of single- and multimode fiber cores.

11.11. How does an optical detector indicate the difference between a logic 1 (high-intensity light) and a logic 0 (low-intensity light)?

11.12. What is the greatest loss in fiber cabling besides absorption attenuation? How is this loss minimized?

11.13. Describe what bends do to the losses in a fiber cable.

11.14. What causes scattering to occur in a fiber core?

11.15. What is the resulting effect of the light signal due to scattering?

11.16. What are the tasks of the FDDI media access control (MAC) protocol?

11.17. What is the primary responsibility of station management (SMT)?

11.18. What are the uses of the physical (PHY) and physical media-dependent (PMD) FDDI protocols?

11.19. Define class A and class B stations in an FDDI SUPERNET network. What is the name given to the class A station used to interface class B stations to the primary ring?

11.20. For SUPERNET, what is the purpose of rotation time and transmission valid time?

11.21. What is an FDDI beacon token? When is it used?

11.22. What application is FDDN designed for?

PROBLEMS

11.1. Calculate the critical and Brewster angles for surfaces that have refractive indexes of 1.25 for n_2 and 1.65 for n_1.

11.2. A ray with an incident angle of 15° approaches a surface with a refractive index of 1.63 from one with a refractive index of 1.15. Is the ray reflected or refracted?

11.3. Compute the acceptance angle of a photodiode with a window diameter of 0.75 mm placed in front of a lens with a focal point of 1.05 mm.

11.4. Compute the refractive variance of a multimode fiber core with a center index of 1.42 and a cladding index of 1.15.

11.5. What is the radius of the tightest bend of a cable whose diameter is 250 μm?

RESEARCH ASSIGNMENT

11.1. Design a fiber-optic system using specifications for light sources, detectors, and the cable itself found in various manufacturers' data manuals. Support your design with the technical information from the data specifications. Your system is to successfully transmit and receive a serial digital stream of information at a data rate specified by your instructor.

11.2. Research current fiber-optic applications. Suggested areas include ISDN and BISDN applications and long-distance telephone companies.

LAB PROJECTS

11.1. Design and construct laser diode and detector circuits that will allow the transmission and correct detection of signals to occur. The laser diode circuit may be modulated with a serial data stream or other method to verify that your circuits operate correctly.

11.2. Design and construct a two-point communications system that uses fiber-optic cables as the connecting medium. Depending on the availability of components and the material previously explored, select the data type (asynchronous or synchronous) and modulation type for your system (FSK, PSK, PCM, etc.). Use modem or CODEC chips as is appropriate.

11.3. Construct and prove your design of research assignment 11.1. Interface your system to an existing serial interface (such as an ACIA or RS232C connector) driven by microprocessor-based trainer or computer.

Answers to Odd-Numbered Questions

11.1. Light is faster and is not affected by most electromagnetic disturbances.

11.3. Light that is reflected is returned toward the source. Light that is refracted is bent but continues away from the source.

11.5. The normal is an imaginary line of reference perpendicular to the core's center. Incident ray is the light ray that enters the core.

11.7. Light-emitting diodes have a wider dispersion pattern and less power out than laser diodes.

11.9. Pulse stretching occurs because of the varying propagation times of rays entering at different incident angles into the fiber core. This requires them to travel different physical distances as they reach the receiving end. Pulse stretching in the extreme causes loss of digital information when two 1s emerge, covering a 0 level between them.

11.11. The electron-hole generation of a photodetector varies directly with the amount of light it senses.

11.13. Bends in the fiber cause reflected and refracted rays to change direction. This can increase the absorption loss and distort the light pattern at the receive end.

11.15. The effect of scattering is to attenuate the light energy as the ray travels through the core.

11.17. It oversees network operation; recovers network ring operation in case of failure of a node; is responsible for initialization, insertion, and removal of stations, and so on.

11.19. Class A stations are connected to both the primary and secondary rings. Class B stations are connected to the primary ring only. Wiring concentrator.

11.21. A beacon token is a token with a nondesignated destination used to recover network operation after the ring has been reestablished following a fault.

Answers to Odd-Numbered Problems

11.1. Critical angle = 310.75°; Brewster angle = 37.15°

11.3. Acceptance angle = 35.54°

11.5. 37.5 mm

Research Assignment and Lab Project Guides

Depend on instructor direction, parts, and reference material available.

Twelve • Data Communications Applications

OBJECTIVES

The purpose of this chapter is to acquaint the student with two widely used data communications applications. Details of the basic concepts behind these applications are coupled with their use. These two areas are:

1. Facsimile, an old technology that has come of age, involving the digitizing of documents to allow their image to be transmitted via telephone lines to a destination

2. Satellite communications that implement frequency division multiplexing to allow a large number of data channels to be transmitted to satellites (and back) at extremely high carrier frequencies (gigahertz range)

ACRONYMS

AM—**A**mplitude **M**odulation

CCD—**C**harge **C**oupled **D**evice

CCITT—International **C**onsultative **C**ommittee for **T**elephony and **T**elegraphy

COMSAT—**COM**unication **SAT**ellite Corporation

DOMSAT—**DOM**estic **SAT**ellite

FAX—**FA**csimile

FCC—**F**ederal **C**ommunications **C**ommission

FDMA—**F**requency **D**ivision **M**ultiplexing **A**ccess

FM—**F**requency **M**odulation

FSK—**F**requency **S**hift **K**eying

INTELSAT—**IN**ternational **TEL**ecommunications **SAT**ellite Organization

PSK—**P**hase **S**hift **K**eying

QAM—**Q**uadrature **A**mplitude **M**odulation

RCA—**R**adio **C**orporation of **A**merica

ROM—**R**ead **O**nly **M**emory

SBS—**S**atellite **B**usiness **S**ystem

TDMA—**T**ime **D**ivision **M**ultiplexing **A**ccess

● 12.1 INTRODUCTION

The process of electronically transferring text messages between two entities has gone on for a long time. The field of data and digital communications has advanced rapidly. Almost lost in the ever-increasing technology are two areas that have become well-established methodologies for moving large quantities of information from a source to its destination. One of these is **FACSIMILE (FAX)**, which allows documents to be copied from the original and sent in that form to a destination. To facilitate this, the original document is converted into a series of transmittable digital codes. At the receive end, the digital information is converted back into a copy of the original document. The other area of technology that has become standard usage is satellite communications.

● 12.2 FACSIMILE—TRANSFERRING WHOLE DOCUMENTS

The idea of scanning a full document and transferring it in its entirety from one location to another is not new. As early as 1842, a patent for such a scheme was awarded to Alexander Bain. Although Bain's machine had very little resemblance to facsimile machines of today, some basic concepts of optical scanning that he introduced are in use today. Practical methods of transmitting pictures over long distances were used during World War II and after by the major newspaper wire services to carry daily photographic records of the events of the day between Europe, Asia, and America. Successful transfer of an image from source to destination requires several functional elements, including optical scanning, data conversion for transmission and reception, and the ability to produce a hard copy of the image at the receiving end.

A number of methods have been used to optically scan a document. For facsimile operation, two basic methods are used, **DRUM SCANNING** and **FLATBED SCANNING.** With a drum scanner, a cylindrical drum, 19 inches in length, is coated with a black opaque finish. A helix groove is cut from one end of the drum to the other as illustrated in Figure 12.1. Mirrors and lenses are placed at the drum's center to direct the light entering through the groove toward a photoelectric cell, which converts that light energy to electrical current.

The drum is rotated at 120 revolutions per minute (rpm). Light projected onto the document face is reflected and directed through a lens into the helix groove as the drum rotates. The document is fed through the scanner by rollers driven at sixty cycles per second as shown in Figure 12.2. The combination of the drum rotation and document feed causes the reflected light to represent a single line of the document. Since the movement is constant, there is no overlapping between lines. The received light that has been directed through the drum's center is converted to electrical current, amplified, and caused to modulate a 2,400-Hz carrier. For an analog system, **FREQUENCY MODULATION (FM)** is used, producing a range of frequencies between 1,500 and 2,300 Hz (lower sideband), depending on the strength of the light.

The earlier analog facsimile machines transferred a document at a rate of ninety-six lines per minute using the drum scanner method. At the receiver side (Figure 12.3), the frequency modulated signal is demodulated and filtered to assure that the 1,500- to 2,300-Hz tones alone are passed through the receiver. A linear slope filter is employed to pass the 1,500-Hz tone at a strength ten times that of the 2,300-Hz tone. The lower frequency represents black and the higher, white. Frequencies in between are varying shades of gray. The stronger signal is required to assure that the black area will be imprinted on the hard copy at the printer. Once again, the hard copy paper is pulled through the printer at a sixty-cycles-per-second

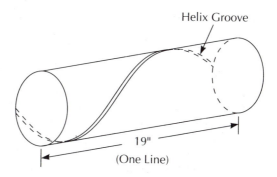

Figure 12.1 Drum Scanner

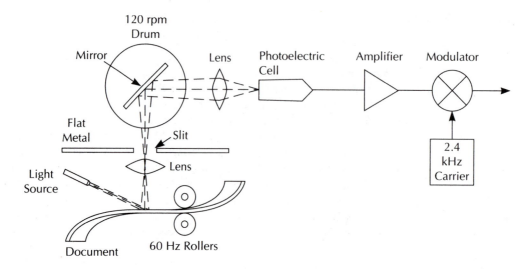

Figure 12.2 Facsimile Drum Scanner

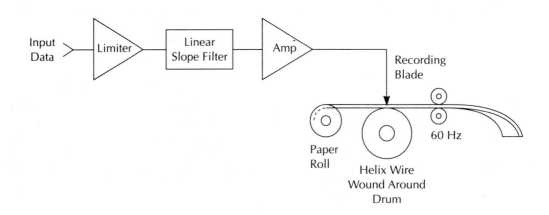

Figure 12.3 Drum Printer

rate. A nonconductive drum with a wire wound in helix fashion around it from end to end rotates at 120 rpm. A recording blade, which is stationary, is positioned on one side of the paper, as shown in Figure 12.3. A voltage between 0V (white) and 60V (black) is placed between the helix wire and the blade. The level of the voltage depends on the strength of the signal from the slope filter. The hard copy paper is made of a special electrosensitive material. The stronger the voltage on either side of the paper, the darker the spot on the paper becomes. The rotation of the helix on the drum assures a left-to-right traveling contact point across the paper. The forward motion of the rollers feeds the paper at the ninety-six-lines-per-minute rate of the scanner.

Drum scanners were developed for analog fax machines, but few of that type are in use today. With the advent of digital fax machines, a second method called flatbed scanning has evolved. The scanned document is pushed or pulled across an aperture slot in a flat metal plate (Figure 12.4). A light source is projected through the slot and is reflected from the document's surface. The reflected light is directed

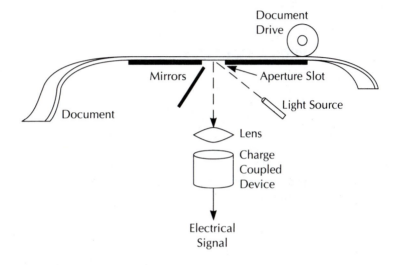

a) Functional Diagram

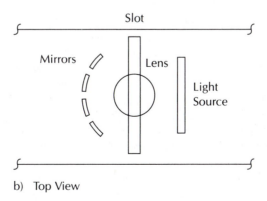

b) Top View

Figure 12.4 Flatbed Scanner

by mirrors and lenses to a **CHARGE COUPLED DEVICE (CCD),** which produces an electrical current in proportion to the amount of light applied to it. Since the scanner is stationary, the number of moving parts is greatly reduced. The analog electrical information is converted into a digital code and sent to its destination. At the destination, the digital codes are reassembled into data required for any one of a number of different printers or terminal displays.

● **12.5 FAX STANDARDS**

The **INTERNATIONAL CONSULTATIVE COMMITTEE FOR TELE-PHONY AND TELEGRAPHY (CCITT)** standards organization has included facsimile requirements as part of its many standards for data communications. These specific fax standards are designated as *RECOMMENDATIONS T.2, T.3, T.4,* and *T.30.* T.2 and T.3 addressed analog fax systems, and T.4 and T.30, digital fax systems. T.2, or Group I, specifications use frequency modulation and a document rate of 6 minutes per page. The resolution of the copy is 96 lines per inch, with a scan rate of 180 lines per minute. The other analog recommendation, T.3 or Group II, uses **AMPLITUDE MODULATION (AM),** decreasing the document rate to 3 minutes per page. Resolution remains the same, but the scan rate is doubled to 360 lines per minute.

Both analog methods are considered slow. At 3 or 6 minutes per page, the time to reproduce several pages would tie up a data system for a considerable length of time. Digital fax systems were developed to decrease document time while increasing resolution. The idea is to convert the analog signals representing the strength of the reflected light into digital pulses. A digital logic level of 1 represents black, and 0 represents white. Thus, a data train of digital pulses is created, representing the light level at a sampled spot. A specific set of consecutive 1s and consecutive 0s represents each gray level of scale between black and white. The method used creates different code lengths (number of bits per sample), each with varying numbers of 1s and 0s. To increase the rate of transmitting this information, the digital pulse trains are encoded using a modified **HUFFMAN CODE** (Table 12.1). The purpose of using this code is to reduce the actual number of bits sent from the source to the destination. Notice on Table 12.1 that for most of the conversions, there are fewer Huffman code bits than the original count of white or black bits.

The basic principle behind the Huffman code is that certain symbols or combinations of symbols appear in text more frequently than others. For instance, the letters E and T are used more frequently than Q and F. The combination of letters TH appears the most frequently out of random letter pairs, and the word THE is the most common word. The process is to assign the shortest codes to the most frequently used symbols. A message coded in a Huffman code would then contain fewer digital bits than any code that uses a fixed number of bits for every symbol.

Table 12.1 Huffman Code Set

	White Bits		Black Bits	
NUMBER OF BITS	HUFFMAN CODE (IN HEX)	NUMBER OF CODE BITS*	HUFFMAN CODE (IN HEX)	NUMBER OF CODE BITS*
0	35	8	037	10
1	07	6	2	3
2	7	4	3	2
3	8	4	2	2
4	B	4	3	3
5	C	4	3	4
6	E	4	2	4
7	F	4	2	5
8	13	5	05	6
9	14	5	04	6
10	07	5		
11	08	5	05	7
12	08	6	07	7
13	03	6	04	8
14	34	6	07	8
15	35	6	018	9
16	2A	6	017	10
17	2B	6	018	10
18	27	7	008	10
19	0C	7	067	11
20	08	7	068	11
21	17	7	06C	11
22	03	7	037	11
23	04	7	028	11
24	28	7	017	11
25	2B	7	018	11
26	13	7	0CA	12
27	24	7	0CB	12
28	18	7	0CC	12
29	02	8	0CD	12
30	03	8		
31	1A	8	069	12

Continued on next page

Table 12.1 Huffman Code Set — Continued

	White Bits		Black Bits	
NUMBER OF BITS	HUFFMAN CODE (IN HEX)	NUMBER OF CODE BITS*	HUFFMAN CODE (IN HEX)	NUMBER OF CODE BITS*
32	1B	8	06A	12
33	12	8	06B	12
34	13	8	0D2	12
35	14	8	0D3	12
36	15	8	0D4	12
37	16	8	0D5	12
38	17	8	0D6	12
39	28	8	0D7	12
40	29	8	06C	12
41	2A	8	06D	12
42	2B	8	0DA	12
43	2C	8	0DB	12
44	2D	8	054	12
45	04	8	055	12
46	05	8	056	12
47	0A	8	057	12
48	0B	8	064	12
49	52	8	065	12
50	53	8	052	12
51	54	8	053	12
52	55	8	024	12
53	24	8	037	12
54	25	8	038	12
55	58	8	027	12
56	59	8	028	12
57	5A	8	058	12
58	5B	8	059	12
59	4A	8	02B	12
60	4B	8	02C	12
61	32	8	05A	12
62	33	8	066	12
63	34	8	067	12

Continued on next page

DIGITAL & DATA COMMUNICATIONS

Table 12.1 Huffman Code Set—Continued

	White Bits		Black Bits	
NUMBER OF BITS	HUFFMAN CODE (IN HEX)	NUMBER OF CODE BITS*	HUFFMAN CODE (IN HEX)	NUMBER OF CODE BITS*
64	1B	5	00F	10
128	12	5	0C8	12
192	17	6	0C9	12
256	37	7	05B	12
320	36	8	033	12
384	37	8	034	12
448	64	8	035	12
512	65	8	006C	13
576	68	8	006D	13
640	67	8	004A	13
704	0CC	9	004B	13
768	0CD	9	004C	13
832	0D2	9	004D	13
896	0D3	9	0072	13
960	0D4	9	0073	13
1,024	0D5	9	0074	13
1,088	0D6	9	0075	13
1,152	0D7	9	0076	13
1,216	0D8	9	0077	13
1,280	0D9	9	0052	13
1,344	0DA	9	0053	13
1,408	0DB	9	0054	13
1,472	098	9	0055	13
1,536	099	9	005A	13
1,600	09A	9	005B	13
1,664	18	6	0064	13
1,728	09B	9	0065	13
SYNC	001	11	001	11

*The number of Huffman code bits plus the code itself yield all unique combinations. Thus, the codes for 1, 2, and 10 white bits and 12 black bits are all different. They are 000111, 0111, 00111 and 0000111, respectively.

E 35%   O 25%   I 15%   A 15%   U 8%   Y 2%

a) Symbols by Percentiles

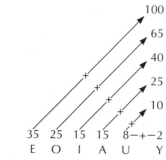

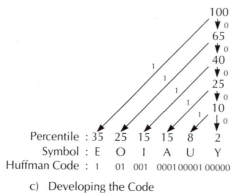

b) Forming Sums

c) Developing the Code

Figure 12.5 Huffman Code Example for Vowels

To illustrate how a Huffman code is developed, assume that a Huffman code is to be developed to represent the vowels A, E, I, O, U, and Y as the only symbols. A statistical analysis is run on a reasonably large sampling of text. Arbitrarily and just for purposes of example, assume that the frequency of the appearance of these vowels is E = 35% of the time, O = 25%, I = 15%, A = 15%, U = 8%, and Y = 2%. Diagrammatically, start by listing the vowels horizontally, sequentially in descending order at the bottom of a page as shown in Figure 12.5a. Add the two lowest percentages (U and Y) and place this sum above the Y. Add the next percentage (A) and place this sum above the sum of U and Y. To the new sum add the next percentage (I) and place this above the last sum. Continue with this until you reach 100% (Figure 12.5b). Draw a line from the 100% to the highest percentage (E). Add another line from the next lowest sum to the next lowest percentage (0). Continue this process to the letters I, A, U, and Y. Finally, draw connecting lines between each sum so the diagram looks like Figure 12.5c. Label each slanted line with a 1 and each vertical line section with a 0. To create the Huffman code, start at 100% and trace your way to each letter. Write down the binary code for each symbol. Note that for this simple example, as the percentage decreases, the binary code gets larger, with the letter E having only 1 bit (1) and the letter Y having 5 bits (00001). In reality, there would be at least two symbols with 1-bit codes (1 and 0) and four with 2 bits (00, 01, 10, 11), and so on. This example is just to give a feel for what a Huffman code is.

Recommendation T.4, or Group III, produces documents at the rate of 1 minute per page; 1,728 points are scanned per line, resulting in a horizontal resolution of 204 lines per inch and a vertical resolution of 98 lines per inch. The transmission rate is specified as 4800 bps but can optionally be operated at 9600 bps given a fairly clean telephone line. **PHASE SHIFT KEYING (PSK)** and **QUADRATURE AMPLITUDE MODULATION (QAM)** techniques are used to transmit and receive the digital data. A secondary handshake channel used for the establishment of a communications dialogue (requests and acknowledgments) operates at 300 bps using **FREQUENCY SHIFT KEYING (FSK)** modulation. Recommendation T.30 defines the handshake procedures to be used by fax systems. It has not yet been completed or implemented.

● **12.6 FAX SYSTEMS**

An overall block layout for a fax system is shown in Figure 12.6. The VOICE/FAX MUX block discerns between a fax signal and regular voice call on the telephone line. In this block, voice communications are directed to a regular telephone handset, while fax data are sent to an appropriate modem. The modem

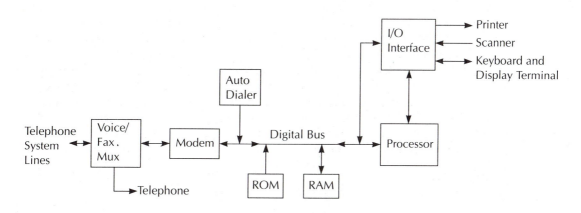

Figure 12.6 Facsimile System

receives and demodulates the fax data, which are decoded and processed by the processor block and sent to a specific peripheral—printer for hard copy and terminal for video display. Fax data can be entered into the system through a scanner or from data previously scanned and stored on disk or in a computer's memory. The data are encoded and sent to the modem. There they are modulated and sent out onto the telephone lines to their destination.

Figure 12.7 shows a Group III fax transmitter. A document is scanned and sampled to produce a 1,728-bit-per-line digital stream. This stream is applied to a binary counter, which counts the bits representing black and the bits representing white. This count is used as an address to the Huffman code **READ ONLY MEMORY (ROM).** The output of the ROM is the compressed Huffman code, which is used to modulate a quadrature amplitude modulated (QAM) modem. Any handshake control data are sent to the frequency shift keying (FSK) modem elements. A 1,650-Hz tone is used for a logic 1 and an 1,850-Hz tone for logic 0. The combined QAM and FSK signals are applied to the telephone lines and sent to their destination.

Received fax data are demodulated using a QAM demodulator. Control data are demodulated using the FSK demodulator. Figure 12.8 shows fax receiver blocks. The digital data from the QAM demodulator are converted from a serial stream to parallel data and are used to address Huffman decompressor (decoder) read only memory. Parallel data from the ROM is converted back to a serial stream to duplicate the information generated by the scanner at the transmitting site. This digital information will be decoded further as required by the printer or video terminal that reproduces the actual image.

A fax machine whose sole purpose is to transmit and receive fax data uses the decoded ROM data directly to reproduce a hard copy of the image. Fax printers used with these machines come in three types—electrothermal, electrostatic, and thermal. An electrothermal printer uses a dry process with two-layer, white-over-black conductive paper. Black bits cause a large voltage to be applied, which burns off the white upper layer to reveal the black layer beneath. Electrostatic printers use a special coated paper that is electrically chargeable. Black bits cause the coating to be charged and attract a black toner. The paper is then passed over a heater, where the toner particles are permanently bonded to the paper. Thermal paper is actually burned black by the application of a black voltage to the stylus.

Facsimile is a method that has been in use for a long time. Its slow acceptance use worldwide was a result of the cumbersome size and process employed by earlier machines. With today's technology, small, compact fax machines at affordable prices are appearing everywhere. Additionally, fax plug-in cards for IBM, Macintosh, and other personal computer systems are bringing fax capabilities into small business and home environments.

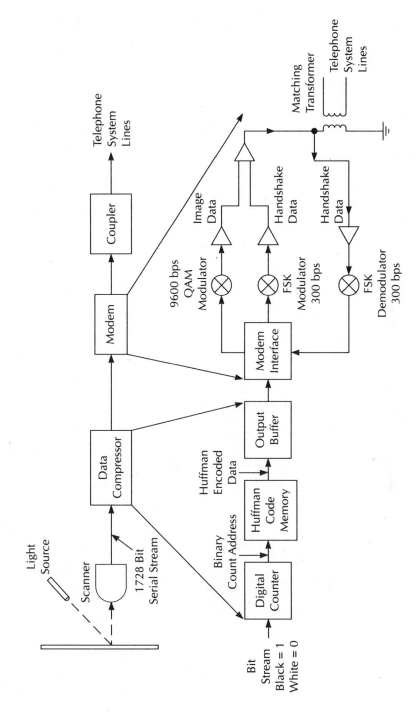

Figure 12.7 Facsimile Transmit Blocks

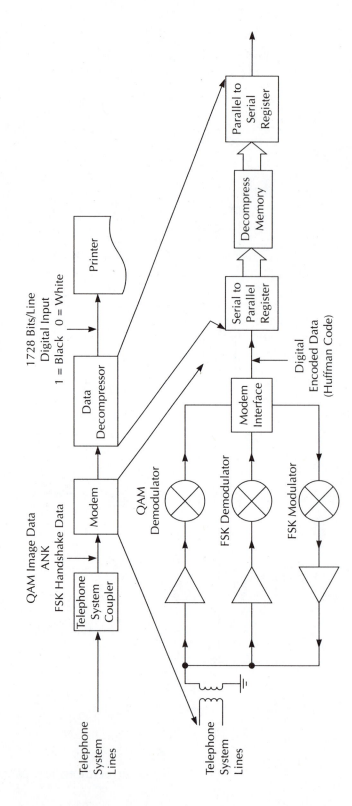

Figure 12.8 Facsimile Receive Blocks

Much of today's communications between distant localities is performed with the assistance of communications satellites. When the Union of Soviet Socialist Republics launched the first successful satellite, the *SPUTNIK,* into orbit in 1957, the Soviets, as well as everyone else, had no idea to what extent satellites would affect our daily lives. Sports and news coverage is instantaneous because of satellite communications. Business deals and up-to-the-minute information concerning the world brings forth quicker and stronger changes in the world's political, financial, and social communities.

Communications satellites are essentially electronic repeaters located many miles above the earth. They are wideband **TRANSPONDERS,** receiving transmissions from one earth station and sending the data received to another earth station. The satellite receives transmissions at one carrier frequency, amplifies the weakened signals, and retransmits the information at a different lower carrier frequency. Satellites receive their operating power from solar cells, attached either to fins or wings connected to the satellite's body or to solar cells mounted directly on the satellite's body. Figure 12.9 illustrates the *Mariner 10* satellite, which uses wing-type extensions to contain the solar cells needed to power the satellite. Also illustrated in Figure 12.9 is an uplink and downlink path between the satellite and two earth stations.

Today's communications satellites are launched into a **GEOSYNCHRO- NOUS ORBIT** 22,282 miles above the equator. They travel at a rate of approximately 6,879 miles per hour, which synchronizes their orbital motion with the rotation of the earth. In this way the satellite appears to remain stationary at one spot above the earth. Because of the apparent motionless aspect of the satellite in respect to earth stations, tracking the satellite is easy. Earlier low, elliptical-orbit satellites required earth station antennas to track them as they came into view until they disappeared over the horizon. This also limited communications with them to a short period of time, usually half an hour.

Geosynchronous satellites are capable of receiving and transmitting to earth stations that are in line of sight to the satellite. In other words, if a person were to take a panoramic picture of the earth from the satellite, earth stations that are located in the area shown in the picture could use that satellite. The *Jupiter* satellite in Figure 12.10 illustrates an earth coverage beam. It should be noted that the *Pioneer* satellite is an exploratory satellite and is not parked in a geosynchronous orbit above the earth. A single satellite using an earth coverage beam can blanket about 42.4% of the earth's surface if the satellite is placed in a geosynchronous orbit as described. Three satellites circling the earth, equally spaced above the equator, can cover 90% of the world. The chief areas not covered by these satellites are the northern and southern arctic regions. These areas are sparsely populated, so it is difficult to actually cite this omission as a "disadvantage." It should be noted that different signal strengths are received by the earth stations in different sections of the beam's

Mariner 10

AUSTRALIA

ASIA

AFRICA

Earth
Station

Earth
Station

NORTH
AMERICA

SOUTH
AMERICA

Figure 12.9 Satellite Communications

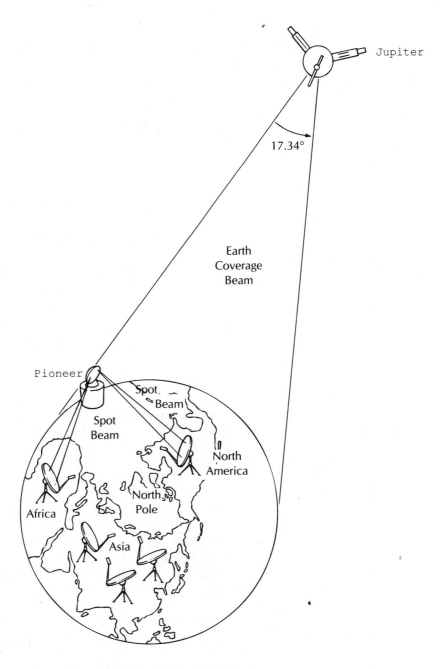

Figure 12.10 Spot and Earth Coverage Beams

coverage. One additional disadvantage of satellites orbiting at such a high distance from earth is the amount of time it takes signals to be received by the destination earth station once they have been sent from an originating station. Typical round-trip propagation delays are approximately 270 ms.

Another type of beam sent by satellite antennas is the spot beam, which concentrates signal strength into a narrow beam. Spot beams emanating from the *Pioneer* satellite are illustrated in Figure 12.10. Spot beams have the advantage of increasing signal strength received by an earth station because of the concentration of beamed power into a narrow corridor. Spot beams are commonly employed for satellite transmissions to areas such as Hawaii and Puerto Rico, where signal strength is concentrated to the islands and not lost in the surrounding oceans. Some satellites employ steerable spot beams, which allow the beam to be directed to a different area on the earth's surface.

Transmissions between earth stations and satellites are allocated by various licensing and regulatory agencies. One such agency is the **FEDERAL COMMU-NICATIONS COMMISSION (FCC)** in the United States. The FCC has assigned specific frequency ranges for commercial and military satellites. International satellite frequency assignments are made to all countries with satellite services. Figure 12.11 charts the FCC and international frequency assignments for United States commercial communications satellites. The chart also defines the frequency range for standardized bands so that the allotted frequencies can be related to the appropriate band.

The transmission from earth to the satellite is referred to as an **UPLINK,** and the return transmission to earth as a **DOWNLINK.** A short form reference to a satellite's operating frequencies is to generalize the uplink and downlink frequency allocation. For example, **SATELLITE BUSINESS SYSTEM (SBS)** satellites used for business communications receive uplink transmissions from 14 to 14.5 GHz in the K band. Downlink frequencies depend on the version of the SBS in orbit. There are three sets, all in the K band: 10.95–12.2 GHz; 12.45–12.7 GHz; and 12.7–13.2 GHz. SBS satellites that use the first or second set of downlink frequencies are referred to as operating at 11/14 GHz. The third set of downlink frequencies cause the satellites to be designated as 12/14-GHz units. As you may surmise, the 11 or 12 GHz refers to the downlink frequency, and the 14 GHz to the uplink. At other times these same satellites are called 14/12- or 14/11-GHz sets, where the uplink frequency precedes the downlink. It actually does not matter which designation is used because the uplink frequency is always higher than the downlink frequency.

The choice of 12/14 GHz for business satellites evolved from the need to locate earth stations close to the businesses they service. This could mean that an earth station antenna might be located on top of a building in the middle of a city. Concern about interference with microwave and other transmissions in a city environment led to the 12/14-GHz allocation for business satellite communication. Another consideration was the volume of traffic expected for business satellites versus telephone communications and entertainment satellites. The 12/14-GHz

Band	Frequency Range	FCC Terrestrial Common Carrier Band (GHz)	International Satellite Frequency Bands		Band-width (MHz)
			Downlink (GHz)	Uplink (GHz)	
P	225–390 MHz				
J	350–530 MHz				
L	390–1,550 MHz				
S	1.55–5.2 GHz	2.11–2.13			20
		2.16–2.18			20
			2.5–2.535	2.655–2.69	35
			3.4–3.7		300
		3.7–4.2	3.7–4.2	4.4–4.7	500
C	3.9–6.2 GHz				
X	5.2–10.9 GHz	5.925–6.425		5.925–6.425	500
			7.25–7.75	7.9–8.4	500
K	10.9–36.0 GHz	10.7–11.7	10.95–11.2		500
			11.45–11.7		500
			11.7–12.2	14–14.5	500
			17.7–21.2	27.5–31	3500
Ku	15.35–17.25 GHz				
Q	36–46 GHz				
V	46–56 GHz				
W	56–100 GHz				

Figure 12.11 Frequency Allocation by Band; FCC Frequency Allocation for Communications Satellites; International Allocation for U.S. Communications Satellites

systems can transmit signals occupying one-half the beam width required for satellites using 4/6-GHz frequencies. On the down side, the higher transmitting frequencies of the business satellites are more susceptible to interference from weather conditions, particularly those resulting in heavy moisture in the air (heavy rains, clouds, or fog). These conditions tend to attenuate 12/14-GHz signals more than those at lower frequencies. The most common use of satellite communications today is long-distance telephone connections, television transmissions, teleconferencing, worldwide fax, Telex, and electronic mail (E-Mail) services.

Satellite communications were preceded by the launching, in late 1958, of a satellite that broadcast a Christmas message recorded by President Eisenhower. The first actual two-way communications came into being in 1960 with the passive *ECHO* satellites. Unlike today's satellites, these earlier models merely reflected the signal sent to them. This was achieved because the satellites were parked into a low orbit due to the limitations of the rocket technology of the day. These low orbits decayed comparatively rapidly, giving these early satellites a low life span.

Two breakthroughs in technology led to the present communications technology. The first involved improved rocket technology, including booster engines that could launch satellites into higher, nondecaying orbits. The second was the development of efficient solar energy panels to supply power to these satellites. This changed them from passive reflectors to active transponders. The first transponder-type satellite, *COURIER,* was sent into orbit in late 1960 by the U.S. Department of Defense. The first commercial communications satellite, *TELSTAR* (which spurned a pop music instrumental record), came on-line in July of 1962. From that time, many additional communications satellites have been placed into orbit. The **INTERNATIONAL TELECOMMUNICATIONS SATELLITE ORGANIZA-TION (INTELSAT)** series began with the launch of the *EARLY BIRD* satellite in 1965. INTELSAT arose from a United Nations resolution to develop worldwide satellite communications on a nondiscriminatory basis. INTELSAT I, better known as *Early Bird,* had two transponders capable of handling 240 voice channels and one television signal within a 25-MHz bandwidth. It was placed in a geostationary orbit and radiated using a 360° omnidirectional antenna. Unfortunately, much of the power transmitted by *Early Bird* went into space instead of to earth. Later versions of the INTELSAT satellites incorporated all the upgrades technology could provide, including earth coverage beams and an increase to 1,200 voice channels using two transponders. The latest version of the INTELSAT satellite *(INTELSAT V)* is a multichannel (12,500 channels), high-bandwidth (2,300-MHz) system. It consists of twenty-seven transponders and weighs 950 kilograms. That is a lot of communications in a relatively small package.

The United States **COMMUNICATION SATELLITE CORPORATION (COMSAT),** although owned through stockholder shares, regulates the use and operation and sets tariffs for U.S. satellites. It operates as a monopoly and sells user time on satellites to many diverse users. Presently, the United States uses the largest share of satellite time (24%), followed by Great Britain (13%) and France (6%). Other users include Germany, Italy, Canada, Iran, and South American countries. COMSAT has a counterpart in Russia called *INTERSPUTNIK,* which controls all aspects of satellite use for Russia and its east European allies.

Western Union's *WESTAR,* begun in 1974, today incorporates twelve transponders retransmitting 1,200 voice channels and one color television channel at a digital data rate of 50 Mbps. Two of the transponders are reserved as backup

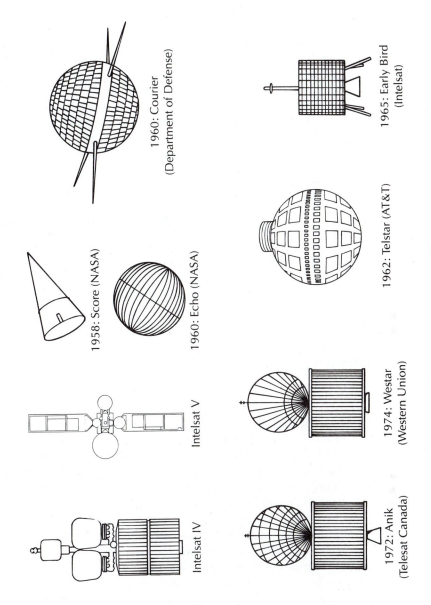

1960: Courier
(Department of Defense)

1965: Early Bird
(Intelsat)

1958: Score (NASA)

1960: Echo (NASA)

1962: Telstar (AT&T)

Intelsat V

1974: Westar
(Western Union)

Intelsat IV

1972: Anik
(Telesat Canada)

Figure 12.12 Early Satellites

in case of a failure of one of the other ten. **RADIO CORPORATION OF AMERICA (RCA)** launched *SATCOM* satellites in 1975, and in 1976 AT&T joined the satellite communications family with its *COMSTAR*. The first commercial satellites were launched by Canada starting in 1972 and were designated by the name *ANIK* which is Eskimo for "Little Brother." some of these satellites are illustrated in Figure 12.12.

The Union of Soviet Socialist Republics launched the first set of **DOMESTIC SATELLITES (DOMSATs),** the *Molniya* (Lightning), in 1966. These satellites beamed television and radio coverage to remote areas within the vast Soviet Republics territory. Four of these satellites were originally spaced equidistant in a nonstationary orbit. As one satellite lost contact with earth stations, the next in line took over.

Currently, the commercial and communications arenas are populated with satellites owned and operated by large companies such as RCA, MCI Communications, AT&T, IBM, and Western Union. Other commercial satellites are managed by individual and combined national governments. Norway, Sweden, and Finland have a joint satellite communications venture called *TELE-X*. France controls a series designated *TELECOM,* and China has one called simply *STW*. There are many others, too numerous to include here.

• 12.9 METHODS OF SATELLITE COMMUNICATIONS

The main technologies for sending and receiving data using satellites incorporate **FREQUENCY DIVISION MULTIPLEXING ACCESS (FDMA) or TIME DIVISION MULTIPLEXING ACCESS (TDMA).** FDMA methods, as explained in chapter 7, illustrate how multiple channels are developed and used to modulate increasingly higher-frequency carriers until the transmission frequencies reach the gigahertz range required for satellite transmissions.

A form of time division multiplexing called time division multiplex access (TDMA) is used as an alternate method to FDMA. As with standard time division multiplexing principles, channels are assigned time slot allocations. These time slots are separated by headers and guard bands. A typical TDMA frame is shown in Figure 12.13. A supervisor burst of information precedes the transmission of channel information. This burst is used to establish symbol and burst timing. The first guard band follows, and then the first channel's transmission. At the beginning of the channel block is a header (or preamble), which repeats the supervisor burst and adds source and destination station addresses. Channel information completes the channel's block. Channels are separated by guard bands to prevent overlapping between channels. A closing burst indicates the end of the transmission and reports status information about the transmission capacity and possible error occurrences.

Figure 12.14 shows a functional block diagram of a TDMA communications system. On the transmit side, the channels are multiplexed into their time slots and

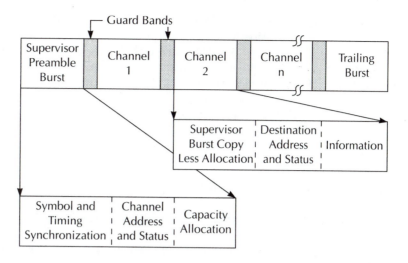

Figure 12.13 TDMA Frame

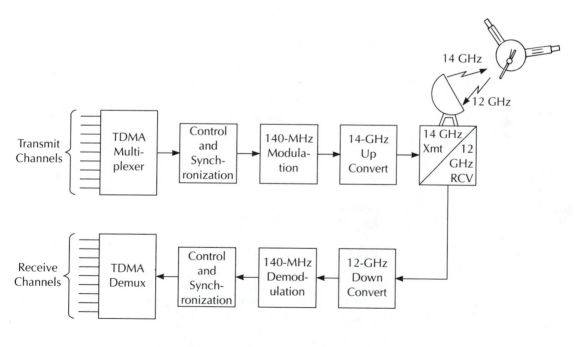

Figure 12.14 TDMA Communications System

the control supervisor burst is added. This data stream modulates a 140-MHz intermediate carrier that is up-converted onto a 14-GHz carrier. The 14-GHz signal is sent to the satellite, arriving with a power of approximately 1 picowatt. The satellite transponder demodulates the 14 GHz back to the 140-MHz intermediate level. It then up-converts it to 12 GHz for transmission back to earth. The 12-GHz signal is received by an earth station, and is down-converted to 140 MHz and demodulated. Finally, the demodulated information is demultiplexed to the various destination stations.

• 12.10 EARTH STATION ANTENNAS

One additional aspect of satellite links is the earth station's capacity to receive high-frequency, low-power signals from the satellite and to transmit high-power, high-frequency signals to the satellites. These are accomplished using large aperture or dish antennas, which are generally parabolic to allow the maximum gathering of the low-power signals sent by the satellite. Some form of a feed horn is used to direct transmitted signals from the earth station to the parabolic reflector, so that for the transmit function signals can be directed in a more directional and concentrated form as they are sent to the satellite. Figure 12.15 illustrates the physical geometry of a typical parabolic reflector antenna.

The feed horn, for transmit purposes, is located at a radius (r) from the reflector portion of the antenna. Using the central axis (c) as a reference line, signals leave the horn at angle (θ). The geometry of the antenna is designed so that the reflected signal leaves in a direction parallel to the central axis. To assure this, the following relationship between the geometric elements of the antenna has to be met:

$$\tan (\theta/2) = d/(4r) \qquad (12.1)$$

where d is the distance between the reflector's edges as shown in Figure 12.15.

Example 12.1 What is the maximum angle a signal can leave a horn for a reflector measuring 3 meters from edge to edge if the horn is placed 1.5 meters from the reflector?

Solution: Substituting the values given into equation 12.1 produces this result:

$$\tan (\theta/2) = (3 \text{ m})/(4)(1.5 \text{ m}) = .5$$

Taking the arc tangent of .5 yields an angle of 26.56°, and thus: $\theta/2 = 26.56°$. This results in a maximum angle of exit from the horn of 53.12°.

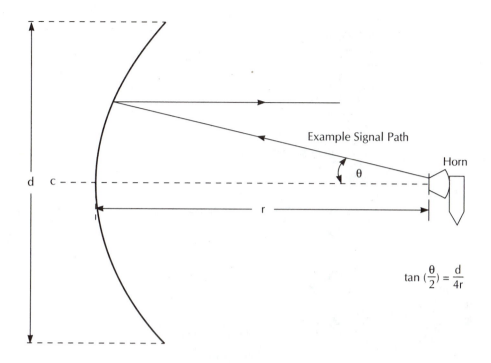

$$\tan\left(\frac{\theta}{2}\right) = \frac{d}{4r}$$

Figure 12.15 Parabolic Reflector Antenna

Other antenna configurations with varying geometries are in use today. Despite their different configurations, the methods of designing them are identical. A relationship between the exiting signal, the reflector geometry, and any intermediate devices (mirror reflectors, corner horns, etc.) exists, which determines the antenna's actual dimensions based on its applications.

12.11 SATELLITE POLARIZED TRANSMISSIONS

Besides the type of beam (spot, steerable, or earth coverage), discussed earlier, different manners of transmitting signals are used with satellites. To expand the number of channels transmitted at one time, polarized transmission was developed. Electromagnetic radiations, such as those emanating from an antenna, have a polarizing quality. Antennas can be made to radiate signals that are oriented 90° apart. Referencing one signal as horizontal, the other would be vertical as in Figure 12.16. This means that two signals of the same frequency can be transmitted simultaneously without interfering with each other if one is polarized horizontally

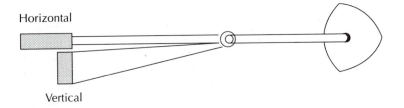

Figure 12.16 Horizontal and Vertical Polarization

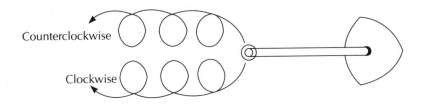

Figure 12.17 Circular Polarization

and one vertically. One example of the use of polarized transmissions is RCA's *SATCOM* satellite, which manages twenty-four transponders, each with a bandwidth of 36 MHz. Half of the transponders transmit signals polarized horizontally, while the other half use vertical polarization.

Another, less-common form of polarization is called **CIRCULAR POLARIZATION.** Signals are radiated in spiral form (Figure 12.17) from the radiating antenna. Reference the direction of a signal as clockwise, a second signal, at the same frequency, can be radiated in a counterclockwise spiral. An important point about polarization is that while it allows additional transmissions to occur simultaneously, it also divides the radiated power between each pair of polarized radiations.

SUMMARY

Satellite communications and facsimile have been with us for quite some time now. Satellite systems have provided rapid and reliable worldwide communications. Facsimile transmissions have provided a means to send actual copies of documents between users through the telephone network. Applications within the data communications field are continually expanding, bringing about new ideas on a weekly basis. For example, a comparatively recent development, the integrating of voice, digital data, video and teleconferencing on a single network is quickly becoming a reality. As the need for larger and faster communication networks and applications increases so will the technology needed to accommodate the need.

GLOSSARY

Charge coupled device (CCD)—Converts light to electrical energy.

Circular polarization—Radiating signals in a spiral form.

Downlink—Transmission from satellite to earth.

Drum scanning—Form of optical scanning that uses a helix groove in a drum to detect reflected light from a document.

Facsimile (FAX)—Image detection and transmission.

Flatbed scanning—Image is scanned as it is moved across a flat surface.

Geosynchronous orbit—A satellite orbit characterized by motion and placement in orbit to be synchronized with the earth's rotation.

Huffman code—Data transmission efficiency code that uses shorter codes for more frequently transmitted bit combinations.

INTELSAT—An international commercial satellite organization.

Transponder—Satellite repeater that receives at one frequency and transmits at another.

Uplink—Transmission from earth to a satellite.

QUESTIONS

12.1. What is facsimile? How long has it been in use?

12.2. What is the main difference between drum and flatbed scanning?

12.3. For analog drum scanners, what type of modulation is used after the received light has been converted into electrical energy?

12.4. What is the line rate for an analog drum scanner?

12.5. Give two differences between a flatbed and a drum scanner besides the physical motion between paper and light.

12.6. Contrast CCITT T.2 and T.3 recommendations for analog fax systems.

12.7. What is the purpose of using the Huffman code?

12.8. Which types of modulation are employed with the CCITT T.4 recommendation?

12.9. What basic function does a satellite perform?

12.10. Describe a geosynchronous orbit. Include the prime advantages when compared with an elliptical orbit.

12.11. Where do satellite transponders get their operating power?

12.12. List the main functions of a satellite transponder.

12.13. Give a good reason why uplink and downlink frequencies are not the same.

12.14. What is the difference between a spot beam and an earth coverage beam? Which would be used to transmit data to Easter Island?

12.15. Give one advantage of using 12/14 GHz for business applications in a large city.

12.16. Which two forms of multiplexing are used for multichannel satellite communications?

12.17. What is gained by using horizontal and vertical polarized radiations? What is the main drawback to polarized radiations?

12.18. How do the earlier *ECHO* satellites differ from today's transponder satellites?

12.19. What is the capacity of the *ITELSAT V* satellite? What is the bandwidth of the transmitted signal?

12.20. For the TDMA system described in the text, what are the frequencies of the uplink and the downlink?

12.21. Why are earth dish antennas generally parabolic in shape?

PROBLEMS

12.1. Following the example in the text, create a Huffman code for the following characters:

TH 22%   ER 18.5%   CH 13%   ST 12%   OUS 10%

ION 10%   CL 9%   MI 5.5%

12.2. What is the radius from a parabolic antenna's horn to its reflector if the reflector measures 3.5 meters from edge to edge and the maximum signal angle from the horn is 60°?

RESEARCH ASSIGNMENT

12.1. Research the latest developments in satellite technology. Include how each advancement in technology improves this form of communication.

12.2. Write a report on the number of different satellites in use today and what their prime functions are. This would include communications (telephone, fax, data, other), television, ISDN, defense, ''spy,'' and business applications, and so on.

12.3. Research and report on facsimile cards used with personal computers. Include how they function in contrast to standalone fax machines.

Answers to Odd-Numbered Questions

12.1. Facsimile is the sending and receiving of digitized images. In analog form it has been around since World War II (1940s).

12.3. Frequency modulation

12.5. Flatbed is used for digital fax machines and has fewer moving parts than analog drum scanners.

12.7. The purpose of the Huffman code is to reduce the number of bits sent for a given pattern of data.

12.9. Repeater function

12.11. From solar cells

12.13. Possibilities include noninterference of uplink and downlink transmissions; allow multiple receptions and transmissions

12.15. It avoids interference with other transmitted signals

12.17. They yield an increased channel capacity by allowing the transmission of two signals simultaneously. Radiated power is divided between both signals.

12.19. 12,500 channels with a bandwidth of 2,300 MHz

12.21. To maximize the gathering of lower-power signals

Answers to Odd-Numbered Problems

12.1. TH 1; ER 01; CH 001; ST 0001; OUS 00001; ION 000001; CL 0000001; MI 0000000

Appendix
Abbreviations and Acronyms

The following is a listing of most of the abbreviations and acronyms used throughout the text and in the data and digital communications field. This list is made to provide the student with an available source for what an abbreviation or acronym stands for. It is not intended to define the terms shown.

A

ACIA	Asynchronous Communications Interface Adapter
ACK	Positive Acknowledgment
A/D	Analog to Digital converter
ADCCP	Advanced Data Communications Control Procedure
ADC	Analog to Digital Converter
ADM	Adaptive Delta Modulation
ADM	Asynchronous Disconnect Mode
AlGaAs	Aluminum Gallium Arsenide
AM	Amplitude Modulation
AMI	Alternate Mark Inversion
ANI	Automatic Number Identification
ANSI	American National Standards Institute
ARM	Asynchronous Response Mode
ARQ	Automatic Request for retransmission or Automatic Repeat reQuest
ASCII	American Standard Code for Information Interchange
ASK	Amplitude Shift Keying
ATDM	Asynchronous Time Division Multiplexing
ATM	Asynchronous Transfer Mode

B

BCC	Block Check Character
BCD	Binary Coded Decimal
B8ZS	Binary 8 Zero Suppression

BDT	Broadband Distant Terminal
BERT	Bit Error Rate Tester
BIOS	Basic Input/Output System
BISDN	Broadband Integrated Services Digital Network
BISYNC	Bisynchronous Protocol
BNT	Broadband Network Termination
BPV	Bipolar Violation
BPS	Bits Per Second
BRI	Basic Rate Interface
BRPRZ	Bipolar Return to Zero
BS	Back space
BTI	Broadband Terminal Interface
BW	Bandwidth

C

CAN	Cancel
CCD	Charge Coupled Device
CCITT	International Consultative Committee for Telegraphy and Telephony
CD	Carrier Detect
CMOS	Complementary Metal Oxide Semiconductor
CODEC	Coder/Decoder
COMPAND	Compress/Expand or Compression/Expansion
COMSAT	Communication Satellite Corporation
CPU	Central Processing Unit
CR	Carriage Return
C/R	Command/Response
CRC	Cyclic Redundancy Check
CRT	Cathode Ray Tube (Monitor)
CSMA/CD	Carrier Sense Multiple Access with Collision Detection
CTS	Clear To Send
CWK	Continuous Wave Keying

D

DAA	Data Access Arrangement
DAC	Digital to Analog Converter
dB	Decibel ratio
dBm	Decibel level (milliwatt reference)
DC	Device Control
DCD	Data Carrier Detect
DCE	Data Communications Equipment or Data Circuit Terminating Equipment
DDCMP	Digital Data Communications Message Protocol
DDD	Direct Distance Dialing
DECNET	Digital Equipment Corporation Network

DISC	Disconnect
DM	Disconnect Mode command
DLC	Data Link Control
DLE	Data Link Escape
DMA	Direct Memory Access
DNA	Digital Network Architecture
DOMSAT	Domestic Satellite service
DOS	Disk Operating System
DPSK	Differential Phase Shift Keying
DSR	Data Set Ready
DTE	Data Terminal Equipment
DTMF	Dual Tone Multiple Frequency (Touch Tone)
DTR	Data Terminal Ready
DUV	Data Under Voice
DUX	Duplex

E

EBCDIC	Extended Binary Coded Decimal Interchange Code
EIA	Electronic Industries Association
EFS	Error-Free Second
EMI	Electromagnetic Interference
EMP	Electromagnetic Pulse
ENQ	Enquiry
EOT	End of Transmission
ESC	Escape
ESF	Extended Superframe
ET	Exchange Termination
ETB	End of Transmission Block
ETX	End of Text

F

FATS	File Allocation Tables
FAX	Facsimile
FCC	Federal Communications Commission
FCS	Frame Check Sequence
FDDI	Fiber Distributed Data Interface
FDDN	Fiber Distributed Data Network
FDM	Frequency Division Multiplexing
FDMA	Frequency Division Multiple Access
FDX	Full Duplex
FEC	Foward Error Correction
FF	Form Feed
FM	Frequency Modulation
FNE	Frame Number Expected

FNS	Frame Number Sent
FRMR	Frame Reject
FSK	Frequency Shift Keying
FT1	Fractional T1

G

GaAs	Gallium Arsenide
GFI	Group Format Identifier
GOSIP	Government Open Systems Interface Protocol
GS	Group Separator

H

HDLC	High-level Data Link Control
HDX	Half Duplex
Hz	Hertz —cycles per second

I

IBM	International Business Machines
IC	Integrated Circuit
IEEE	Institute of Electrical and Electronic Engineers
INTELSAT	International Telecommunications Satellite Organization
INTERNET	International Network
I/O	Input/Output (peripheral) device or operation
IRQ	Interrupt Request
ISDN	Integrated Services Digital Network
ISO	International Standards Organization
ITB	End of Intermediate Transmission Block

L

LAN	Local Area Network
LANCE	Local Area Network Controller
LAP	Link Access Protocol
LAPB	Link Access Protocol for B channels or Balanced
LAPD	Link Access Protocol for D channels
LCGN	Logical Channel Group Number
LCI	Logical Channel Identifier
LCN	Logical Channel Number
LCU	Line Control Unit
LED	Light-Emitting Diode
LF	Line Feed
LLC	Logical Link Control
LRC	Longitudinal Redundancy Check

LSB	Least Significant Bit
LSI	Large Scale Integration
LT	Line Termination
LU	Logical Unit

M

MAC	Media Access Control
MAN	Metropolitan Area Network
MAP	Manufacturing Automation Protocol
MFOTS	Militarized Fiber-Optic Transmission Network
MODEM	Modulator/Demodulator
MOS	Metal Oxide Semiconductor
MSB	Most Significant Bit
MSI	Medium Scale Integration
MUX	Multiplexer

N

NAK	Negative Acknowledgment
NAU	Network Addressable Unit
NETBIOS	Network Basic Input Output Services Interface
NL	New Line
NR	Frame Number expected to be Received Next
NRZ	Non-Return to Zero
NRZ/AMI	Non-Return to Zero Alternate Mark Inversion
NRZB	Non-Return to Zero Bipolar
NRZI	Non-Return to Zero Mark Inversion
NS	frame Number being Sent
NT, NT1, NT2	Network Termination, 1, 2

O

| ODI | Open Data Link Interface protocol |
| OSI | Open Systems Interconnection model |

P

PABX	Private Automatic Branch Exchange
PAD	Packet Assembler/Disassembler
PAM	Pulse Amplitude Modulation
PBX	Private Branch Exchange
PC	Personal Computer
PCM	Pulse Code Modulation
PCN	Path Control Network
PHY	Physical protocol

PDN	Packet or Public Data Network
PLL	Phase Lock Loop
PM	Phase Modulation
PMD	Physical Media-Dependent protocol
POTS	Plain Old Telephone System
PRI	Primary Rate Interface
PSK	Phase Shift Keying
PU	Physical Unit
PVC	Permanent Virtual Circuit

Q

QAM	Quadrature Amplitude Modulation
QPSK	Quadrature Phase Shift Keying

R

RAM	Random Access Memory
RCA	Radio Corporation of America
RCV	Receiver
RD	Received Data
RI	Ring Indicator
RIM	Request Initialization Mode
RLSD	Receive Line Signal Detector
ROM	Read Only Memory
RS	Recommended Standard
RTS	Request To Send
RZ	Return to Zero
RZB	Return to Zero Bipolar
RZI	Return to Zero Mark Inversion

S

SAP	Service Access Point
SBS	Satellite Business System
SCC	Satellite Communications Controller or Specialized Common Carrier
SDA	Station Device Address
SDLC	Synchronous Data Link Control
SF	Superframe
SIA	Serial Interface Adapter
SIM	Set Initialization Mode
SIMP	Satellite Information Message Protocol

SIO	Serial Input/Output interface
SMT	Station ManagemenT protocol
SNA	Systems Network Architecture
SNR	Signal-to-Noise Ratio (also S/N)
SNRM	Set Normal Response Mode
SOH	Start of Heading
SONET	Synchronous Optical Network
SP	Space
SPA	Station Polling Address
SPS	Symbols Per Second
SPN	Subscriber's Premises Network
SQD	Signal Quality Detect
SSA	Station Selection Address
SSAP	Source Service Access Point
SSCP	Source Service Control Point
SSI	Small Scale Integration
STACO	Station Controller
STATDM	Statistical Time Division Multiplexing
STATMUX	Statistical Multiplexer
STDM	Synchronous Time Division Multiplexing
STX	Start of Text
SVC	Switched Virtual Circuit
SYN, SYNC	Synchronization
SYSCON	System Configuration

T

TA	Terminal Adapter
TANSTAAFL	There Ain't No Such Thing As A Free Lunch
TCP/IP	Transmission Control Protocol/INTERNET Protocol
TD	Transmitted Data
TDM	Time Division Multiplexing
TDMA	Time Division Multiple Access
TE, 1, 2	Terminal Equipment, 1, 2
TELCO	Telephone Company
TELENET	Telephone Public Data Network
TELEX	Teleprinter Exchange service
TEP	Terminal End Point
TOP	Technical Office Protocol
TTL	Transistor-Transistor Logic
TTY	Teletypewriter
TWX	Teletypewriter Exchange Service
TYMNET	Timeshare Incorporated Network

U

UA	Unnumbered Acknowledge frame
UART	Universal Asynchronous Receiver/Transmitter
UDS	User Datagram Service
UI	Unnumbered Information frame
USART	Universal Synchronous/Asynchronous Receiver/Transmitter

V

VAN	Value Added Network
VC	Virtual Circuit
VCO	Voltage Controlled Oscillator
VGM	Voice Grade Medium
VLSI	Very Large Scale Integration
VRC	Vertical Redundancy Check

W

WAN	Wide Area Network
WATS	Wide Area Telecommunications Service
WPM	Words Per Minute
WYSIWYG	What You See Is What You Get

X

XFR	Transfer
XID	Transfer ID

Bibliography

Texts and Magazine Articles

Annamalai, K., et al. "FDDI Chips: The Dawn of a New LAN." *ESD*, October 1987.

ANSI/IEEE Committee *802.2 Local Area Networks—Logical Link Control*. IEEE, 1984.

Barnoski, M. *Fundamentals of Optical Fiber Communications*. 2nd ed. Academic Press, 1981.

Bartee, T. C., ed. *Data Communications Networks and Systems*. Howard W. Sams, 1985.

Bellamy, J. C. *Digital Telephony*. Wiley, 1982.

Berry, P. *Operating the IBM PC Networks*. Sybex, 1986.

Black, U. D. *Data Communications, Networks and Distributive Processing*. Reston Publishing, 1983.

Bleazard, G. B. *Introducing Satellite Communications*. National Computing Centre, 1985.

Bousquet, M. and Maral, G. *Satellite Communication Systems*. Wiley, 1986.

Brodsky, I. "Tapping into ISDN." *Data Communications* 19, no. 4 (April 1990).

Brown, J. K., and Lew, P. M. "Critical User Issues for Fiber Backbones." *Telecommunications* 24, no. 5 (May 1990): 27–30.

Byrd, M., and Drefler, F. J., Jr. "WAITS." *P.C. Magazine,* March 1990.

Byrne, W. R., et al "Broadband ISDN Technology and Architecture." *IEEE Network* 13, no. 1 (January 1989): 7–12.

Callahan, P. "Getting More Fiber." *LAN Magazine* 5, no. 7 (July 1990): 93–102.

Campbell, G. M., and Morreale, P. A. "Metropolitan Area Networks." *IEEE Spectrum* 14, no. 5 (May 1990): 40–42.

Carr, J. "Promises, Promises." *LAN Magazine* 5, no. 6 (June 1990): 113–120.

Chew, J., and Duncanson, J. "The Ultimate Link?" *Byte* 13, no. 7 (July 1988): 278–286.

Chorafas, D. N. *Designing and Implementing Local Area Networks*. McGraw-Hill, 1984.

Telephony Today and Tomorrow. Prentice-Hall, 1984.

Das, J. *Review of Digital Communications*. Wiley, 1987.

Deasington, R. J. *X.25 Explained*. 2nd ed. Ellis Horwood, 1986.

DeFrance, J. *Communications Electronic Circuits*. 2nd ed. Rinehart Press, 1972.

Dixon, R. C. "Lore of the Token Ring." *IEEE Network* 1, no. 1 (January 1987): 11.

Douglas, J. L. *Applied Data Communications Handbook*. Universal Data Systems, 1987.

Durfey, D. C. "Functionality Overview of NetWare APIs." *NetWare Technical Journal*, October 1988.

Falek, J. I., and Johnston, M. "Standards Makers Cementing ISDN Subnetwork Layers. *Data Communications* (October 1987): 237.

Foldvik, R. G. *The Evolutionary Path to Broadband ISDN*. Phoenix Conference on Computers and Communications, IEEE Computer Society Press, March 1990.

Glass, B. "The Token Ring." *Byte* (January 1989): 363.

"Understanding NETBIOS." *Byte* (January 1989): 301.

Greenfield, D. "Bridge Compression Taps Unused Bandwidth." *Data Communications* 19, no. 7 (July 1990): 54–59.

Greenfield, D., and Heywood, P. "FDDI: Just Say Not Yet." *Data Communications* 19, no. 4 (April 1990).

Halsall, F. *Introduction to Data Communications and Computer Networks*. Addison-Wesley, 1985.

Handel, R. "Evolution of ISDN Towards Broadband ISDN." *IEEE Network* 3, no. 1 (January 1989): 7–12.

Henshall, J., and Shaw, S. *OSI Explained*. Wiley, 1988.

Hildebrand, D. "OSI: Not Ready for Prime Time?" *Data Communications* 19, no. 3 (March 1990).

Hindin, E. M. "Easing LAN Application Development." *Data Communications* 19, no. 2 (February 1990).

"Hayes Unveils ISDN API Which Could Become Standard." *Data Communications* 19, no. 2 (February 1990).

"IBM 8209 LAN Bridge Links Ethernet to Token Ring." *Data Communications* 19, no. 3 (March 1990).

"LAN APIs—Getting in Touch with Network Resources." *Data Communications* 19, no. 2 (February 1990).

"OSI Links to SNA." *Data Communications* 19, no. 7 (July 1990): 59–66.

Hornbach, B. H. "The 5ESS Switch as International Gateway." *AT&T Technology* 5, no. 2 (1990).

Howard, M. "The OSI Switch Is On." *LAN Times* 7, no. 8 (July 1990): 49–64.

Hurwicz, M. "Whys and Whiches of Brouters and Bridges." *Computer World* 24, no. 6 (February 5, 1990).

Johannessen, P. "Implementation of NetWare LU6.2." *NetWare Technical Journal*, October 1988.

Johnson, J. T. "Scheme to Avert Ethernet Packet Collisions." *Data Communications* 19, no. 3 (March 1990).

"When Is a Ring Not a Ring?" *Data Communications* 19, no. 4 (April 1990).

Johnson, S. "NetWare Drivers Pack It In." *LAN Times* 7, no. 5 (May 1990): 97–100.

Keiser, G. *Optical Fiber Communications*. McGraw-Hill, 1983.

Kelly, P. "Connecting LANs with Bridges." *Telecommunications* 24, no. 6 (June 1990): 31–38.

Kingdon, K. "The NetWare Requester: Functionality Overview." *NetWare Technical Journal,* October 1988.

Krall, G. "SNMP Opens New Lines of Sight." *Data Communications* 19, no. 3 (March 1990).

Lacy, E. *Fiber Optics*. Prentice-Hall, 1982.

Lehmann, J. N. "What ISDN Will Mean for Australia." *Data Communications* (October 1987): 208.

McCool, J. F. "The Emerging FDDI Standard." *Telecommunications* 21, no. 5 (May 1987).

"FDDI: Getting to Know the Inside of the Ring." *Data Communications* (March 1988).

McCool, J. F., and Ullal, J. V. *Fiber Optic Network Standard Delivers Speed and Reliability*. Computer Design Publishing, 1987.

McNamara, J. *Technical Aspects of Data Communication*. Digital Equipment Corporation, 1977.

Malvino, A. *Electronic Principals*. 3rd ed. McGraw-Hill, 1984.

Mank, S. "Myths and Misconceptions about OSI." *Telecommunications* 24, no. 5 (May 1990): 27–30.

Markley, R. W. *Data Communications and Interoperability*. Prentice-Hall, 1990.

Marks, H. "No Fault Insurance." *LAN Magazine* 5, no. 7 (July 1990): 53–60.

Martin, J. *Data Communication Technology*. Prentice-Hall, 1988.

Local Area Networks. Prentice-Hall, 1989.

Mier, E. E. "PBX Upgrades: Forget the Forklift." *Data Communications* 19, no. 3 (March 1990).

Miller, K. "Fractional T1: An Idea Whose Time Has Come." *LAN Times* 1, no. 6 (June 1990): 47.

Miller, M. A. *The 68000: Architecture, Programming and Applications*. Merrill, 1988.

Mohanty, M. "Fiber on a Chip." *LAN Magazine* (February 1988).

Mokhoff, N. "Five Chip Token Ring Passing Set Operates LANs at 100 Mbps." *Electronic Design* 36, no. 21 (September 1987).

Murphy, R. J. *Telecommunication Networks—A Technical Introduction*. Howard W. Sams, 1987.

Noordsij, K. M., and Rajala, D. W. "AT&T's Integrated Access and Cross Connect System Goes International." *AT&T Technology* 5, no. 2 (1990).

Osmundsen, S. "Blueprints for the 1990s." *Byte* 15, no. 3 (March 1990).

Palais, J. *Fiber Optic Communications*. Prentice-Hall, 1984.

Pitt, D. "Standards for the Token Ring." *IEEE Network* 1, no. 1 (January 1987): 19.

Presuhn, R. "Considering CMIP." *Data Communications* 19, no. 3 (March 1990).

Prittchard, W., and Scivlli, J. *Satellite Communication Systems Engineering*. Prentice-Hall, 1986.

Purton, P., and Tate, P. "The ISDN Ingredient." *Datamation*, January 1, 1987, 79.

Reddy, S. "Pathfinders." *LAN Magazine* 5, no. 6 (June 1990): 101–111.

"Seeing the Light." *LAN Magazine* 5, no. 6 (June 1990): 123–128.

Reynolds, G. *Introduction to Business Telecommunications*. Merrill, 1984.

Rickert J. B., Jr. "Evaluating MAC Layer Bridges—Beyond Filtering and Forwarding." *Data Communications* 19, no. 5 (May 1990).

Ross, F. E. "Rings are 'Round for Good!" *IEEE Network* 1, no. 1 (January 1987): 31

Routt, T. J. "SNA Network Management: What Makes IBM Netview Tick?" *Data Communications* (June 1988): 203.

Rux, P. T. and Lew, P. M. "T1 Standards—A Dynamic World." *Telecommunications* 24, no. 5 (May 1990): 67–72.

Schoenbeck, R. *Electronic Communications, Modulation and Transmission*. Merrill, 1988.

Schwartz, M. *Information Transmission, Modulation and Noise*. 2nd ed. McGraw-Hill, 1970.

Schweber, W. *Data Communications*. McGraw-Hill, 1988.

Scott, K. "Commercial Interworking Services for LAN to LAN Connections." *Data Communications* 19, no. 2 (February 1990).

"SNMP Brings Order to Chaos." *Data Communications* 19, no. 3 (March 1990).

"Taking Care of Business with SNMP." *Data Communications* 19, no. 3 (March 1990).

Siegel, J. A. "Using Novell's 3270 APIs." *NetWare Technical Journal*, October 1988.

Sinnema, W. *Digital, Analog and Data Communication*. Reston Publishing, 1982.

Stallings, W. *Data and Computer Communications*. Macmillan, 1985.

"Managing the Well Tempered LAN." *Byte* 15, no. 4 (April 1990).

Stein, D. H. *Introduction to Digital Data Communications*. Delmar Publishers, 1985.

Stephenson, P. "Taking a Look at 10-BASE-T Ethernet." *LAN Times* 5, no. 6 (May 1990): 138–153.

Straus, P. R. "Standard Bridges, MANs, Routers: Tools for Linked LANs." *Data Communications* (October 1988): 49.

Strole, N. C. "The IBM Token Ring Network: A Functional Overview." *IEEE Network* 1, no. 1, (January 1987): 23.

Tolly, K. "Opening the Gateways to SNA Connectivity." *Data Communications* 19, no. 3 (March 1990).

"Playing to Win with IBM's LAN Gateways." *Data Communications* 19, no. 2 (February 1990).

Tomasi, W. *Advanced Electronic Communications Systems*. Prentice-Hall, 1987.

Electronic Communications Systems, Fundamentals Through Advanced. Prentice-Hall, 1988.

Tomasi, W., and Alisouskas, V. *Telecommunications, Voice/Data with Fiber Optic Applications*. Prentice-Hall, 1988.

Udell, J. "NetWare 386: Less Pain, Grest Gain." *Byte* 15, no. 3 (March 1990).

Ullal, J. V. "The Dynamics Behind the Uses of Fiber Optics." *Electronic Engineering Times*, November 1987.

Vinzant, D. R. "SQL Database Servers." *Data Communications* 19, no. 1 (January 1990).

Yoneda, S. "Broadband ISDN ATM Layer Management . . . et al." *IEEE Network Magazine* 14, no. 3 (May 1990): 31–36.

Weissberger, A. J. "The Evolving Versions of ISDN's Terminal Adapter." *Data Communications* 18, no. 3 (March 1989): 133.

Specification Manuals

Advanced Micro Devices. *AMLINK3 ISDN Communication Protocol Software*. AMD, 1988.

Advanced Micro Devices. *AM79C30A/32A Digital Subscriber and ISDN Data Controllers and AM79C401 Integrated Data Protocol Controller Data Books*. AMD, 1988.

Advanced Micro Devices. *ISDN AmLINK*. AMD, 1989.

Advanced Micro Devices. *SuperNet Family for FDDI*. AMD, 1989.

Advanced Micro Devices. *World Network Catalogue*. AMD, 1989/90.

Advanced Micro Devices. *Z85C30 Enhanced Serial Communication Controller*. AMD, 1988.

Hewlett Packard. *X.25 the PSN Connection*. Hewlett Packard Company, 1988.

Intel. *Telecommunications Products Handbook*. Intel Corp., 1984.

Intel. *Microcommunications Handbook*. Intel Corp., 1987.

Motorola. *8-Bit Microprocessor and Peripheral Data*. Motorola, Inc., 1983.

Novell. *Distributed Application Processing in the NetWare Environment*. Novell, Inc., 1988.

Novell. *NetWare 386 Technical Overview*. Novell, Inc., 1989.

RCA. *QMOS Integrated Circuits Databook*. RCA Corp., 1985.

Signetics. *Logic TTL Specifications Data Manual*. Signetics Corp., 1978.

Western Digital. *Communication Products Handbook*. Western Digital Corp., 1984.

Western Digital. *Network Products Handbook*. Western Digital Corp., 1983.

Standards

Electronics Institutes Association:
> *EIA-232-C* published in August 1969
> *EIA-232-D* published in November 1986
> *EIA-422-A* published December 1978
> *EIA-449* published November 1985

Application Notes

Fronheiser, K. *AN-754 Device Operation and System Implementation of the Asynchronous Communications Interface Adapter (MC6850)*. Motorola, Inc., 1987.

Index

voltage controlled oscillator (VCO), 168, 195

W

WAN (wide area network), 4, 272, 341, 348

X

X.25, 273, 291
XMODEM protocol, 122